Fundamentals of Mobile
and Pervasive Computing

Fundamentals of Mobile and Pervasive Computing

Frank Adelstein

Sandeep K. S. Gupta

Golden G. Richard III

Loren Schwiebert

McGraw-Hill

New York   Chicago   San Francisco   Lisbon   London   Madrid
Mexico City   Milan   New Delhi   San Juan   Seoul
Singapore   Sydney   Toronto

The McGraw·Hill Companies

CIP Data is on file with the Library of Congress

3 4 5 6 7 8 9 0 IBT / IBT 0 1 0 9 8 7

ISBN 978-0-07-141237-7
ISBN 0-07-141237-9

The sponsoring editor for this book was Stephen S. Chapman and the production supervisor was Sherri Souffrance. It was set in Century Schoolbook by International Typesetting and Composition. The art director for the cover was Anthony Landi.

Printed and bound by IBT Global.

McGraw-Hill books are available at special quantity discounts to use as premiums and sales promotions, or for use in corporate training programs. For more information, please write to the Director of Special Sales, McGraw-Hill Professional, Two Penn Plaza, New York, NY 10121-2298. Or contact your local bookstore.

This book is printed on recycled, acid-free paper containing a minimum of 50% recycled, de-inked fiber.

To my friends and family who kept asking, "Is it done yet?"
(Frank)

To Poonam, Ayush, and Animesh
(Sandeep)

To Daryl, Q, and David Byrne
(Golden)

To Pam, Joel, and Kyle
(Loren)

Contents

Preface   xv
Acknowledgments   xix

Chapter 1. Mobile Adaptive Computing   1

1.1 What Is Mobile Computing?   1
1.2 Adaptability—The Key to Mobile Computing   3
     1.2.1 Transparency   4
     1.2.2 Constraints of mobile computing environments   5
     1.2.3 Application-aware adaptation   6
1.3 Mechanisms for Adaptation   8
     1.3.1 Adapting functionality   8
     1.3.2 Adapting data   9
1.4 How to Develop or Incorporate Adaptations in Applications   11
     1.4.1 Where can adaptations be performed?   12
1.5 Support for Building Adaptive Mobile Applications   19
     1.5.1 Odyssey   19
     1.5.2 Rover   22
1.6 Summary   24

Chapter 2. Mobility Management   27

2.1 Mobility Management   27
2.2 Location Management Principles and Techniques   30
     2.2.1 Registration area–based location management   33
2.3 Location Management Case Studies   48
     2.3.1 PCS location management scheme   49
     2.3.2 Mobile IP   50
2.4 Summary   53

Chapter 3. Data Dissemination and Management   55

3.1 Challenges   59
3.2 Data Dissemination   61
     3.2.1 Bandwidth allocation for publishing   63
     3.2.2 Broadcast disk scheduling   65

3.3 Mobile Data Caching   67
     3.3.1 Caching in traditional distributed systems   68
     3.3.2 Cache consistency maintenance   69
     3.3.3 Performance and architectural issues   70
3.4 Mobile Cache Maintenance Schemes   72
     3.4.1 A taxonomy of cache maintenance schemes   72
     3.4.2 Cache maintenance for push-based information dissemination   74
     3.4.3 Broadcasting invalidation reports   75
     3.4.4 Disconnected operation   77
     3.4.5 Asynchronous stateful (AS) scheme   78
     3.4.6 To cache or not to cache?   84
3.5 Mobile Web Caching   86
     3.5.1 Handling disconnections   87
     3.5.2 Achieving energy and bandwidth efficiency   87
3.6 Summary   88

Chapter 4. Context-Aware Computing   91

4.1 Ubiquitous or Pervasive Computing   92
4.2 What Is a Context? Various Definitions and Types of Contexts   94
     4.2.1 Enumeration-based   94
     4.2.2 Role-based   96
4.3 Context-Aware Computing and Applications   96
     4.3.1 Core capabilities for context awareness   97
     4.3.2 Types of context-aware applications   98
     4.3.3 Developing context-aware applications   100
4.4 Middleware Support   102
     4.4.1 Contextual services   103
     4.4.2 Actuator service   104
     4.4.3 An example: context toolkit   104
     4.4.4 Providing location context   105
4.5 Summary   106

Chapter 5. Introduction to Mobile Middleware   109

5.1 What Is Mobile Middleware?   109
5.2 Adaptation   110
5.3 Agents   111
5.4 Service Discovery   111

Chapter 6. Middleware for Application Development:
Adaptation and Agents   113

6.1 Adaptation   113
     6.1.1 The spectrum of adaptation   114
     6.1.2 Resource monitoring   114
     6.1.3 Characterizing adaptation strategies   115
     6.1.4 An application-aware adaptation architecture: Odyssey   117
     6.1.5 A sample Odyssey application   119
     6.1.6 More adaptation middleware   120

6.2 Mobile Agents   123
     6.2.1 Why mobile agents? And why not?   125
     6.2.2 Agent architectures   127
     6.2.3 Migration strategies   130
     6.2.4 Communication strategies   131
6.3 Summary   133

Chapter 7. Service Discovery Middleware: Finding Needed Services   137

7.1 Common Ground   140
7.2 Services   142
     7.2.1 Universally unique identifiers   142
     7.2.2 Standardization   144
     7.2.3 Textual descriptions   145
     7.2.4 Using interfaces for standardization   149
7.3 More on Discovery and Advertisement Protocols   150
     7.3.1 Unicast discovery   150
     7.3.2 Multicast discovery and advertisement   151
     7.3.3 Service catalogs   155
7.4 Garbage Collection   156
     7.4.1 Leasing   156
     7.4.2 Advertised expirations   158
7.5 Eventing   159
7.6 Security   163
     7.6.1 Jini   163
     7.6.2 Service location protocol   164
     7.6.3 Ninja   165
7.7 Interoperability   166
     7.7.1 Interoperability success stories   167
7.8 Summary   167

Chapter 8. Introduction to Ad Hoc and Sensor Networks   171

8.1 Overview   171
     8.1.1 Outline of chapter   172
8.2 Properties of an Ad Hoc Network   175
     8.2.1 No preexisting infrastructure   175
     8.2.2 Limited access to a base station   175
     8.2.3 Power-limited devices   177
     8.2.4 No centralized mechanisms   178
8.3 Unique Features of Sensor Networks   178
     8.3.1 Direct interaction with the physical world   178
     8.3.2 Usually special-purpose devices   179
     8.3.3 Very limited resources   180
     8.3.4 Operate without a human interface   181
     8.3.5 Specialized routing patterns   181
8.4 Proposed Applications   182
     8.4.1 Military applications   182
     8.4.2 Medical applications   185
     8.4.3 Industrial applications   186
     8.4.4 Environmental applications   187
     8.4.5 Other application domains   188

Chapter 9. Challenges

191

9.1 Constrained Resources   191
    9.1.1   No centralized authority   192
    9.1.2   Limited power   193
    9.1.3   Wireless communication   196
    9.1.4   Limited computation and storage   197
    9.1.5   Storage constraints   198
    9.1.6   Limited input and output options   199
9.2 Security   200
    9.2.1   Small keys   200
    9.2.2   Limited computation   201
    9.2.3   Changing network membership   201
    9.2.4   Arbitrary topology   202
9.3 Mobility   203
    9.3.1   Mobility requirements   204
    9.3.2   Loss of connectivity   205
    9.3.3   Data loss   206
    9.3.4   Group communication   207
    9.3.5   Maintaining consistent views   208
9.4 Summary   209

Chapter 10. Protocols

213

10.1 Autoconfiguration   213
    10.1.1   Neighborhood discovery   214
    10.1.2   Topology discovery   215
    10.1.3   Medium access control schedule construction   216
    10.1.4   Security protocol configuration   220
10.2 Energy-Efficient Communication   221
    10.2.1   Multihop routing   222
    10.2.2   Communication scheduling   225
    10.2.3   Duplicate message suppression   226
    10.2.4   Message aggregation   228
    10.2.5   Dual-radio scheduling   230
    10.2.6   Sleep-mode scheduling   232
    10.2.7   Clustering   232
10.3 Mobility Requirements   235
    10.3.1   Movement detection   235
    10.3.2   Patterns of movement   236
    10.3.3   Changing group dynamics   237
    10.3.4   Resynchronization   239
10.4 Summary   240

Chapter 11. Approaches and Solutions

245

11.1 Deployment and Configuration   245
    11.1.1   Random deployment   246
    11.1.2   Scalability   246
    11.1.3   Self-organization   247
    11.1.4   Security protocol configuration   247

11.1.5   Reconfiguration/redeployment   249
11.1.6   Location determination   249
11.2   Routing   252
11.2.1   Event-driven routing   253
11.2.2   Periodic sensor readings   254
11.2.3   Diffusion routing   260
11.2.4   Directional routing   265
11.2.5   Group communication   268
11.2.6   Synchronization   271
11.3   Fault Tolerance and Reliability   273
11.3.1   FEC and ARQ   273
11.3.2   Agreement among sensor nodes (Reliability of
         measurements)   274
11.3.3   Dealing with dead or faulty nodes   278
11.4   Energy Efficiency   279
11.4.1   Uniform power dissipation   280
11.4.2   Sensor component power management   281
11.4.3   MAC layer protocols   282
11.4.4   Tradeoffs between performance and energy efficiency   283
11.5   Summary   284

Chapter 12. Wireless Security   287

12.1   Traditional Security Issues   287
12.1.1   Integrity   287
12.1.2   Confidentiality   288
12.1.3   Nonrepudiation   288
12.1.4   Availability   288
12.2   Mobile and Wireless Security Issues   290
12.2.1   Detectability   290
12.2.2   Resource depletion/exhaustion   291
12.2.3   Physical intercept problems   291
12.2.4   Theft of service   291
12.2.5   War driving/walking/chalking   292
12.3   Mobility   293
12.4   Problems in Ad Hoc Networks   293
12.4.1   Routing   294
12.4.2   Prekeying   294
12.4.3   Reconfiguring   295
12.4.4   Hostile environment   295
12.5   Additional Issues: Commerce   295
12.5.1   Liability   296
12.5.2   Fear, uncertainty, and doubt   296
12.5.3   Fraud   296
12.5.4   Big bucks at stake   297
12.6   Additional Types of Attacks   297
12.6.1   "Man in the middle" attacks   297
12.6.2   Traffic analysis   298
12.6.3   Replay attacks   298
12.6.4   Buffer-overflow attacks   298
12.7   Summary   299

Chapter 13. Approaches to Security   301

13.1   Limit the Signal   301
    13.1.1   Wire integrity and tapping   301
    13.1.2   Physical limitation   301
13.2   Encryption   302
    13.2.1   Public and private key encryption   302
    13.2.2   Computational and data overhead   303
13.3   Integrity Codes   304
    13.3.1   Checksum versus cryptographic hash   304
    13.3.2   Message authentication code (MAC)   305
    13.3.3   Payload versus header   306
    13.3.4   Traffic analysis   307
13.4   IPSec   307
    13.4.1   Authentication header (AH)   307
    13.4.2   Encapsulating security payload (ESP)   308
13.5   Other Security-Related Mechanisms   308
    13.5.1   Authentication protocols   308
    13.5.2   AAA   313
    13.5.3   Special hardware   315
13.6   Summary   315

Chapter 14. Security in Wireless Personal Area Networks   317

14.1   Basic Idea   317
    14.1.1   Bluetooth specifications   317
    14.1.2   Bluetooth network terms   318
    14.1.3   Bluetooth security mechanisms   320
14.2   Bluetooth Security Modes   320
14.3   Basic Security Mechanisms   321
    14.3.1   Initialization key   322
    14.3.2   Unit key   322
    14.3.3   Combination key   322
    14.3.4   Master key   323
14.4   Encryption   323
14.5   Authentication   325
14.6   Limitations and Problems   325
14.7   Summary   327

Chapter 15. Security in Wireless Local Area Networks   329

15.1   Basic Idea   329
15.2   Wireless Alphabet Soup   331
15.3   Wired-Equivalent Privacy (WEP)   333
    15.3.1   WEP goals   333
    15.3.2   WEP data frame   334
    15.3.3   WEP encryption   334
    15.3.4   WEP decryption   335
    15.3.5   WEP authentication   335
    15.3.6   WEP flaws   336
    15.3.7   WEP fixes   338

15.4 WPA   340
15.5 802.11i   340
      15.5.1 Encryption protocols   340
      15.5.2 Access control via 802.1x   342
15.6 Fixes and "Best Practices"   344
      15.6.1 Anything is better than nothing   344
      15.6.2 Know thine enemy   344
      15.6.3 Use whatever wireless security mechanisms are present   345
      15.6.4 End-to-end VPN   345
      15.6.5 Firewall protection   346
      15.6.6 Use whatever else is available   346
15.7 Summary   348

Chapter 16. Security in Wireless Metropolitan Area Networks (802.16)   349

16.1 Broadband Wireless Access   349
16.2 IEEE 802.16   350
16.3 802.16 Security   350
      16.3.1 Key management   351
      16.3.2 Security associations   351
      16.3.3 Keying material lifetime   352
      16.3.4 Subscriber station (SS) authorization   353
      16.3.5 Encryption   353
16.4 Problems and Limitations   354
16.5 Summary   354

Chapter 17. Security in Wide Area Networks   357

17.1 Basic Idea   357
17.2 CDMA   359
17.3 GSM   360
      17.3.1 GSM authentication   360
      17.3.2 GSM encryption   360
17.4 Problems with GSM Security   361
      17.4.1 Session life   362
      17.4.2 Weak encryption algorithm   362
      17.4.3 Encryption between mobile host and base station only   363
      17.4.4 Limits to the secret key   363
      17.4.5 Other problems   364
17.5 The Four Generations of Wireless: 1G–4G   364
17.6 3G   364
17.7 Limitations   366
17.8 Summary   366

Appendix A Brief Introduction to Wireless Communication
            and Networking   369
Appendix B Questions   375
Index   389

Preface

Mary sits with her PDA in a coffee shop in Woods Hole reviewing her appointments for the day and notes that the location for her lunch meeting with a CEO is marked TBD. After reviewing menus for a few nice restaurants, she selects an appropriate location. As she gets into her car, with notes for the lunch meeting she printed at the coffee shop, she quickly checks her credit card limit with her PDA, and then sends the CEO a note on the suggested location. Suddenly, 9000 miles to the west, a sperm whale surfaces in the Pacific Ocean. The whale carries a small, low-power radio transmitter that has been recording its vital signs for the last 2 months. Now above the water, the transmitter establishes contact with a low-earth-orbit satellite and uploads its data. It also uploads vital signs for other similarly equipped whales it passed since the radio last contacted the satellite. The data indicate a growth in the pod population. Scientists halfway around the world are thrilled with the new data and news filters up the chain to the company CEO, who can now approve a research grant for the chief scientist of a marine biology institute he is meeting for lunch. As she leaves the restaurant, Mary shares the good news with members of her research group using her cell phone.

A common thread in the above scenario is the use of mobile and pervasive computing. Mobile computing is characterized and driven by portable, lightweight hardware, wireless communication, and innovations in application and system software. The technology drives new applications which in turn fuel the demand for the new technology. Pervasive computing uses small, battery-powered, wireless computing and sensing devices embedded in our environment to provide contextual information to new types of applications. The trend of technology getting smaller and more portable is likely to continue for the foreseeable future. Bandwidth available to a home user through a broadband Internet connection exceeds the bandwidth available to all but the largest organizations several years ago. The bandwidth in low-priced network connections continues to increase at a similar rate. Wireless and cellular technologies allow users to connect to networks from practically any location and to remain connected in transit. This is the age of mobile and pervasive computing!

This book provides a focused look at important topics in *Mobile and Pervasive Computing*. Although the underlying technology is essential, we avoid being completely bound to it. Technology will change and evolve, and in fact *has* changed even during the course of writing this book. Data rates increase. Protocols evolve. More efficient algorithms are created. But the underlying techniques, the essentials of how computers share, protect, and use data in an environment in which the location is not fixed, remain relatively stable. This book is intended to address these fundamental points, the ones that continue to be significant long after this book is published.

Audience

This book has been designed for two audiences. The first group contains professionals seeking to get a solid understanding of mobile computing—the problems and solutions, the protocols, and the applications. The second group comprises students in a senior-undergraduate or graduate-level course. To help the second group, Appendix B contains both short questions, to help review the chapters, and longer questions and project ideas that instructors may select as starting points for lab assignments or for further research.

Book Organization

Conceptually, the book is organized into four parts:

Part One, which comprises the first four chapters, covers issues related to mobile and pervasive computing applications. This includes not only disseminating and caching data, but also routing and location management to determine *where* to send data in the first place. Context aware computing is the final chapter of the first part.

Part Two, which comprises Chapters 5 to 7, focuses on middleware, the layer that bridges mobile computing applications to the underlying systems that support this mobility. Middleware topics include agents and service discovery, as well as techniques and methods.

Part Three, which comprises Chapters 8 to 11, covers an important enabling networking technology for pervasive computing: adhoc and wireless sensor networks. Adhoc networks form when needed, on-the-fly, without central management or infrastructure. Wireless sensors record some type of data, such as temperature and humidity, and form an adhoc network to efficiently disseminate these readings.

Part Three discusses the applications, problems, approaches, and protocols for wireless sensor networking.

Part Four, which comprises Chapters 12 to 17, covers security in wireless environments. After describing common security problems, these chapters discuss how security is handled in personal, local, metropolitan, and wide area networks, as well as current research work and future trends.

How to Use This Book

Since this book provides the coverage of almost all the important aspects of mobile and pervasive computing, it can be used in several ways. Two of the authors have used this book in semester-long courses. Since each part of the book is almost self-contained, an instructor can easily select the parts most suitable for his or her course, permitting the book to be used in quarter-long courses or even short introductory courses. The book provides two appendices to assist in teaching this course. Appendix A provides a brief introduction to wireless communication and networking that can be used by instructors to quickly cover the essentials of wireless communication and networking. Appendix B provides questions for each chapter of the book. These questions can be used by an instructor to encourage students to further explore the topics covered in this book.

Acknowledgments

Over the entire process of the development of this book, many people have helped us; without them, this book would not have been possible.

First we thank the many people who reviewed drafts of this book: John Bodily, Michel Gertraide, Manish Kochhal, Fernando Martincic, John Quintero, Georgios Varsamapoulos, and Yong Xi.

We are grateful to the National Science Foundation (NSF) and the Center for Embedded Networking Technologies (CEINT) for supporting our research in the areas related to this book.

We thank our employers, the Arizona State University, the University of New Orleans, the Wayne State University, and the Odyssey Research Associates/ATC-NY, for providing the working environment that made this book possible. We also thank the Ohio State University, where all of us obtained our doctoral degrees, and the Buckeye Spirit. Go Bucks, Go!

Our heartfelt thanks go to our original sponsoring editor Marjorie Spencer; our editor Steve Chapman, who provided the needed level of support and encouragement to see this book through from conception to completion; Priyanka Negi, who oversaw the copyediting process; and the entire crew of McGraw-Hill.

Last but not least, we thank our family members and friends who accepted the time away from them spent writing this book. Without their support, this book could never have been finished.

Fundamentals of Mobile and Pervasive Computing

Mobile Adaptive Computing

1.1 What Is Mobile Computing?

What is mobile computing? What are its different aspects? Basically, mobile computing systems are distributed systems with a network to communicate between different machines. Wireless communication is needed to enable mobility of communicating devices. Several types of wireless networks are in use currently, and many books have been devoted to mobile communication. In this book we cover only the basic aspects of mobile communication. However, this book is not devoted to wireless communication and networks. Readers not familiar with wireless communication and networking can refer to Appendix A for a short description. In this book, we are concerned mostly with the logical aspects of mobile communication.

What is the difference between mobile computing and mobile communication? Communication is necessary for distributed computing. Many mobile computing tasks require mobile communication. But that is not the end of it. Mobile communication does not solve all the problems. As we will see in this book, many issues need to be resolved from a higher-level perspective than just being able to exchange signals/packets.

What interests us the most about mobile computing is what we can do with this new facility of mobile communication. What new applications can be enabled? Think about the applications we currently use. Many of these applications do not adapt very well to our needs. For example, suppose that we are browsing the World Wide Web and are interested in going to an Italian restaurant for lunch. What information will we get when we do a search for "Italian restaurant" through search engines such as Google? Usually we get a lot of information that we do not require. The search engines of today do not care about who is invoking the search or the location from where the search is being invoked. You can type "Italian restaurant in Tempe" or "Italian restaurant near Arizona State University." You can make your query as specific as possible. But wouldn't it be nice if the applications were aware of your location or, more generally, your context and if the response from the search engine were ordered according to your context? Surely, such context-aware applications will make us more productive.

Let us consider another application, such as video streaming over the Internet. Suppose that you are on the move, watching a movie. Wireless communication is different from wired communication. When you have a wire, you usually have a fixed bandwidth available to you and your application. Once the application is started and the movie starts, you can watch the movie at a good *quality of service* (QoS). (Currently, this is not really true if streaming is done across the Internet. Nevertheless, it is expected that this will become a reality in the near future.) In wireless communication, however, the bandwidth is mostly shared among several users in a dynamic fashion. This is to say that no dedicated bandwidth is available. Even if your application can reserve certain wireless bandwidth, the usable bandwidth fluctuates owing to the nature of the wireless medium. There are both short- and long-term fluctuations. The question is, "How should the application respond to these fluctuations?" One way that the application can respond is in a uniform manner, irrespective of what you are watching. The other approach is to respond based on the type of movie you are watching. For example, suppose that you are watching an action movie. The application can reduce the bandwidth requirement by switching from full-color video to black and white or by reducing the resolution. On the other hand, if you are watching an interview of Bill Gates, the application may switch simply to audio streaming. There are several important points to note here. These decisions are based on the content of the video, and the decision making may involve both the client and the server (the client needs to inform the server that it no longer wants the video frames, only the audio).

In this chapter we will reexamine computer (software) system design to see how it needs to be changed to accommodate mobility. As we will see, many of the changes are concerned with providing mechanisms for adapting to changing environmental and system conditions, such as the location and the available resources. Mobile computing is about providing information anytime anywhere or, more generally, computing anytime and anywhere.

Mobile computing is also about dealing with limitations of mobile computing devices. For example, *personal digital assistants* (PDAs) and laptops have small interfaces and are powered by batteries. One of the major issues is how to do computation in an energy-efficient manner. Battery technology is not advancing at the same pace as processor technology, so it is not expected that battery capacity will double at a fixed rate, as is true about processor speeds, according to Moore's law. One does not have to deal with these issues when designing systems for stand-alone or distributed systems. One may have to deal with issues of fault tolerance in distributed systems, such as server crashes or network link failures. However, energy usually is not an issue. In mobile systems, energy becomes a resource like processing time or memory space. Therefore, one now has to design resource management techniques for energy, just as traditional operating systems deal with process and memory management.

Different computers in a mobile computing environment may have different capabilities. Any collaborative activity between these devices needs an underlying software entity to deal with the heterogeneity of these devices. This software entity is called a *middleware layer*. A middleware layer also may allow a mobile device and a wired device to interact. When mobile clients move from one administrative domain to another, they may want to know what new services are available.

How about security or privacy? Wireless communication takes place over an "open" wire and is relatively easy to tap. It may seem that traditional techniques of cryptography can be used to secure wireless communication. However, the main problem is that these secure techniques are designed for wired networks and are computation- and communication-intensive. Attempts to reduce these overheads lead to security schemes that are relatively easy to break. We will look into security schemes for mobile networks in later chapters.

1.2 Adaptability—The Key to Mobile Computing

Humans have evolved to be on the top of the food chain. It is probably undeniable that we have done so by being able to adapt quickly and effectively to varying situations. It is only because of our ability to adapt that we can be found from Arctic regions to the Sahara Desert. What are the techniques by which we are able to adapt to such diverse environments? Can we incorporate some of these techniques into our computing systems? Anyone who has used computers seriously would like them to be more resilient and adaptive to our needs and circumstances. Computing systems and applications fail for various reasons. What is most frustrating is when they fail for no apparent reason. You install a new application, and some other apparently unrelated application stops functioning. And sometimes we just want our computers to learn from our past actions and act proactively and appropriately. Making systems resilient and adaptive is not a trivial task. It took us millions of years to rise to the top of the food chain. Computers as we know them today are not even a hundred years old.

The vision of mobile computing is to be able to roam seamlessly with your computing devices while continuing to perform computing and communication tasks uninterrupted. Many technological advances at various fronts such as security, privacy, resource allocation, charging, and billing have to be made to make this feasible. A quintessential characteristic of any solution to mobile computing problems is its ability to adapt to dynamic changes in computing and communication environments. A system's agility to react to changes in the computing environment and continue its computing tasks uninterrupted is a new measure of performance in mobile computing environments.

Consider a scenario in which you move from one coverage area of an access point to another while a video streaming application is running on your computer. To continue receiving the video stream uninterrupted and without deterioration in video quality, the video stream packets now should be routed automatically through the new access point. In an Internet Protocol (IP)–based network this may involve the mobile client obtaining a new IP address in the new access point's IP network and informing the server of the new address so that it can now send the packets to the new address. Many more sophisticated solutions to this problem have been developed. The point here is that the underlying system has to take many actions automatically to ensure continued connectivity, in this case uninterrupted viewing of the video stream. In essence, the system has to adapt to the changes in the environment, such as the network configuration and the availability of communication and computation resources and services. However, is this enough? More specifically, the preceding adaptation scheme does not take into account the applications requirements, and the applications themselves did not play any part in the adaptation. Does this application-transparent way of adapting suffice to meet the goals of mobile computing?

1.2.1 Transparency

Transparency is the ability of a system to hide some characteristics of its underlying implementation from users. Much of the research effort in distributed computing has been devoted to developing mechanisms for providing various forms of transparency. Examples of these include the following:

- *Access transparency* is the ability of a system to hide the differences in data representation on various machines and the mode of access of a particular resource.

- *Location transparency* is the ability of a system to conceal the location of a resource. Related to location transparency are *name transparency* (which ensures that the name of a resource does not reveal any hints as to the physical location of the resource) and *user mobility* (which ensures that no matter which machine a user is logged onto, she should be able to access resources with the same name).

- *Failure transparency* is the ability of the system to hide failure and recovery of a system component.

Mobile computing systems can be viewed as a form of distributed system, and attempts can be made to provide "mobility transparency," which would encompass the transparencies just mentioned. This would, in essence, support application-transparent adaptation. But is this an achievable, or even desirable, goal for building mobile computing systems and applications? Let us look closely at the characteristics of the mobile computing environment and their implications in these regards.

1.2.2   Constraints of mobile computing environments

Mobile computing has many constraints that distinguish a *mobile computing environment* (MCE) from the traditional desktop workstation/PC–based distributed computing environment. Notable among these are the following (Satyanarayanan, 1996a, 1996b):

1. *Mobile computers can be expected to be more resource-poor than their static counterparts.* With the continued rapid improvement of hardware technology, in accordance with Moore's law, it is almost certain that a laptop computer purchased today is more powerful that the desktop computer purchased just a year or even a few months ago. However, mobile computers require a source of electrical energy, which is usually provided by battery packs. Since batteries store a finite amount of energy, they need to be replaced or recharged. The first option costs money, and the second option, although cheaper in terms of money expended, requires plugging in the computer for recharging, restricting mobility. This has an impact on the design of mobile computers— all the hardware and software components in mobile computers are designed to reduce energy consumption and to increase the lifetime of the batteries. For example, processors on mobile computers are designed to consume less energy and, consequently, achieve a lower computation performance.

2. *Mobile computers are less secure and reliable.* Since mobile computers accompany their users everywhere, they are much more likely to be lost or stolen. Furthermore, they are more likely to be subjected to hostile or unfriendly use, e.g., a child throwing his daddy's PDA in a tantrum.

3. *Mobile connectivity can be highly variable in terms of its performance (bandwidth and latency) and reliability.* Disconnections, both

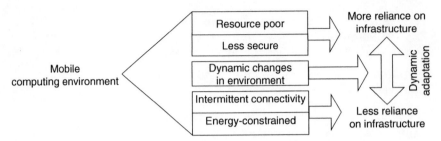

Figure 1.1   Need for dynamic adaptation in mobile computing environments.

voluntary and involuntary, are common. The link bandwidth can vary by orders of magnitude over time and space.

Thus, in general, resource availability and quality vary dynamically.

The preceding characteristics of a mobile computing environment require rethinking about how mobile applications and systems should be designed. Resource paucity and the lower reliability of mobile devices point toward designing systems in such a manner that more reliance is placed on the static infrastructure. On the other hand, the possibility of disconnections and poor connectivity point toward making systems less reliant on the static infrastructure. Further, as mobile devices are moved around (or even if they are not), their situations keep changing over time. Mobile devices should change their behavior to be either more or less reliant on static infrastructure depending on current conditions. Figure 1.1 illustrates this need for dynamic adaptation in a mobile computing environment.

1.2.3   Application-aware adaptation

Who should be responsible for adaptation—the application, the system, or both? There are two extreme approaches to designing adaptive systems: application-transparent (the system is fully responsible for adaptation) and laissez-faire (the system provides no support at all) (Satyanarayanan, 1996a). Obviously, the laissez-faire approach is not desirable because it puts too much burden on the application developer. Further, no support from the underlying system restricts the types of adaptations that can be performed. However, as the following example points out, the application-transparent approach is not sufficient either. Consider two different multimedia applications. In one you are video-conferencing using a mobile device, and in the other you are watching a live video stream from a remote server on your mobile device. Now consider the following scenarios. In one, you move from an area with sufficient bandwidth for your application to an area where the amount of bandwidth is less than that needed by your application. In the other, your laptop's battery power level drops considerably. Both scenarios deal with changes in availability of resources. How would you like your system/application to behave under each scenario?

In the application (user)–transparent approach, the system/application may behave the same irrespective of the application running. However, different responses may be suitable depending on the type of application that is running. For example, in the first scenario, a non-adaptive system may just do nothing and let the audio/video quality drop. In the second scenario, the system may just give a warning to the user without any assistance on how to deal with the situation. In an adaptive system, various behaviors can be envisioned. For example, the system may try to do its best in both situations. However, the system's adaptation does not take into account the kind of application that is running. For example, in the first scenario, the system may try to adapt by requesting that the server or other peers start to send lower-quality video, in effect requiring lower bandwidth. In the second scenario, the system may try to conserve energy by reducing the intensity of the backlight of the display (besides warning the user of the lower battery power level). A still more adaptive approach is possible in which the system interacts with the user/application in deciding how to adapt.

In the application-transparent approach, the responsibility of adaptation lies solely with the underlying system. On the other hand, in the application-aware approach, the application collaborates with the underlying system software. Here the system provides status information about the available resources. The application uses this information to make decisions on how to adapt to changes in the resource availability. Each application can adapt in its own way. Figure 1.2 illustrates the spectrum of adaptation strategies that are possible (Satyanarayanan, 1996a).

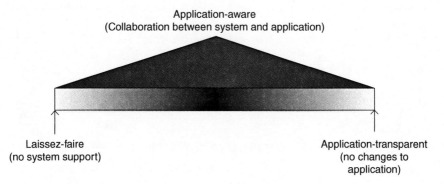

Figure 1.2 The spectrum of adaptation strategies (Satyanarayanan, 1996a).

1.3 Mechanisms for Adaptation

What can be adapted? As we will see in this section, both the functionality of various components in the mobile application and the data that are delivered to the application can be adapted. The next question is, "How to adapt?" In the context of the *client-server* (CS) model, functionality can be adapted by varying the partition of duties between the client and the server; e.g., during disconnection, a mobile client works autonomously, whereas during periods of strong connectivity, the client depends heavily on the fixed network, sparing its scarce local resources. Next we look at these approaches in more detail.

1.3.1 Adapting functionality

The first approach is to change dynamically the functionality of the computational entities involved in response to the change in the operating conditions. An example of this approach is the extended CS model (Satyanarayanan, 1996a). The CS paradigm is the most widely used architecture for distributed computing. In the standard CS model, the roles of the client and server are defined usually at design time, and these remain fixed during operation of the system (run time). Typically, a small number of servers provide some services such as access to databases, Web pages, allocation of temporary IP addresses, and name translation to a usually larger group of clients. A client, or the underlying system—middleware—may select dynamically the server from which to request the service. A server may or may not maintain information, or state, about the clients to which it is providing service. The state information may be maintained as soft or hard. Soft state information, once installed, has to be updated periodically to avoid automatic deletion by the state maintainer (in our case, a server), whereas hard state information, once installed, requires explicit deletion. Soft state is useful in systems with very dynamic configurations, such as mobile systems. This is so because soft state requires no explicit action to make the state information consistent with dynamic changes in the system. Soft state is used in various protocols, such as the Resource Reservation Protocol (RSVP) and the Internet Group Management Protocol (IGMP) to adapt gracefully to dynamic changes in the system state. Specifically, in the case of data servers (such as file servers), the CS model (as implemented by Coda) has the following characteristics (Satyanarayanan, 1996a):

- A few trusted servers act as the permanent safe haven of the data.
- A large number of un-trusted clients can efficiently and securely access the data.

- Good performance is achieved by using techniques such as caching and prefetching.
- Security of data is ensured by employing end-to-end authentication and encrypted transmission.

Developers of the Coda file system point to the following advantages of the CS model: It provides good scalability, performance, and availability. "The CS model decomposes a large distributed system into a small nucleus that changes relatively slowly, and a much larger and more dynamic periphery of clients. From the perspective of security and system administration, the scale of the system appears to be that of the nucleus. From the perspective of performance and availability, a client receives service comparable to stand-alone service" (Satyanarayanan, 1996a).

Impact of mobility on the CS model. The CS model permits a resource-poor mobile client to request a resource-rich server to perform expensive computations on its behalf. For example, the client can send a request to the server, go to sleep to conserve energy, and later wake up to obtain the result from the server. For the sake of improved performance and availability, the boundary between the clients and servers may have to be adjusted dynamically. This results in an *extended CS model*. In order to cope with the resource limitations of clients, certain operations that normally are performed at the client may have to be performed by resource-rich servers. Such lean clients with minimal functionality are termed as *thin clients*. Conversely, the need to cope with uncertain connectivity requires the clients sometimes to emulate the functions of the servers. This results in a short-term deviation from the classic CS model. However, from the long-term perspective of system administration and security, the roles of servers and clients remain unchanged.

1.3.2 Adapting data

Another way to adapt to resource availability is by varying the quality of data (fidelity) that is made available to the application running on the mobile client. *Fidelity* is defined as the "degree to which a copy of data presented for use at the client matches the reference copy at the server" (Noble et al., 1997). This kind of adaptation is extremely useful in mobile information access applications. The QoS requirements for such applications are

- *Information quality.* "Ideally, a data item being accessed on a mobile client should be indistinguishable from that available to the application if it were to execute on the server storing the data." (Noble et al., 1997).

- *Performance*
  - *From the mobile client's perspective.* Latency of data access should be within tolerable limits.
  - *From the system's perspective.* Throughput of the system should be maximized.

In general, it is difficult to provide both high-performance and highest-quality information in a mobile computing environment. In some cases, information quality can be traded for increased performance. The basic idea behind data adaptation is as follows: Assume that any data item accessed by a mobile client has a *reference copy* which is maintained at a remote server. The reference copy of a data item is assumed to be complete and up-to-date. At times when resources are plentiful, the mobile client directly accesses and manipulates the reference copy. However, whenever resources are scarce, the mobile client may choose to access or manipulate a data item of lower fidelity, and consequently, consume fewer resources.

Fidelity and agility.   Data fidelity has many dimensions depending on its type. Consistency is a dimension which is shared by all data types. The other dimensions are type-dependent (Noble et al., 1997):

- *Video data*—frame rate and image quality
- *Spatial data such as topographic maps*—minimum feature size
- *Telemetry data*—sampling rate and timeliness

Various dimensions of data fidelity can be exploited for adapting to mobility. For example, a mobile client can choose to use the locally cached stale copy when it is disconnected from the server and a possibly more current copy when the server is accessible. Fidelity of data can be changed in several ways. This requires knowledge of data representation. For example, a video stream can be degraded by reducing the frame rate, reducing the quality of individual frames, or reducing the size of individual frames. Another point to note is that different applications using the same data may exploit different trade-offs among dimensions of fidelity. For example, a video editor may choose to slow the frame rate, whereas a video player may choose to drop frames. When developing different strategies for trading off data fidelity dimensions against performance, an issue that arises is how to determine which strategy is better. Developers of the Odyssey system (a middleware for application-aware adaptation developed at Carnegie Mellon University) have evaluated their system using agility as a metric.

Agility is defined as the speed and accuracy with which an adaptive application detects and responds to changes in its computing environment, e.g., change in resource availability (Noble et al., 1997). The larger the change, the more important agility is. For example, an adaptive system that tries to adapt to the availability of connection bandwidth can try to determine how well the system reacts to sudden changes in bandwidth. One issue is how to model changes in the environment. The developers of Odyssey have used reference waveforms—step-up, step-down, impulse-up, and impulse-down—to model variation in wireless bandwidth availability. These waveforms were generated using a trace modulation technique that emulates a slower target network over a faster wired local area network (LAN)—in this case, a wireless LAN over a faster wired LAN. In general, the results obtained from such studies should be interpreted by keeping in mind that an adaptation strategy is strictly better than another if it provides better fidelity with comparable performance or better performance with comparable fidelity. Further, the comparison must take into account the application's goals. The interested reader should refer to Noble et al. (1997) for a detailed performance study of an adaptive system.

1.4 How to Develop or Incorporate Adaptations in Applications

In general, it is difficult to enumerate all the mechanisms that can be employed to construct adaptive programs. However, it should be clear intuitively that all adaptive programs must adapt to some detectable *change* in their *environment*. Either a program can implement its own mechanisms to detect the changes, or mechanisms may be provided by some other entity—a middleware layer or operating system—to make the program aware of these external changes. In general, we can view these entities as software sensors (as opposed to hardware sensors, which we will come across later in this book). For example, a Transmission Control Protocol (TCP) client adapts its transmission window size by indirectly monitoring the congestion level in the network. Conceptually, it maintains a software timer for each packet sent, and as long as it receives an acknowledgment for a packet before its timer expires, it keeps increasing the size of its transmission window up to a maximum allowable window size. However, in the event of a loss—timeout or receipt of triple acknowledgment for a packet—it assumes that the loss is due to a buildup of congestion in the network, so it backs off by reducing its transmission window size. As a side note, this behavior is not suitable for wireless networks because the packet loss may be due to the high error rate in a wireless link on the delivery path or may be because an endpoint has moved. In such cases, the TCP client should not back off but instead should continue trying to push the packets through the network. Many techniques have been developed to "adapt" TCP for wireless networks, e.g., TCP-snoop (Balakrishnan, Seshan, and Katz, 1995).

In the state-based approach, changes in mobile computing are viewed as state transitions. Irrespective of how the state of the environment is sensed, the adaptation of function and/or data can be performed when a state transition occurs. Logically, each system state corresponds to an *environmental state*. Each system state is associated with some appropriate functionality. As long as the environment remains in a particular state, the system behaves according to the functionality associated with that state. When the environment's state changes, the system may have to perform some bookkeeping functions associated with state transition before assuming the functionality associated with the new state. In order to perform these operations, the system may have some additional states.

For example, consider the functionality adaptation in the Coda (Continued data availability) distributed file system developed at Carnegie Mellon University. Coda is designed to maximize the availability of data at the expense of possible access to stale data. Each Coda client (called *Venus*) maintains a local cache. Venus adapts its functionality based on the state of the connectivity between the client and the server. Venus uses the following four states:

- *Hoarding.* Venus is in the hoarding state when it has *strong connectivity* with the server. In this state, the client aggressively prefetches files from the server to store locally. Files to be prefetched are decided on the basis of user preference and access pattern.

- *Emulating.* Venus is in the emulating state when it is *disconnected* from the server. In this state, the client emulates the server by optimistically allowing both read and write access to local files. To later update the primary copy of the files on the server and to detect any conflicting updates, the client maintains a log of all file operations.

- *Write-disconnected.* Venus is in the write-disconnected state when the client has *weak connectivity* to the server. In this state, a Coda client decides whether to fetch files from the server or to allow local access.

- *Reintegration.* Venus enters this state when the connectivity improves to *strong connectivity*. In this state, Venus resynchronizes its cache with the accessible servers. The log of operations is used for this purpose. If any conflict is detected, then user assistance may be required. On completion of resynchronization, Venus enters the hoarding state.

Note that the first three states correspond to some environmental state but that the last state corresponds to an environmental state transition. The state-transition diagram for Venus is shown in Fig. 1.3.

1.4.1 Where can adaptations be performed?

In a distributed application, in particular, a CS application, the adaptation can be performed at the client, at the server, or at both the client and the server. Further, there are additional possibilities. The adaptation

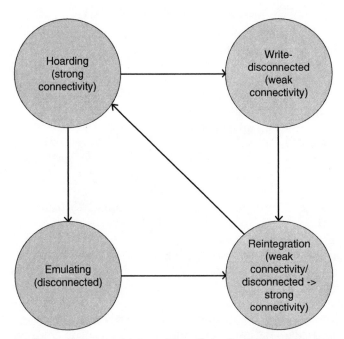

Figure 1.3 State transition used by a Coda client (Venus) to adapt its functionality to changes in connectivity to a server (Venus states hoarding, write disconnected, emulating and reintegration correspond to strong connectivity, weak connectivity, and no connectivity (disconnected), respectively). The reintegration state corresponds to transition to strong connectivity from no/weak connectivity link state.

also can be performed within the network, say, at an intermediate software entity called a *proxy*. For example, consider a typical CS application. Several different adaptations may be performed on different components located at different points in the data and control path between the client and the server:

- *Adapting to the hardware / software capabilities of the mobile device*—in the proxy and/or at the server

- *Adapting to the connectivity of the mobile device*—at the server and/or the client

- *Adapting to the resource availability at the mobile device*—at the client

Let us look at some concrete examples to get a better understanding of incorporating adaptations in mobile applications.

Proxies.   Proxies have been used by many applications to perform various tasks, such as filtering data and connections (e.g., security firewalls) and modifying control data (e.g., *network address translators* [NATs] change the IP fields). Of particular interest to data adaptation are transcoding proxies. Transcoding is the process of converting data objects from one representation to another. Transcoding proxies can be used to adapt to various situations dynamically, such as the availability of bandwidth and capabilities of the end device. For example, if the end device is not capable of handling full-motion video, a transcoding proxy may convert it to a form that can be displayed on the end device (Han et al., 1998).

Conceptually, a transcoding proxy may be viewed as consisting of three modules: (1) an *adaptation-policy module* (2) a *data (content) analysis module*, and (3) a *content-transformation module*. Figure 1.4 shows the architecture of the transcoding proxy developed at the IBM T. J. Watson Research Center (Han et al., 1998). The adaptation-policy module can take as input information such as the server-to-proxy bandwidth (B_{sp}), proxy-to-client bandwidth (B_{pc}), client device capabilities (e.g., video display capabilities), user preferences, and content characteristics (provided by content-analysis module). Based on these inputs, it can decide on whether and how to modify the content. The content-transformation module performs the actual modification.

Figure 1.5 presents an example of logic that can be implemented in a transcoding policy module. The *transcoding threshold* is the document size beyond which transcoding becomes beneficial. To get an understanding of this parameter, consider the following simple cost-benefit analysis. Assume that the proxy uses the store-and-forward mechanism for delivering documents to the client. This is to say, the request from the client first is submitted to the proxy, the proxy then obtains the entire document from the server, and finally, it forwards the document to the client. Further, assume that the goal is to minimize the latency of document retrieval. Following the analysis in Han et al. (1998), the total delay for document retrieval without the proxy (D_{sc}) and with the proxy (D_{spc}) is

$$D_{sc} = 2 \times \text{RTT}_{pc} + 2 \times \text{RTT}_{sp} + S/\min(B_{pc}, B_{sp})$$

$$D_{spc} = 2 \times \text{RTT}_{pc} + 2 \times \text{RTT}_{sp} + D_p(S) + S/B_{sp} + S_p(S)/B_{pc}$$

where RTT_{pc} is the round-trip delay between the proxy and the client, RTT_{sp} is the round-trip delay between the server and the proxy, S is the document size, $D_p(S)$ is a proxy delay function that relates the proxy processing delay to the document size, and $S_p(S)$ is an output size function that relates the transcoded document size to the input document size.

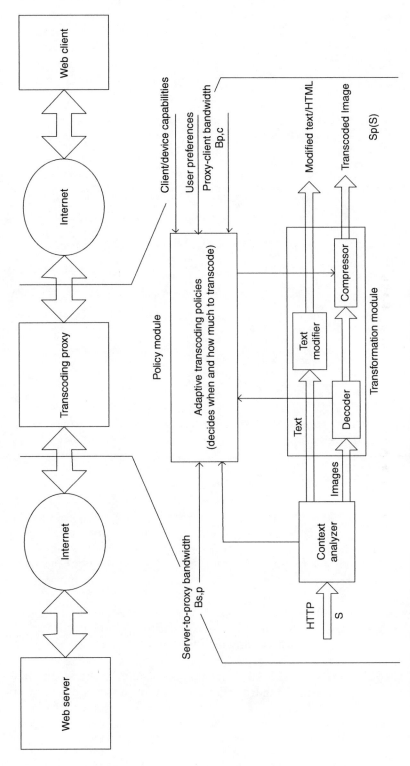

Figure 1.4 Dynamic adaptation in IBM's transcoding proxy (Han, Bhagwat, and LaMaire, 1998).

```
Switch(Client_Type)
 Case Laptop/PC:
  If (Input_Size > TranscodingThreshold)
   Switch(Input_type)
       Case GIF:
     If(Input is well-compressed GIF)
       TranscodedImage = T(Input, GIF, GIF, user preference)
     Else
       TranscodedImage = T(Input, GIF, JPEG, user preference)
     End If
       Case JPEG:
           TranscodedImage = T(Input, JPEG, JPEG, user preference)
   End Switch
  End If
  If (Output size > input Size)
           Send Input Image to Client
  Else
           Send TranscodedImage to Client
  End If
 Case Palm PDA:
       TranscodedImage = T(Input, Any, 2-bit_grayscale, user preferences);
       Send TrancodedImage to Client
End Switch
```

Figure 1.5   Example of a transcoding policy module (The example is based on the code fragment of a policy module in [Han transcoding proxy 1998]). T (input object, input object type, output object type, user preferences) performs data transformation on the <input object> of type <input object type> into <output object> of type <output object type>.

Obviously, using transcoding is better when $D_{spc} < D_{sc}$. From this, a lower bound (*transcoding threshold*) on the input document size can be obtained to be

$$S > [D_p(S) + S_p(S)/B_{pc}]/(1/B_{pc} - 1/B_{sp})$$
$$= \text{transcoding threshold} \tag{1.1}$$

Note that the server-to-proxy bandwidth and the proxy-to-client bandwidth correspond with the bottleneck bandwidth along the data path from the server to the proxy and the proxy to the client, respectively. These bandwidths can vary dramatically as the user moves and with fluctuations in the available wireless bandwidth. Equation (1.1) can be used to adapt dynamically to changes in available bandwidth. However, for this to work, the transcoding proxy would need a good bandwidth estimator of the bandwidth that might be available in the near future.

Transcoding proxies also can work in the streamed mode. In this mode, the data stream is modified and passed on to the client as it is obtained from the server. For an analysis of when adaptive-streamed transcoding is beneficial and other details about transcoding proxies, see Han et al. (1998).

WebExpress: an example of adaptation using proxies. Browsing over wireless networks can be expensive and slow owing to pay-per-minute charging (in cellular networks) and the characteristics of wireless communication. Additionally, the Hyper-Text Transport Protocol (HTTP) was not designed for wireless networks and suffers from various inefficiencies—connection overhead, redundant transmission of capabilities, and verbosity. WebExpress (Housel, Samaras, and Lindquist, 1998), developed by researchers at IBM, significantly reduces user cost and response time in wireless communication by intercepting the HTTP data stream and performing various optimizations on it. It is aimed at enabling routine commercial applications on mobile computers.

WebExpress uses proxies called *intercepts* that allow WebExpress to be used with any Web browser and any Web server. They enable WebExpress to intercept and control communications over the wireless link for the purpose of reducing traffic volume and optimizing the communication protocol, reducing data transmission. As shown in Fig. 1.6, the WebExpress architecture consists of two components that are inserted into the data path between the Web client and the Web server—*client-side intercept* (CSI), also known as *client-side proxy*, and *server-side intercept* (SSI), also known as *server-side proxy*. CSI is a process that runs in the end-user client mobile device, whereas SSI is a process that runs

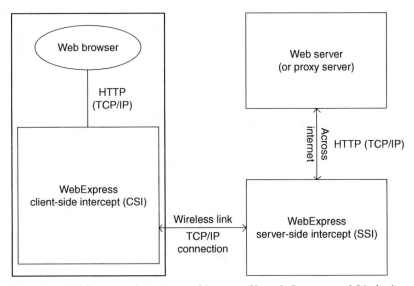

Figure 1.6 WebExpress adaptation architecture (Housel, Samara, and Lindquist, 1998).

within the wireline network. One of the features of this client-proxy-server model, also called the *intercept model*, is that the proxies are transparent to both Web browsers and servers. This makes this adaptation technique insensitive to the evolution of technology. This is a very important advantage because HTML/HTTP technology was (and still is) maturing rapidly when WebExpress was developed. Another advantage is the highly effective data reduction and protocol optimization without limiting browser functionality or interoperability. WebExpress employs several optimization techniques such as *caching, differencing, protocol reduction,* and *header reduction.*

■ *Caching.* WebExpress supports both client and server caching using the *least recently used* (LRU) algorithm. Cached objects persist across browser sessions. Caching reduces the volume of application data transmitted over the wireless link through cross-browser sessions. We will look at the details of this in Chap. 3.

■ *Differencing.* Caching techniques do not help in common graphic interface (CGI) processing, where each request returns a different result, e.g., a stock-quote server. However, different replies from the same program (application server) are usually very similar. For example, replies from a stock-quote server contain lots of unchanging data such as graphics. For each dynamic response from a CGI (HTML file) cached at the SSI, the SSI computes a base object for the page before sending it to the CSI. If the SSI receives a response from the CGI server and the cyclic redundancy check (CRC) received does not match the CRC of the base object, the SSI returns both the difference stream and the base object. This is called a *basing operation* in WebExpress parlance. *Rebasing* is carried out in the same fashion when the SSI detects that the difference stream has grown beyond a certain threshold.

■ *Protocol reduction.* Repeated TCP/IP connections and redundant header transmissions present additional overhead. The WebExpress system uses two techniques to reduce this overhead and optimize browsing in a wireless environment:

■ *Reduction of TCP/IP connection overhead using virtual sockets.* WebExpress establishes a single TCP/IP connection between the CSI and the SSI. The CSI sends requests over this connection. The SSI establishes a connection with the destination server and forwards the request. Thus overhead is incurred between the SSI and the Web server but not over the wireless link. *Virtual sockets* are used to provide multiplexing support. Virtual sockets are implemented in the following manner: Data sent are prefixed by a virtual socket ID, command byte, and a length field. At the CSI, the virtual socket ID is associated with a real socket at the browser. At the SSI, the virtual socket ID is mapped to a socket connection at an HTTP server. This mechanism permits efficient transport of HTTP requests and responses while maintaining protocol transparency.

- *Reduction of HTTP headers.* HTTP request headers containing lists of Multipurpose Internet Mail Extensions (MIME) content types can be hundreds of bytes in length. The CSI allows this information to flow in the first request and then saves the list. On subsequent requests, the CSI compares the received list with the saved one. If the two lists match, the MIME content-type list is deleted from request. The SSI inserts the saved one if none is present.

The transparent proxy–based architecture of WebExpress allows the operation of commercial Web applications on wireless networks. Differencing and virtual sockets offer the most critical optimizations in the WebExpress system.

1.5 Support for Building Adaptive Mobile Applications

Adaptations should be customized to the needs of individual applications. We have argued that applications are in a better position to perform application-specific adaptations than the operating system alone. However, what does this mean with regard to where adaptations can be performed? Applications not only make local adjustments, but they also should collaborate with other adaptation technologies that are available in other components of the system. For example, in the CS scenario, both the client and the server may need to adapt. The advantage of application-aware adaptation is that the application writer knows best how the application should adapt. However, does the application writer know the underlying system as well as the application? Furthermore, application-aware adaptation tends to work only at the client or perhaps at the server. If this is the only mechanism for adaptation, without any monitoring on resource usage by each application, this may result in selfish behavior by the applications. In the following sub-sections we look at details of some current efforts to developing adaptive applications. Here, we focus only on the adaptation mechanisms of these systems. We will re-examine some of these systems later in this book from the perspective of the middleware services they provide to mobile applications.

1.5.1 Odyssey

Odyssey (Noble et al., 1997) aims to provide high fidelity and to support concurrent mobile applications with agility. It emphasizes collaboration between applications and the operating system in performing adaptation to handle constraints of the mobile computing environment, especially those imposed by the presence of wireless links.

The following case is a motivating example from the Odyssey group. Imagine a user with a lightweight/wearable mobile computer with ubiquitous wireless access to remote services, an unobtrusive heads-up display, a microphone and earphones, and speech recognition for computer interaction with online language translation. The user has ubiquitous connectivity, for example, owing to an overlay network, but the quality varies as he moves and as different networks are accessed. The user simultaneously gets voice, video, and other data sent to him. When the user moves to a relatively shadowed area and the network bandwidth drops dramatically, Odyssey informs the video, audio, and other applications of these changes, allowing them to make the proper adaptations in their network usage and their behavior. Why use application-aware adaptation here? The presumption is that for this environment, only the application knows what to do because the operating system does not have the application-level knowledge. If the operating system makes the decision, it may do the wrong thing. However, the operating system must be involved to ensure fairness among competing applications. In essence, the basic Odyssey adaptation model is as follows: The operating system support on the portable machine monitors network conditions. Each application interacts with the operating system tools to negotiate services. When network conditions change, the operating system notifies the applications of what has happened.

As shown in Fig. 1.7, the Odyssey architecture consists of two main components—*viceroy* and *warden*. Viceroy performs centralized resource management and monitors the availability of resources, notifying applications of changes. Wardens provide data-type-specific operations, namely, tsop() functions, to change the fidelity. A tsop() function is similar to the transcoding proxies we saw earlier but specific to a data type. Wardens are also responsible for communicating with servers and for caching data.

In the Odyssey application-adaptation model, applications do not interact directly with their remote servers. Applications talk to their wardens, and wardens talk to the servers. Applications tend to have limited roles in actually adapting transmissions. They may know about different formats and tolerances and accept data in their different adapted versions.

Applications interact with Odyssey to adapt their behavior. All data to and from the server flow through Odyssey. Applications must register their preferences and needs with Odyssey in the form of requests. A request specifies that an application needs a particular resource within certain limits, e.g., between 100 kbps and 1 Mbps of bandwidth. If the request can be satisfied currently, it is. If things change later, the

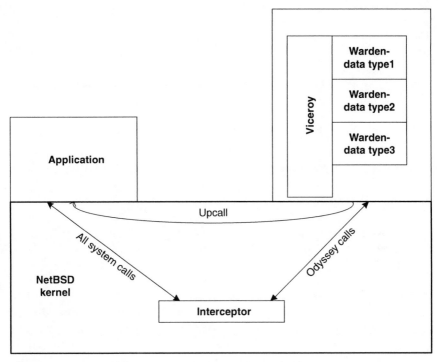

Figure 1.7 Odyssey architecture (Satyanarayanan, 1996b).

application is notified using an upcall to the applications. In response, the application can adjust itself and make another request. Note that upcalls can occur because resource availability became either worse or better.

The following example illustrates a typical interaction between an application and Odyssey. A video application requests enough bandwidth to receive a color video stream at 20 frames per second (fps). Odyssey responds, "No way, try again." The application retries and requests bandwidth sufficient for 10 fps in black and white, specifying the minimum and the maximum needed for this application so that it can improve the quality if it is worthwhile. The channel subsequently gets noisy, and the bandwidth drops. Odyssey performs an upcall informing the application that the bandwidth is outside the specified limits. The application requests a lower bandwidth suitable for a lower fps rate.

Odyssey wardens mediate server-application interaction. A warden is a data-type-specific module capable of performing various adaptations on that data type. Additionally, wardens do caching and prefetching. If Odyssey needs to handle a new data type, not only does the application

need to be altered, but a new warden also has to be written. The better
the warden understands the data type, the better is the potential adap-
tivity. The Odyssey viceroy is the central controlling facility that handles sharing of resources. The viceroy notices changes in resource
conditions. If these changes exceed preset limits, the viceroy informs the
affected applications using the upcall mechanism. Developers of
Odyssey have evaluated their system to answer the following questions:

How agile is Odyssey in the face of the changing network bandwidth?

How beneficial is it for applications to exploit the dynamic adaptation
made possible by Odyssey?

How important is centralized resource management for concurrent
applications?

Interested readers should refer to Noble et al. (1997) for details.

1.5.2 Rover

Rover is an object-based software toolkit for developing both *mobility-
aware* and *mobility-transparent* CS distributed applications (Joseph
et al., 1995, 1997). It provides application developers with two programming and communication abstractions specifically designed for
assisting applications in harsh network environments such as mobile
computing—*relocatable dynamic objects* (RDOs) and *queued remote procedure calls* (QRPCs). RDOs can be used to reduce interaction between
two weakly connected entities, such as a client on the mobile device
and a server in the wireline network. Rover RDOs are objects with
well-defined interfaces and are loadable dynamically from the server to
the client. This, in essence, moves objects to the client machine, avoiding the client having to communicate with the object at the server.
QRPCs can be used to handle disconnections. Rover QRPCs are essentially nonblocking *remote procedure calls* (RPCs) that support *split-
phase operations*. That is, they allow an application to make an RPC
without worrying about whether the destination is currently reachable.
If the destination of the RPC is not reachable at the time of the call, the
call is queued. On reconnection to the RPC's destination, the RPC is performed. The result of the RPC is delivered asynchronously to the application. Figure 1.8 illustrates the various components in the distributed
CS object system of the Rover toolkit and the control flow within the
toolkit.

- How does one use Rover? By writing and perhaps rewriting the application using the toolkit—invoking these tools when network connectivity is poor. However, this requires a good understanding of network
  programming and user mobility. To use RDOs, applications—usually

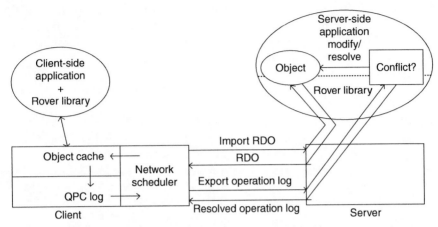

Figure 1.8 Adaptation mechanisms provided in Rover Toolkit (Joseph et al., 1995).

clients—import RDOs from the server. They then invoke methods on imported RDOs. When finished, RDOs are exported back to the server. If needed, Rover also can cache copies of objects. This allows clients to use the cached copy instead of fetching the original from the server. Updates to objects are handled by an optimistic CS replication method. RPCs are queued only at the client, where they are stored until the client can handle them. When the appropriate level of connectivity is established, Rover clears the queue intelligently using an RPC prioritization mechanism. It also can batch related requests. If the client is not available when the response comes back, the server drops the response. The rationale is that queued requests at the client eventually will be replayed. This may cause some inefficiency, but it simplifies application design. A mobile application can employ QRPCs in various situations dynamically to optimize cost to the user or the performance of the application (Reiher, 1998).

- *Optimize use of expensive links.* Consider a situation in which the mobile user pays for wireless connectivity based on the duration of usage, e.g., pay-per-minute plans for cellular phones. To optimize the monetary cost to the mobile user, QRPC can batch several requests and then disconnect from the network after invoking all the batched QRPC calls in a single connection.

- *Make use of asymmetric links.* The queued RPC requests are not associated with a particular network interface. Thus responses can be obtained over any network device. This permits, for example, an application to launch requests over expensive links and to receive responses over cheaper links. This is beneficial for situations where the request size is much smaller than the expected response size (e.g., in Web browsing) and when the response can be used incrementally, as it arrives.

■ *Stage messages near their destination.* RPC queuing can be arranged to occur just before a "bad" link. If the link quality improves, the RPC queue will then be cleared out. Meanwhile, the transmitter does not get blocked on the bad link.

1.6 Summary

In this chapter we discussed the limitations of mobile computing environments. In order to cope with these limitations, mobile applications have to be adaptive. Adaptations can be performed either by changing functionality or by changing the data provided to the application. The kinds of applications we considered in this chapter mostly fall under the domain of mobile information access. Many of these applications involve downloading different kinds of information. For example, Web browsing involves downloading HTML multimedia documents, database access involves accessing different kinds of databases, and video streaming applications involve downloading of continuous video and audio streams. We saw what kind of adaptations can be performed to enable these applications to continue working in mobile computing environments. Furthermore, many of these applications are based on the CS model. In order to enable adaptation, this model is extended.

Once we established the need for adaptation and the mechanisms that can be used for performing adaptations, we looked at what should be responsible for it—the system or the application itself. Approaches to developing adaptive applications can be categorized as completely internal to the applications, layered outside the application, using special operating system features and libraries, and interacting with other mechanisms such as intelligent use of proxies. We studied various approaches to incorporating adaptation in an application, such as using ad hoc methods to achieve the needs of the application, using tools specific to applications (e.g., Web browser proxy method), and using general adaptation tools such as toolkits and operating system features. However, in the current state of the art, most mechanisms only work at the client side, some work at the server, and a few help with anything in between. The main question for future work in this direction is, "How can an adaptive application obtain finer control over the entire path between its distributed components?" (Reiher, 1998).

1.7 References

Balakrishnan, H., S. Seshan, and R. H. Katz, "Improving Reliable Transport and Handoff Performance in Cellular Wireless Networks," *ACM Wireless Networks* 1(4):469, 1995.
Han, R., P. Bhagwat, R. LaMaire, et al., "Dynamic Adaptation in an Image Transcoding Proxy for Mobile Web Browsing Personal Communications," *IEEE Personal Communications* 5(6):8, 1998 (see also *IEEE Wireless Communications*).

Housel, B., G. Samaras, and D. Lindquist, "WebExpress: A Client/Intercept Based System for Optimizing Web Browsing in a Wireless Environment," *Mobile Networks and Applications* 3:419, 1998.

Joseph, A., A. F. deLapinasse, J. A. Tauber, et al., "Rover: A Toolkit for Mobile Information Access," Proceedings of the fifteenth ACM symposium on Operating Systems Principles SOSP'95, Cooper Mountain, Colorado, December 1995.

Joseph, A., A. F. deLapinasse, J. A. Tauber, et al., "Mobile Computing with Rover Toolkit," *IEEE Transactions on Computers* 46(3):337, 1997.

Noble, B., M. Satyanarayanan, D. Narayanan, et al., "Agile Application-Aware Adaptation for Mobility," Proceedings of the sixteenth ACM symposium on Operating Systems Principles SOSP'97, December 1997.

Reiher, P., Lecture notes CS239 Hot Topics in Operating Systems, UCLA, 1998.

Satyanarayanan, M., "Fundamental Challenges in Mobile Computing," *Proceedings of the Fifteenth Annual ACM Symposium on Principles of Distributed Computing (PODC),* Philadelphia, PA, 1996a.

Satyanarayanan, M., "Mobile Information Access," *IEEE Personal Communications* 3(1): XX, 1996b.

Mobility Management

In modern times, people are much more mobile—relocate more often—than in olden days, yet we are expected to be always reachable. The average American family moves once every four to five years. In addition, most of us have many credit cards in our wallets. To ensure that we continue to receive our bills on time (and other letters, such as junk mail), we do the following: (1) we inform the post office of our new address, and (2) we inform the credit card companies of our new address. The post office forwards letters arriving at the old address to our new address for a period of one year (only 60 days for periodicals), so any credit card bills sent to you before you inform the credit card company of your new address will reach you. The credit card companies send any future bills to your new address. And in time, you manage to inform all the important senders of your new address, and the forwarding service ends up being used mainly for junk mail. This example highlights various tasks that need to be performed when a receiving end changes its address.

We now begin our journey into mobility management schemes for the mobile computing world in which a mobile node changes its physical location (address) at a much smaller timescale. It is assumed that you are familiar with the basics of wireless communication and networks. Readers not familiar with the basics of wireless communication and networks can refer to Appendix A for a brief overview.

2.1 Mobility Management

To ensure that a mobile node m is able to communicate with some other node n in a network, the networking infrastructure has to ensure that (1) m's location (e.g., its access point [AP] in wireless local area networks [WLANs] and base stations in cellular networks) can be determined so that a route can be established between m and n, (2) when m moves out of the range of the current AP (henceforth, we use the two terms, access point and base station, interchangeably), it establishes a connection with another AP, and (3) the connection/data packets are rerouted correctly to the new AP. The first task, which requires maintaining the current location of every mobile node in the network, is known as *location management*. Conceptually, any location management scheme consists of two operations: *search* and *update*. The search operation is invoked by a node that wants to establish a connection with a mobile node whose location it currently does not know. The update operation, also known as a *registration operation*, is performed to inform the system of the mobile node's current location. The update operation helps in making the search operation more efficient. For example, if location updates are never performed for a mobile node m, then locating m may involve paging all the mobile nodes in the network with the message, "If you are m, then please report your location." Paging can be very expensive both for the network and for the mobile nodes. However, if the location update operations are performed very frequently, then the volume of these operations may overwhelm the location management system. In general, the overhead (cost) of search operations depends on the granularity and currency of location information, the structure of the database that stores the location information, and the search procedure.

Location information can be maintained at various granularities. In a cellular system, for example, the finest granularity at which location can be (and needs to be) maintained is a *cell*. This would require a mobile node to update its location whenever it moves from one cell to another. However, if the location information is maintained at a coarser granularity, say, in an area consisting of certain number of contiguous cells, then the search cost increases because a larger number of cells need to be paged to obtain the exact location (cell) of the mobile node each time a call needs to be established. Thus the granularity of location information maintained for a mobile node by the system has an impact on the performance of the location management scheme.

Another important aspect is the organization of the location registrars, databases that store the location information of the mobile nodes. In order to achieve good performance, scalability, and availability in the face of an increasingly mobile population and the need to maintain location information at finer granularities (owing to shrinking of cell sizes to accommodate more users), several database organizations and techniques have been developed to maintain fast-changing location information efficiently and reliably. In Section 2.4 we will describe location management schemes used by Global System for Mobile (GSM) and Mobile Internet Protocol (Mobile IP). We will motivate these schemes by first presenting several location management schemes with different desirable characteristics and organization. We also will discuss some interesting location management schemes that have been proposed to handle the fast-growing mobile population.

Another important mobility management task, which is known as *handoff*, is concerned with ensuring that the mobile node remains connected to the network while moving from one cell to another. This is especially important when a mobile node has several active connections that have in-transit packets. Handoff conceptually involves several subtasks: (1) deciding when to hand off to a new AP, (2) selecting a new AP from among several APs in the vicinity of the mobile node, (3) acquiring resources such as channels, (4) informing the old AP so that it can reroute the packets it gets for this mobile node and also transfer any state information to the new AP. The decision to initiate a handoff (which can be taken either by the mobile node, i.e., *mobile-controlled handoff* [MCHO], or by the AP, i.e., *network-controlled handoff* [NCHO]) may depend on several factors such as (1) the quality of the wireless communication between the mobile node and the AP (as indicated by the *signal-to-noise ratio* [SNR]) and (2) the load on the current AP (if the current base station is running out of communication channels, it may want to switch a mobile node to a neighboring lightly loaded AP). Access technologies such as Code Division Multiple Access (CDMA) permit smooth handoffs (as opposed to hard handoffs) when a mobile node can be in communication with several base stations simultaneously before selecting a base station to which to hand off. The choice of the base station to which to hand off may depend on such factors as (1) the SNR of the beacon signals from these APs, (2) the region the mobile node is expected to move to in the near future, and (3) the availability of resources at the AP. The main resources that need to be acquired in the new cell are the uplink and downlink channels in a connection-oriented circuit-switched network and the address (such as an IP address) in a packet-switched network. The task of allocating (managing) channels is handled by channel allocation schemes. Great care needs to be taken in allocating the channels because this can have an impact on ongoing communications and the potential to satisfy future requests for communication channels. Several sophisticated *channel allocation schemes* have been developed because this is a problem of both considerable complexity and considerable revenue-generation potential (for the mobile telecommunications industry).

Location management assists in establishing new connections to a mobile node, whereas handoffs ensure that the mobile node remains connected to the network. However, there may be several in-transit packets when a mobile node moves from one AP to another. The old AP can forward some of these packets to the new AP for a short duration after the handoff. However, in the long term it may be better to route the packets directly to the new AP. Thus the third major task is to ensure that the packets or connection are routed to the new AP. For connection-less traffic, such as IP datagrams on the Internet, this may just involve informing the sender to use the IP address of the new AP as the destination address. For connection-oriented communication, such as Transmission Control Protocol (TCP) connections on the Internet or communication circuits in Asynchronous Transfer Mode (ATM) or Public Switched Telephone Network (PSTN) networks, this operation becomes much more involved. Since a TCP connection is identified by quadtuple (source IP address, source port number, destination IP address, destination port number), an ongoing TCP connection may break when a mobile node moves and acquires a new IP address. In the case of an ATM or a PSTN circuit, the circuit may have to be reestablished to the new AP. The combined task of handoff and connection rerouting is referred to as *handoff management.*

In summary, *mobility management* consists of: location management and handoff management (Akyildiz, 1998). In the case of mobile telephony, location management is needed to ensure that the mobile node can be located quickly when a new call arrives so that a connection can be established. A call to a mobile node is dropped if the mobile node cannot be reached within a certain time. Location management plays a crucial role in minimizing the number of calls that are dropped. Handoff management is needed to ensure that ongoing calls continue with minimal degradation in quality of service (QoS) irrespective of the mobility of the endpoints (caller/callee) of the connection. For packet communication, location management is needed to inform the sender of the new address of the mobile node so that future packets can be addressed to the current address of the mobile node. Handoff management ensures that the mobile node always remains reachable to receive (or send) any packets and to forward the packets from the old address to the new address.

2.2 Location Management Principles and Techniques

Location management schemes use several databases called *location registrars* to maintain the location and other information, such as preferences and service profile, of mobile nodes. To understand why more than one location registrar may be helpful, let us consider a simple location management scheme that uses a *single-location registrar*, called the *home location registrar* (HLR), to maintain the location information of all the mobile nodes in the network. In this simple location management scheme, the search and update operations are performed as follows (Fig. 2.1):

- The location of a mobile node is maintained at the granularity of a cell, i.e., which cell the mobile node was in when it last registered. For each

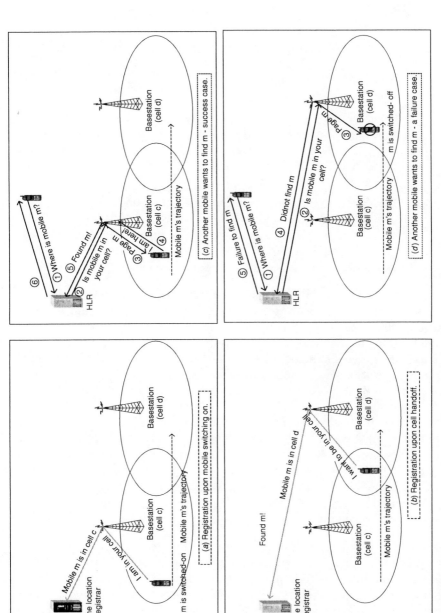

Figure 2.1 Search and registration in basic location management technique.

mobile node m, the HLR maintains a *mobility binding* (m, c), where c is the latest cell (location) of m known to the HLR. The location information of m in the HLR is updated as follows:

- When a mobile node is switched on, the HLR is notified of the current location of m (the cell in which the mobile node is located). As illustrated in Fig. 2.1*a*, the mobile node m's location is sent to the location server. The registration message travels via the base station of the cell to the location server.

- Whenever handoff occurs, the HLR is notified of the cell ID to which m is handing off to. As illustrated in Fig. 2.1*b*, when the mobile node moves to cell d from cell c, the mobile node may decide to register its location to be cell d.

- To find a mobile node m's current location, first the HLR is contacted. The HLR contacts the base station of cell c in the mobility binding for m. The base station pages for mobile m in its cell. If m is in cell c and is switched on, then it can respond to the page message, and connection can be established. Figure 2.1*c* illustrates the messaging between various entities in this location management scheme.

Obviously, if the mobile node is not switched on, the call cannot be established (Fig. 2.1*d*)]. Another scenario in which the system may be unable to establish a call is when the location information provided by the HLR is not the most recent location information for mobile node m. This can happen if m handed off to another cell between the time its location information was obtained from the HLR and cell c paged for it. For example, this would be the case if in the scenario illustrated in Fig. 2.1 mobile node m hands off to cell d just after the mobile node attempting to contact it obtains its location information. As the average time a mobile node stays in a cell before moving to another cell, called the *cell residency time*, decreases, this situation can occur with increasing frequency. In general, the average cell residency time depends on the cell size and the mobility pattern of the mobile node.

In order to decrease the probability of failure in locating a mobile, the preceding location management scheme can be enhanced as follows:

- The mobility binding of a mobile node has two additional pieces of information: t_u and ttl, where t_u is the time when the binding was last updated, and ttl is the *time-to-live* value, which determines how long the binding is valid. The time-to-live entry reduces the chance of trying to contact a mobile that is currently powered off. However, this requires that the mobile node updates its location information periodically, every t_p seconds, where t_p is less than ttl. A side effect of this is that the number of updates performed at the HLR per mobile node is increased dramatically. On the positive side, the search operation can use the fact that the mobile node updates its location every t_p time units (whenever it is on) to reduce the search cost.

- When the mobile node is not found in cell c, a set of cells around cell c is paged. These cells can be paged simultaneously, or an *expanded ring search* can be performed for a maximum of k rings centered at cell c—*the last known location of the mobile node*. Increasing the paging area increases the chance that the search operation will succeed at the expense of using more network resource such as wireless bandwidth and consumption of battery power for processing paging messages. For example, if the speed of the mobile node m is a maximum of v_m cells per second, then k can be set to $v_m \times t_p$.

Periodic updates, also know as *time-based updates*, are an example of *dynamic update schemes* (as opposed to *static update schemes*, an example of which include updates done whenever a mobile node crosses a registration area boundary). We will discuss other dynamic location update schemes later in this chapter.

As mentioned earlier, the periodic updates done by mobile nodes increase the number of updates that a single HLR has to handle. With this increase in the volume of updates, the HLR may become a bottleneck, and the latency of both the search and update operations would increase. Further, if the lone HLR fails, then all the mobile nodes become unreachable. Thus, from the perspective of both performance of search and update operations and resiliency to failures of the location registrar; it is desirable to have multiple location registrars. How should the location information be distributed among these multiple location registrars? Should the location information of a mobile node be replicated among several location registrars? If so, what should be the degree of replication? How should these location registrars be organized? Should a flat (single-tier) organization or a hierarchical (multitier) organization be used? Where should these location registrars be placed in the network? Many of these are challenging optimization problems. For example, where to place location registrars can be formulated as a facilities location problem. In the following, we present some of the solutions to these challenging problems.

2.2.1 Registration area–based location management

Personal communication service (PCS) networks such as GSM use a registration area–based mobility management scheme. The *service area* of a PCS network consists of the set of all the cells (the union of the coverage area of all the cells) belonging to the PCS network. Location and other services are provided only within the service area. The service area is partitioned into several *registration areas* (RAs). Each RA consists of several contiguous communication cells. In the GSM standard, RAs are called *location areas* (LAs). We will use the terms *registration area* and *location area* interchangeably. Figure 2.2 illustrates location update and search operations in the single-location registrar system. Cells c and d are in registration area RA_1 and cell e is in registration area RA_2. Compare this figure with Fig. 2.1. Note that the average update cost has decreased because the HLR is not informed when handoff involves cells belonging to same RA, e.g., when mobile node m moves from cell c to cell d. However, the search cost has increased because all the cells in the registration area have to be contacted when the exact location (cell) of the mobile node needs to be obtained to establish a call.

Cellular systems such as GSM use an RA location management scheme that employs a two-level hierarchy of location registrars to avoid contacting all the cells in the RA to locate a mobile node. Conceptually, a location registrar is associated with each RA. In practice, a location registrar may be in charge of several RAs. Unless mentioned otherwise, we will assume that each location registrar is in charge of a single RA. In the following, we present a model for a two-level location management scheme and discuss some optimizations that can be used to optimize the location management cost by taking into account the call and mobility rates of mobile nodes.

Consider that there are n registration areas $(RA_1, RA_2, \ldots, RA_n)$ in the service area, n location registrars $(LR_1, LR_2, \ldots, LR_n)$, and LR_i is associated with RA_i, where $1 \leq i \leq n$, is registration area i. We will refer to LR_i as the *local location registrar* of RA_i and all other location registrars as *remote location registrars* of RA_i. The rationale for this classification is that accessing a local location registrar is less expensive than accessing a remote location registrar. We will denote the cost of accessing a local location registrar to be A_l and that to access a remote location registrar to be A_r. Consider the hypothetical case where the entire United States is the service area, and each state is an RA. In addition, assume that a location registrar keeps the current mailing address of the residents in its state. Now suppose that Bob, who is a resident of Arizona, wants to find the mailing address of another resident of Arizona. He can simply ask the local Arizona location registrar. However, if Bob wants to locate another person, say, Alice, who is a resident of New York, then he would have to contact New York's registrar, which would take much longer. The farther away Alice is from Arizona, the longer it will take Bob to discover her location. Now suppose that Alice temporarily moves to Texas. An easy way for her to ensure that all her correspondents, including Bob, are able to reach her is by informing her HLR, the New York registrar, that she is currently in Texas. In this way, when Bob

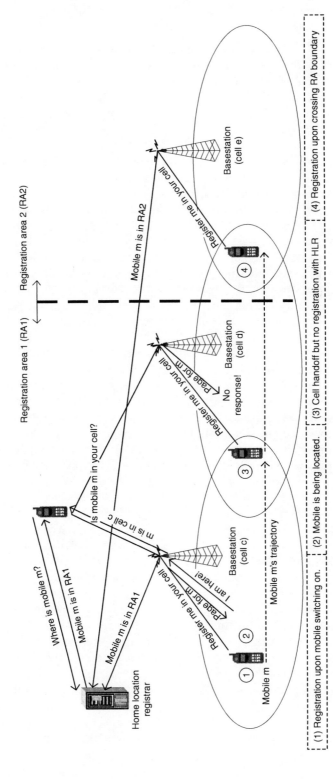

Figure 2.2 Registration area based location management.

contacts the New York registrar for Alice's address, he will be asked to contact her *visitor location registrar*, i.e., the Texas registrar. Bob now will have to contact the Texas location registrar to get Alice's current address. We next discuss the advantage of keeping a forwarding pointer to the Texas registrar at the New York registrar rather than the actual address of Alice.

Forwarding pointers.   In order to highlight the difference between maintaining the actual address of the mobile node at its HLR as opposed to a pointer, we consider the scenario in which Alice's current job in Texas requires her to move quite often within Texas. This would mean that every time she moves, she has to inform the New York registrar (in addition to the Texas registrar) of her new address in Texas. Won't it be easier (less burdensome) for Alice to tell the New York registrar that she is currently in Texas and that the person who wishes to find her current address should contact the Texas registrar for her current address? Maintaining a forwarding location pointer at the New York registrar instead of the actual address reduces the burden on Alice (update cost) but increases the burden on Bob (search cost) because now he has to contact the New York registrar first and then the Texas registrar. Which scheme is better, maintaining the actual address at the home registrar or the location pointer? Assuming a cooperative society in which we want to reduce the overall burden, irrespective of who is burdened, the answer depends on whether Alice moves more often than she is being contacted by some other person. Obviously, if Alice never changes her address while in Texas, it is best to maintain the exact address in the New York registrar. However, if Alice moves every week—the first week in Dallas, the second week in El Paso, the third week in Austin, and the fourth week in Houston (maybe she is an investigative reporter for the *New York Times* on assignment in Texas and living at various places in the state)—and she is contacted only by Bob (from Arizona) once a month, as illustrated in Fig. 2.3, it is better to maintain a location pointer at the New York registrar (the average total cost per month is 4 × local update cost + 3 × remote search cost for the location pointer scheme versus 4 × [local update cost + remote update cost] + remote search cost = 4 × local update cost + 5 × remote search cost for the actual address scheme). Note in this case that if Bob remembers that Alice is currently in Texas, then he simply can contact the Texas registrar for subsequent location queries. We will discuss location caching later in this chapter.

Continuing with this example, we motivate some other popular location management optimization techniques. We will see that techniques that reduce the search cost tend to increase the update cost, and vice versa. Consider now that Alice, who is on some investigation assignment,

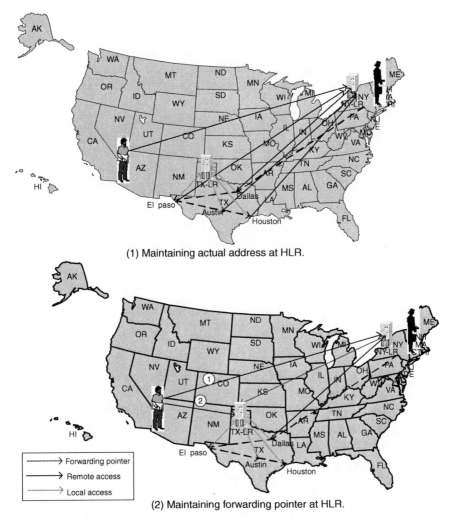

(1) Maintaining actual address at HLR.

(2) Maintaining forwarding pointer at HLR.

Figure 2.3 Maintaining actual address versus forwarding pointer at HLR.

has to visit several states within the United States. Her investigation assignment takes her from New York to Texas to Alaska to Alabama. As illustrated in Fig. 2.4, forwarding pointers can be maintained at registrars in New York, Texas, and Alaska, forming a *chain of forwarding pointers:* New York → Texas → Alaska → Alabama. Suppose that Bob is trying to locate Alice. He will have to start with the New York registrar and follow the forwarding pointers at each intermediate location registrar to finally reach the Alabama location registrar. The use of forwarding pointers

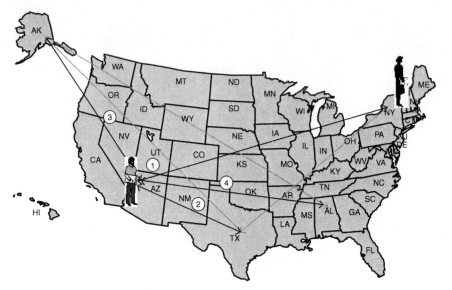

Figure 2.4   Location management using a chain of forwarding pointers.

reduces the update cost (two local updates for each move as opposed to one remote and one local update) but increases the search cost ([1 + number of links in the chain] × remote access cost as opposed to 2 × remote access cost). For example, as illustrated in Fig. 2.4, Alice's chain of forwarding pointers is of length three, and Bob has to contact the following four location registrars: New York, Texas, Alaska, and Alabama (in that order).

Analyzing cost versus benefit of location management optimization techniques.   We have seen that there is an intrinsic trade-off between the cost of update and the cost of search. So how do we evaluate whether an optimization proposed for location management is beneficial or not? In the following, we illustrate how to determine this analytically. Let f_s and f_u be the number of searches per unit time for a mobile node. Further, assume that $C_{s,u}$ and $C_{u,u}$ are the respective search and update costs before optimization that and $C_{s,o}$ and $C_{u,o}$ are the respective search and update costs after optimization. Then the optimization in question is worthwhile if the average cost of location management before the optimization is more than the average cost after the optimization is performed, that is,

$$f_s C_{s,u} + f_u C_{u,u} > f_s C_{s,o} + f_u C_{u,o} \rightarrow f_s(C_{s,u} - C_{s,o}) > f_u \, (C_{u,o} - C_{u,u}) \quad (2.1)$$

Now, if the optimization reduces the search cost, then $C_{s,u}$ is greater than $C_{s,o}$, and $C_{s,u} - C_{s,o}$ is positive. For such optimizations, we can write the preceding inequality as

$$\rightarrow f_s/f_u > -[(C_{u,u} - C_{u,o})/(C_{s,u} - C_{s,o})] \tag{2.2}$$

The fraction on the left-hand side of the equation is referred to as the *call-to-mobility ratio* (CMR). For the optimization that reduces the update cost, $C_{s,u} - C_{s,o}$ will be negative, so the inequality becomes

$$\rightarrow f_s/f_u < -[(C_{u,u} - C_{u,o})/(C_{s,u} - C_{s,o})] \tag{2.3}$$

Let us apply Eq. (2.3) to the forwarding-chain optimization scheme to determine the maximum chain length for a mobile device m with a fixed CMR. For a forwarding chain of length k, the increase in search cost is $k \times A_r$. Further, the update cost is reduced by A_r. According to Eq. (2.3), forwarding pointers are beneficial as long as the CMR is less than $1/k$. Thus, the chain length should be no more than the integer value of $1/$CMR. For a more detailed analysis of efficacy of forwarding pointers in the context of location management for PCS networks, interested readers should see the research paper by Jain and Lin (1995).

Dynamic update schemes. RA-based location update is an example of a static update scheme. Such update schemes do not take into account the dynamic mobility behavior of mobile nodes. Boundaries of RAs are determined taking into account the aggregate mobility pattern of the mobile nodes. Such static boundaries can lead to lot of location updates owing to mobile nodes moving between two adjacent RAs in quick succession. Such ping-pong effects are eliminated in dynamic update schemes. Periodic updates, also known as *time-based updates*, are an example of a dynamic update scheme. Other examples of dynamic update schemes include

- *Movement-based updates.* A mobile node updates its location whenever it *crosses* a certain number of cell boundaries M since it last registered.
- *Distance-based updates.* A mobile node updates its location whenever it moves a certain number of cells D away from the last cell at which it last updated its location.

The maximum search area that needs to be paged depends on the mobility pattern of the mobile node and the update scheme used. Dynamic update schemes have the potential to reduce the maximum search area (search operation latency) over static schemes. Which of the dynamic schemes performs best? This depends on the mobility and call patterns. Among the three schemes, the time-based scheme is the simplest to implement. However, it is not suitable for stationary users. A movement-based scheme requires keeping a count of the number of handoffs since the last update and is suitable for mostly stationary users. A distance-based scheme requires knowing the topology of the cellular network and thus is more difficult to implement than the other two dynamic update schemes. This scheme is suitable for a mobile user who moves within a locality. As opposed to static update schemes, dynamic update schemes can be adapted to the mobility pattern of a user by appropriately choosing the parameters of the dynamic update schemes.

Different dynamic update schemes have different update rates and search costs associated with them. Assume that each update has a unit cost and that the search cost is proportional to the number of cells searched. If f_s is the average call frequency to a mobile node, then $1/f_s$ (also equal to t) is the mean time between two calls to a mobile node. A time-based scheme performs, on average, $t/T = 1/(Tf_s)$ updates between two calls. If v is the maximum speed of a mobile node (in terms of cells per unit time), then the maximum area to be searched is a circle of radius $v \times \min(t, T)$ cells centered on the last known location of the mobile node. Assuming that an expanding ring search is performed to locate the mobile node, the average cost of the search operation can be taken as $v \times \min(t, T)$ assuming the system knows the exact location of the mobile node at the completion of a search operation. Also, for simplicity, assume that the update cost is 1 unit per update operation. The average cost of location management with a time-based scheme is $\text{cost}_{\text{tbus}}(T) = f_s \times v \times \min(1/f_s, T) + 1/T$, for $T \geq 1/f_s$ and $\text{cost}_{\text{tbus}}(T) \approx v$. On the other hand, if $T < 1/f_s$, then the optimal value of T is $T_{opt} = 1/(\sqrt{(v \times f_s)}) \min \text{cost}_{\text{tbus}}(T)$, and the optimal value of $\text{cost}_{\text{tbus}}(T)$ is $2\sqrt{v}$. Therefore, the optimal value of T is $1/\sqrt{(v \times f_s)}$. This implies that as the average velocity increases, the mobile node should update its location more often. Also, when the call frequency to the mobile node increases, it should update its location more frequently. A similar analysis of a distance-based scheme shows that the optimal value of $D = \sqrt{v/f_s}$, and the optimal value of $\text{cost}_{\text{dbus}}(D)$ is $2\sqrt{v}$. Interested readers are encouraged to see the research paper by Bar-Noy and colleagues (1995) for a more detailed analysis of these schemes that takes into account the mobility pattern of the mobile nodes and different search schemes.

Per-user location caching.   The technique of *per-user location caching* can be used to avoid accessing a roaming mobile node's location frequently (Jain, Lin, Ho, and Mohan, 1994). For example, the first time Bob contacts the New York registrar and finds out that Alice is in Texas, he can record that information locally and use it later if needed. However, after Alice moves from Texas to Alaska, Bob's cached location information becomes incorrect. Now, if Bob tries to reach Alice in Texas after she has moved to Alaska, his first attempt to contact her in Texas will fail. And he will have to contact the New York registrar to find out that she is in Alaska (assuming that a forwarding pointer is not used). Let's determine when it is beneficial for Bob to cache Alice's information. Assume that the frequency with which Bob tries to reach Alice is f_s and that the frequency with which Alice changes her location (the state she is in) is f_u. Conceptually, CMR (f_s/f_u) gives the average number of search operations between two consecutive update operations. As illustrated in Fig. 2.5, the first search after an update operation will result in a cache miss, after which the remaining (CMR $-$ 1) search operations will result in cache hits. Thus the cache hit ratio p_h = (CMR $-$ 1)/CMR = (1 $-$ 1/CMR). What should the value of p_h (CMR) be for the caching scheme to be beneficial? Since caching reduces the average search cost without affecting the update cost, Eq. (2.2) cannot be used. However, we just need to determine the values of CMR for which the average search cost with caching ($C_{s,o}$) is less than the average search cost without caching ($C_{s,u}$). If we let A_l be the local access cost and A_r be the remote access cost, then $C_{s,u} = 2A_r$ and $C_{s,o} = p_h(A_l + A_r) + (1 - p_h)(A_l + 2A_r)$. This implies that $C_{s,o} < C_{s,u}$ when $p_h > A_l/A_r$ or, equivalently, when CMR $> [A_r/(A_r - A_l)]$.

In the per-user location-caching scheme, the cache is maintained with location registrars. In this case, the call frequency f_s is the aggregate frequency of calls originating from a registration area for a particular mobile node because once the location of a particular mobile node is cached, it can be used to eliminate (reduce) the search cost for any subsequent search operation before the next move. Thus the search latency of all the calls for a particular mobile node m originating from the RA whose location registrar caches m's location is reduced. Let RCMR$_{i,m}$ be the *regional call-to-mobility ratio*, which is defined as the ratio of the frequency of calls originating from RA_i to the mobility rate of mobile node m. Then the location information for mobile node m can be cached at location registrar i if the RCMR$_{i,m} > [A_r/(A_r - A_l)]$. In the preceding case, a *lazy cache maintenance scheme* was used because the stale cached information was not updated until it was required. However, note that the latency of the location search operation after a cache miss has increased. This can be avoided if somehow the cache entry is invalidated in all the location caches containing location information about mobile node m whenever m changes its RA or, better, if cache information can be updated. *Eager cache maintenance* achieves this by maintaining a list of registrars that are currently caching the location information of mobile node m. This *update list* of m can be maintained at the current local registrar of m. Whenever m moves to another RA, all the registrars in the update list for m are notified of the new location of m. Eager caching decreases the search cost at the expense of increasing the update cost. The average search cost for the registrars in the update list is

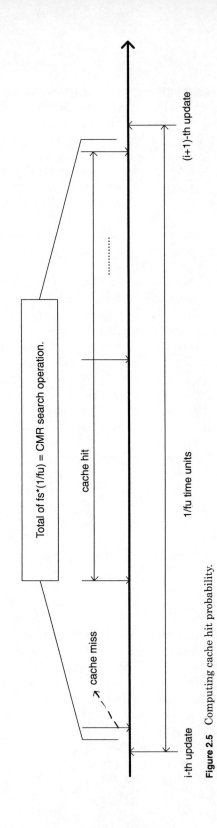

Figure 2.5 Computing cache hit probability.

reduced because the probability of a cache hit (p_h) increases (p_h becomes almost 1). If k is the size of the update list for m, then the update cost now includes the cost to updates of k cached entries.

For efficient implementation of an eager caching scheme, two issues need to be resolved: (1) At which location registrars should the information be cached? and (2) When should the location information be cached? Taking into account the mobility and call patterns of users helps in this regard. Most people have different mobility rates at different times during the day. For example, on weekdays, people drive to work in the morning, stay at work during working hours, and return home. Further, many people are called regularly by only a few people. This set may change over time, but it is considerably small for most people. The *working-set approach* exploits this fact to adaptively maintain cache at only those registrars where it will help in reducing the overhead of location management (Rajagopalan and Badrinath, 1995).

In this approach, the *working set* of mobile node m for time duration $(t, t + dt)$ refers to the set of registrars where it is beneficial to maintain location information about m during this duration. Let $f_{s,i}$ be the frequency of calls for mobile node m originating from registration area i, and let f_u be the mobility rate of mobile node m during duration $(t, t + dt)$. However, at any instant t, it generally does not know what the future call and mobility pattern will be for a mobile node. Assuming that there are fairly long periods of stable behavior in the call and mobility patterns of mobile users, the call and mobility behavior in the recent past can be used as an approximation of behavior in the near future. These are the actual values of $f_{s,i}$ and f_u during $(t - dt, t)$ and are taken as the approximate values of $f_{s,i}$ and f_u as the instantaneous value of search and move frequency. In fact, for sake of efficient implementation, the working-set approach maintains a *sliding window* of the last w operations along with the time each operation was performed. When a new operation occurs, a search or an update, it is added to the window, and the oldest operation is dropped. If there were u update operations in the window and dt is the difference between the time of the last and the first operation in the window, then $f_u = u/dt$, $f_s = (wi - u)/dt$, and $f_{s,i}$ = (number of search operations in the window that were initiated from users in RA_i)/dt.

We explain next how these values are used to adjust the update list dynamically to approximate its membership to the working set of a mobile node. Whenever a new operation o is initiated for mobile node m, the window is "shifted" to include operation o, and the update list is adjusted in the following manner:

- *Case 1: Operation o is a search operation from RA_i.* If location registrar i is not in the update list already, a decision as to whether or not to include i in the update list is made. Note that inclusion of i in the update list will reduce the search cost for calls originating from RA_i and increase the update cost. Equation (2.2) is used to make this decision. If $f_{s,i}$ × decrease in search cost > f_u × increase in update cost is true, then i is included in the update list.

- *Case 2: Operation o is an update (move) operation.* All the registrars in the update list are reevaluated to determine whether it would be beneficial to keep them in the list. A registrar i for which $f_{s,i}$ × increase in search cost < f_u × decrease in update cost is deleted from the update list.

Computation of the search and update costs can be based on the estimated network communication latencies to accomplish search and update operations. This depends on the search and update procedures being employed. For example, in the case of Mobile IP, the search and packet routings are integrated. The packet from sender s is routed to the home network of the mobile node, and the mobile node's home agent (HA) on the home network then tunnels the packet to the current destination network of the mobile node. In this case, the search cost before caching location information is

$$CC[s, \text{HA}(m)] + CC[\text{HA}(m), \text{CA}(m)] \qquad (2.4)$$

where $CC(a, b)$ denotes communication latency between any two nodes a and b on the Internet, $\text{HAD}(m)$ is the home address of node m, and $\text{CAD}(m)$ is the care-of address (either a foreign agent's address or the mobile node's address on the foreign network) of node m. Whereas the search cost after caching is simply $CC[s, \text{CAD}(m)]$, if the updates are sent individually, then the incremental update cost is $CC[\text{CAD}(m), s]$. If a multicast tree were used, then the cost would depend on the structure of the tree and could be substantially less than $CC[\text{CAD}(m), s]$. Interested readers should see the research paper by Rajagopalan and Badrinath (1995) for a description and analysis of working-set technique in the context of Mobile IP.

Replicating location information. If no replication is used, then mobile node m's location information is maintained at one of the n location registrars. The assignment of mobile nodes to a location registrar can be permanent. Care should be taken to ensure that the average load on each location registrar is balanced. However, if any one of these location registrars fails, then all the mobile nodes handled by that location registrar become unreachable. To ensure that the location management service does not go down just because a location registrar fails, one may want to replicate the information among several location registrars.

One of the issues in this regard is how many replicas should one maintain. Full replication increases the cost of updates because each location registrar still has to handle updates for all the mobile nodes. Thus partial replication is preferable. In the following, we discuss partial replication under two different organizations of the location registrars: flat and hierarchical.

Flat organization.   Consider a system with n location registrars. If we assume that the cost of accessing all the location registrars is approximately the same, the location information for a mobile node can be maintained at any location registrar without any penalty in terms of access cost. Under such a scenario, a flat organization of the location registrars may be of interest. Suppose that the system maintains k ($\leq n$) replicas for each mobile node. For simplicity, we assume that n is divisible by k (i.e., $n \bmod k = 0$). A simple update scheme that also achieves load balancing is as follows: Randomly select the first location registrar LR_i to which to add mobile node m's location information, and then place the mobile node m's location information at $(k-1)$ additional location registrars at a stride of $s = n/k$ in a wrap-around fashion. That is, mobile node m's location information is maintained at location registrars with index i, $(i + s) \bmod n$, $(i + 2s) \bmod n$, . . . , $[i + (k-1)s] \bmod n$. In order to search for the location for a particular mobile node, the following procedure can be used: Start with a randomly selected location registrar and sequentially continue the search (in wrap-around manner) until a location registrar with the location information is found. Note that in this scheme at most n/k location registrars would have to be searched to obtain the location information for a particular mobile node.

Figure 2.6 illustrates the preceding location update and search procedure in a flat organization with 16 location registrars. The location registrars are shown to form a logical ring to conveniently illustrate the wrapping around during search and update procedures. Assume that replication factor $k = 4$ is used. In the figure, an update operation is shown that starts at the randomly selected LR_6 and continues to place location information for the mobile node at LR_{10}, LR_{14}, and LR_2. Similarly, the search for the same mobile node's location information starts at randomly selected location registrar LR_{12} and proceeds sequentially until the mobile node's location information is found in the location registrar LR_{14}.

What is the best (or optimal) value for replication factor k? Obviously, the cost of an update increases with an increase in k. However, the cost of a search operation is inversely proportional to k. Further, it would be nice to adapt the replication factor to call and mobility rates for a mobile node. For a system with n location registrars, the following crude but simple analysis illustrates how the optimal value for replication factor

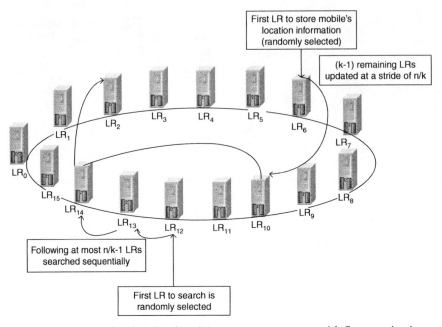

First LR to store mobile's
location information
(randomly selected)

(k-1) remaining LRs
updated at a stride of n/k

Following at most n/k-1 LRs
searched sequentially

First LR to search is
randomly selected

Figure 2.6 Search and update in a location management system with flat organization.

k can be determined for a particular mobile node m with call rate f_s and mobility rate f_u. Obviously, the location update cost is k location registrar accesses per update. Further, under the preceding search scheme, a maximum of n/k location registrars have to be accessed to find a mobile node's location. Thus the normalized location management cost for the mobile node m is $(n/k)f_s + kf_u$. This function is minimized when replication factor $k = \sqrt{(n \times \mathrm{CMR})}$, with $1 \le k \le n$. We can draw several conclusions from this simple analysis:

- As the CMR of the mobile node increases, the number of replicas used for that mobile node also should increase to achieve optimal location management cost.

- When an optimal replication factor is used, the search and update costs are proportional to $\sqrt{n}$.

The second observation has motivated the development of tree organization for location registrars. Tree (or hierarchical) organizations for location registrars promise search and update costs that are $O[\log(n)]$ with a replication factor also of $O[\log(n)]$. Further, they exploit locality in the call and mobility patterns. For optimal location management schemes based on flat (nonhierarchical) organization for location registrars, interested

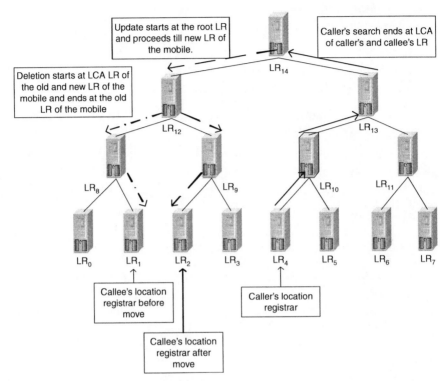

Update starts at the root LR and proceeds till new LR of the mobile.

Caller's search ends at LCA of caller's and callee's LR

Deletion starts at LCA LR of the old and new LR of the mobile and ends at the old LR of the mobile

Callee's location registrar before move

Caller's location registrar

Callee's location registrar after move

Figure 2.7 Update and search in a hierarchical (Tree) location management system.

readers should see the research papers by Krishnamurthi, Azizoğlu, and Somani (2001) and Prakash, Haas, and Singhal (2001).

Hierarchical organization.   Consider a simple multiple-level hierarchical organization (tree) of location registrars. A location registrar that is a leaf node in the tree has information on all the mobile nodes in the RA(s) associated with it. A nonleaf location registrar replicates location information in all the location registrars in the subtree rooted to it. Consequently, the root location registrar stores information on all the mobile nodes in the system. For example, Fig. 2.7 shows a tree hierarchy consisting of 15 location registrars: $LR_0, LR_1, \ldots, LR_{14}$. Assume that there are 8 RAs ($RA_0, RA_1, \ldots, RA_7$) in the service area and that location registrar $LR_i, 0 \le i \le 7$, is associated with RA_i. Consider a mobile node (referred to as the *callee* in Fig. 2.7) that is initially in RA_1. The location information for this mobile node will be maintained in location registrars along the tree path from LR_1 to the root: LR_1, LR_6, LR_{12}, and LR_{14}.

Searches and updates can be performed as follows (Pitoura and Samaras, 1998):

- Let the caller be in RA_i and the callee be in RA_j. The location registrars along the path from the leaf location registrar associated with RA_i to the root are searched until the location information for the callee is found. Let LCA(i, j) denote the location registrar that is the least common ancestor of LR_i and LR_j. Then the search will stop at the location registrar at LCA(i, j).

- If a mobile node moves from RA_i to RA_j, then location information is deleted in all the location registrars along the path from LR_j to LCA(i, j) (except LCA[i, j]), and the location information is updated (or in some cases added) in all the location registrars along the path from root to LR_j.

For example, in the scenario illustrated in Fig. 2.7, the caller mobile node is in RA_4. To locate the callee mobile node, which is in RA_1, the search operation will have to contact LR_4, LR_{10}, LR_{13}, and LR_{14} (in that order) before it can obtain the location information for the callee mobile node. Note that LR_{14} is the least common ancestor of LR_1 and LR_4.

Now suppose that the callee mobile node moves from RA_1 to RA_2. The location information needs to be updated in LR_{14} and LR_{12} and added to LR_7 and LR_2. Further, the location information should be deleted from LR_6 and LR_1.

Assuming a balanced tree, the cost of both the search and update is $O[\log(n)]$, where n is the number of location registrars in the tree hierarchy. Several non-tree hierarchy–based approaches have been developed. For example, see the paper by Awerbuch and Peleg (1995) for a scheme based on regional matching.

2.3 Location Management Case Studies

In this section we present two location management schemes: PCS location management in cellular systems and Mobile IP's location management for the Internet. Both PCS and Mobile IP location management use a two-level hierarchy of location registrars consisting of home location and visitor location registrars for location management. Note that the same location registrar may serve as the home location registrar for some mobile nodes and the visitor location registrar for other mobile nodes. Conceptually, the location registrars have been arranged in a two-level hierarchy: home location registrar in tier 1 and visitor location registrars in tier 2. Each of the several location registrars in the system may serve as the home location registrar for some mobile nodes. The mapping of a mobile node to its home location registrar is usually static (based on the area code in the case of mobile phones or the network address in the case of mobile nodes with IP address).

2.3.1 PCS location management scheme

Two types of location registrars are used: *home location registrars* (HLRs) and *visitor location registrars* (VLRs). The HLR keeps the location and profile information for all the mobile nodes to which the PCS network is supposed to provide service.

Each RA has a VLR associated with it that records the location (cell ID) of all the mobile nodes that currently are in that RA. The HLR for a mobile node records the RA, which is the ID of the VLR associated with the RA in which the mobile node currently is located. When a call needs to be established to mobile node m that is currently located in cell c of RA_c, first the HLR of mobile node m [HLR(m)] is consulted to obtain the ID of the VLR [VLR(n)] that may have information about m; next, VLR(m) is contacted to obtain the current cell in which mobile node m is located.

When a mobile node is switched on, it registers with one of the available access points (base stations). This registration operation also involves updating the VLR and the HLR. While the mobile node remains active, a reregistration is performed (1) when a handoff occurs and (2) periodically.

When a mobile node m moves from cell c to cell d, the following two scenarios are possible:

- *Both cells c and d belong to the same RA.* In this case, only the VLR is updated to indicate in which cell mobile node m is currently located. This helps when mobile node m needs to be located. In this case, there is no need to contact the HLR(m).

- *Cells c and d belong to different registration areas, RA_c and RA_d, respectively.* In this case, the following two actions need to be taken:

- Mobile mode m needs to register with RA_d and deregister with RA_c.

- HLR(m) needs to be notified that mobile node m is now in RA_d.

Now let's examine in more detail what actions need to be performed when mobile node m needs to be located, for example, to establish a connection between mobile nodes n and m. Assume that mobile node n is in cell c and that mobile node m is in cell d, where cell c belongs to RA_c and cell d belongs to RA_d.

- First, VLR(RA_d) is consulted to see whether mobile node m is in RA_d. If so, a search is performed in the vicinity of last reported cell of mobile node m.

- If m is not RA_d, then the HLR(m) is contacted to get the current RA_m.

- VLR(RA_m) is contacted and performs a local search in the last reported cell of mobile node m, and if successful, it returns the current location of mobile node m.

2.3.2 Mobile IP

The Internet is a network of IP networks. In the Internet, a node can be connected to a network via several interfaces. Each interface is assigned an IP address. An IP address, which is 32 bits long, consists of two parts: (1) the host address (x least significant bits) and (2) the network address (32 to x most significant bits). All nodes (interfaces) on an IP network have the same network address but different host addresses. This restricts the IP address that can be assigned to an interface on a given IP network. However, this greatly helps in the amount of routing information that needs to be maintained in the intermediate routers in the core of the network to enable them to forward packets correctly to their intended destination nodes through use of a mechanism called *route aggregation*. A router simply needs to maintain one entry for all the hosts that have a common prefix and are reachable by the same interface of the router.

When a mobile node moves from one IP-network to another IP network, the IP address of the interface connecting it to the new IP network may become invalid on the new IP network. If this is not corrected and the interface is allowed to retain its IP address, then no IP packets can be routed to this interface unless the routing (forwarding) table of all (or, more accurately, a major number of) the routers in the entire network is updated to include an entry for this particular IP address. This solution is not at all desirable for the following reasons: (1) the routing tables will become extremely large and (2) millions of routers on the entire Internet may have to be updated when a mobile node moves. However, if the mobile node (interface) acquires a new IP address every time it moves, then all the nodes (or at least those which want to communicate with the mobile node or want to access the services hosted by the mobile node) will have to be informed about the new IP address. Since the Domain Name System (DNS) maintains the mapping between host domain names and IP addresses, this can be addressed potentially just by updating the DNS entry for the domain name of the mobile node. However, there are several problems with this approach. A source node may use incorrect mapping for a long time after it has been updated in the DNS owing to prevalent use of DNS caching. Datagram packets sent during the period between when the node moves and the source node obtains the new IP address of the mobile node from the DNS server may be lost.

Mobile IP is an extension of Internet Protocol version 4 (IPv4) to support host mobility at the IP layer (Perkins, 2002). In Mobile IP, the operations of location management and packet rerouting are tied closely together. This solves the problem caused by mobile nodes frequently changing their IP network by associating two IP addresses with each node: (1) a *permanent* IP address and (2) a *care-of* IP address. A mobile node has a permanent IP address that is topologically significant on its *home network*. While a mobile node is attached to its home network, the packets destined to it get routed in the normal fashion. When a node moves to a *foreign network*, it acquires another temporary IP address. Further, it *registers* this temporary IP address with its home agent (HA). An HA is a software entity running on some machine attached to the home network and is responsible for ensuring that the IP packets destined for a currently roaming mobile node get rerouted to the node's current location. Among other things, an HA maintains a mobility binding for each roaming mobile node. Simply stated, a *mobility binding* consists of (1) a mapping between the permanent IP address of a mobile node and its current care-of address (or addresses in case of multihomed mobile nodes) and (2) a time-to-live (TTL) value that denotes how long the binding is valid. The mobility binding for a mobile node is established when it sends a registration message to its HA. If this binding is not refreshed (via reregistration) within the time period specified in the TTL field of the binding, the binding becomes invalid. This takes care of situations where the mobile node is no longer connected to the last registered network or has been powered off temporarily, as in the case of a laptop.

When a mobility binding for a mobile node m is active at its home agent [HA(m)], HA(m) intercepts all the packets that arrive for m on its home network. It then forwards these packets to the care-of address (or addresses) of m. This is done by a mechanism called *IP tunneling,* where in the original IP packet is *encapsulated* in another IP packet whose destination address is the care-of-address of the mobile node and the payload is the original IP packet. When this encapsulated packet arrives at the other end of the tunnel, it is decapsulated, and the original packet is delivered to the mobile node. The mobile node uses the source address in the original IP packet to send back any response to the sender of the original IP packet. In the Mobile IP literature, this routing of packets from a sender to a mobile recipient via its HA and that of the response directly from the mobile node to the sender is referred to as *triangle routing.*

Care-of IP addresses are of two types: (1) foreign agent's IP address and (2) colocated IP address. A *foreign agent* (FA) is a mobility agent that provides some services to a roaming mobile node when it is on a foreign network. When a mobile node uses a foreign agent's address, the FA acts as the exit endpoint of the tunnel and performs the decapsulation operation. The original IP packet is then delivered to the destination mobile node through the underlying link layer protocol. Figure 2.8 illustrates how packets from an Internet host (either fixed or mobile) get delivered

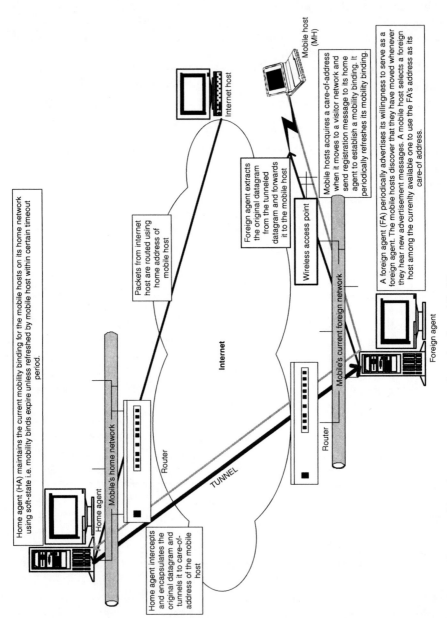

Figure 2.8 Integrated location management and routing in mobile IP.

Home agent (HA) maintains the current mobility binding for the mobile hosts on its home network using soft-state i.e. mobility binds expire unless refreshed by mobile host within certain timeout period.

Packets from internet host are routed using home address of mobile host

Foreign agent extracts the original datagram from the tunneled datagram and forwards it to the mobile host

Mobile hosts acquires a care-of-address when it moves to a visitor network and send registration message to its home agent to establish a mobility binding. It periodically refreshes its mobility binding.

A foreign agent (FA) periodically advertises its willingness to serve as a foreign agent. The mobile hosts discover that they have moved whenever they hear new advertisement messages. A mobile host selects a foreign host among the currently available one to use the FA's address as its care-of address.

Home agent intercepts and encapsulates the original datagram and tunnels it to care-of-address of the mobile host

Internet host

Mobile host (MH)

Wireless access point

Mobile's current foreign network

Internet

Router

Router

TUNNEL

Mobile's home network

Home agent

Foreign agent

to a mobile host via a foreign agent. A colocated IP address is a topologically significant IP address on the foreign network. This address can be statically assigned to a mobile node or can be acquired dynamically by using mechanisms such as Dynamic Host Configuration Protocol (DHCP). In this case, the endpoint of the tunnel is the mobile host itself; i.e., the mobile host is in charge of decapsulating the encapsulated packet.

An interesting issue is how a mobile node detects that it has moved (to a new network). This is achieved by agent advertisement messages periodically broadcasted by foreign agents. Each foreign agent willing to serve as a foreign agent periodically broadcasts an agent advertisement message in networks it wants to serve as a foreign agent. A mobile host detects that it has moved whenever it either stops hearing agent advertisements from a certain foreign agent or starts hearing agent advertisements from a new foreign agent (i.e., a foreign agent currently not on the list of active foreign agents). For more details on Mobile IP, interested readers should consult Perkins (2002).

2.4 Summary

Mobility management ensures that communication can be established and maintained with communication endpoints that are moving as long as they are in the coverage area of some access point or base station and as long as they are switched on. Mobility management is a fundamental task of any mobile computing system. Higher-level services can be built on top of mobility management services, thus enabling novel mobile applications. In this chapter we discussed fundamental aspects of mobility management. We focused mainly on location management aspects of mobility management. Several location management techniques that are essential for achieving scalability, precision, and reliability were discussed. Owing to its importance, mobility management is an active area of research and will continue to be so as long as mobile computing is needed.

2.5 References

Akyildiz, I. F., McNair J., Ho J. et al., "Mobility Management in Current and Future Communication Networks," *IEEE Network Magazine* (July–August):39, 1998.

Awerbuch, B., and D. Peleg, "Online Tracking of Mobile Users," *Journal of the Association for Computing Machinery (JACM)* 42(5):1021, 1995.

Bar-Noy, A., I. Kessler, and M. Sidi, "Mobile Users: To Update or Not to Update," *Wireless Networks* 1(2):175, 1995.

Jain, R., and Y.-B. Lin, "An Auxiliary User Location Strategy Employing Forwarding Pointers to Reduce Network Impacts of PCS," *Wireless Networks* 1(2):197, 1995.

Jain, R., Y.-B. Lin, C. N. Ho, and S. Mohan, "A Caching Strategy to Reduce Network Impacts of PCS," *IEEE Journal on Selected Areas in Communications* 12(8):1434, 1994.

Krishnamurthi, G., M. Azizoğlu, and A. K. Somani, "Optimal Distributed Location Management in Mobile Networks," *ACM/Baltzer Mobile Networks and Applications* 6(2):117, 2001.

Perkins, C. (ed.), IP Mobility Support for IPv4, RFC 3220, January 2002, *http://www.rfc-editor.org/rfc/rfc3220.txt*.

Pitoura, E., and G. Samaras, *Data Management for Mobile Computing*. Norwell, MA: Klewer Academic Publishers, 1998.

Prakash, R., Z. J. Haas, and M. Singhal, "Load-Balanced Location Management for Cellular Mobile Systems Using Quorums and Dynamic Hashing," *Wireless Networks* 7(5):497, 2001.

Rajagopalan, S., and B. R. Badrinath, "An Adaptive Location Management Strategy for Mobile IP," MobiCom, 1995.

Data Dissemination and Management

Timely and accurate information (data) is essential for performing our various day-to-day tasks, including decision making. This information is provided to us from various sources. Some of this information is explicitly "pushed" to us via such mechanisms as television and radio advertisements, telemarketing, electronic billboards, and flyers in the mail. Other information we explicitly seek and "pull" onto our laptops to read. Information that is sent to us, without our explicitly asking, is usually of a general nature. Information we explicitly seek is usually of a personal nature. Obviously, we need to expend more effort in obtaining information that is of particular interest only to us. In the world of computing, there is a similar model of information delivery. Information in which we are interested can be obtained through two modes: on-demand and publish-subscribe.

In the *on-demand mode* (also known as *pull mode*), we (the information sink) send an explicit query every time we need particular information to an information source (a server or a peer). The information source sends back the latest version of the requested (or queried) information. The access latency in the on-demand mode therefore is at least the round-trip latency. In addition, the sender has to expend resources to send the query every time it needs the information.The *publish-subscribe mode* (also known as *push mode*) is used for obtaining information whenever it is available. For example, we can subscribe to a stock ticker, and whenever the stock information is updated, it will be sent to us.

Publish-subscribe mode. In mobile computing, the publish-subscribe mode is combined with the broadcast nature of wireless communication

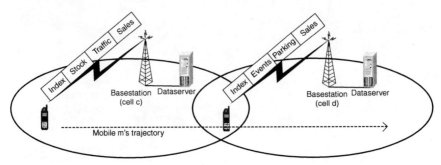

Figure 3.1   Broadcast channel.

to provide resource-efficient and scalable information delivery (Fig. 3.1). Mobility presents some novel challenges in designing services such as e-mail and information broadcasting in a wireless mobile environment. Wireless transmission is characterized by such problems as signal fading, path loss, interference, and time dispersion. These lead to higher error rates and signal distortion. A consequence is that wireless links have a lower bandwidth and a higher communication latency. Providing information services to mobile clients is an important application of the publish-subscribe mode of data dissemination. Examples of information services include financial information services, airport information services, and emergency services for traffic information. Wireless information services can be classified into picoservices, macroservices, and wide area services based on the area of coverage (cell sizes). The following are some issues in providing wireless information services that we will address later in this chapter:

- How to structure an information server to provide wide area services? Some issues in this regard are *Publication content*: Which items to publish?

- *Publication frequency*: How often to publish?

- *Bandwidth allocation*: How to adaptively allocate bandwidth between uplink and downlink channels to reflect the changing usage patterns in a cell?

- How can mobile users access services transparently? Figure 3.1 shows an example of the service handoff that is needed when a mobile client moves from one cell to another. It needs to discover the "address" of the broadcast channel for the new cell. A simple way to accomplish this is to provide the user with a directory channel that has information about all the broacast channels available in the new cell. The user can access this channel to, in turn, find the channel that has the information she is interested in.

When mobile hosts are accessing data, how can their energy consumption be minimized? Some items are requested more frequently than others. These items are called *hot items*. In a mobile computing environment, the on-demand mode requires energy for uplink requests. Furthermore, this architecture is not scalable with respect to the number of mobile clients. Requests have to be made by each mobile client interested in a particular data item. This is especially problematic for hot items. Many modern CPUs offer "doze mode" operations. A low-power radio circuit can be programmed to perform matching for a predefined set of packet addresses, to store data in a low-energy buffer, and to wake up the CPU after a certain time interval. The clock also runs in a low-power mode. The main issue here is how can doze mode operations be used for energy-efficient information services? This approach conserves battery power of mobile hosts because no uplink query is required. In some wireless systems, packet transmission consumes more energy than the reception of packets. Further, the publish-subscribe mode is more scalable because access time for a published data item is independent of the number of mobile hosts requesting that data item. As an additional advantage, this mode is more useful for asymmetric environments in which the downlink bandwidth is substantially greater than the uplink bandwidth.

Information caching.   Most of us maintain a cache of money in our wallets for quick access when needed. Frequently, we withdraw some money from our bank accounts and replenish the supply of money in our wallets. This technique of maintaining a small supply of essential items close to us (money in our wallets, food items in our pantry, and beer in the refrigerator) prevents our making frequent and time-consuming trips, e.g., to a bank or grocery store. We may replenish the supply on an as-needed basis (e.g., when we run out of that item) or in a predictive manner (e.g., if there is a party next Friday, we might buy enough beer in advance). And when we come to know that a certain item, such as gasoline, may not be available in the future, some of us try to hoard that item. Hoarding is a very common practice in places where essential items are in short supply and become available only occasionally. Usually the expertise of deciding when to hoard and what to hoard is acquired by experience. Sometimes, making a mistake in this regard can be fatal. For example, forgetting to hoard water while going on a long trip in a desert can cause a person great discomfort or even cost his precious life. We usually try to hoard only nonperishable items. If perishable items such as vegetables have to be hoarded, we convert them into a nonperishable form, e.g., by canning them. In the process, we give up some flexibility (or preferences) in favor of guaranteed availability in the future.

In the information age, information itself has become an important commodity to cache and hoard. When we go on a trip, we fill our carry-on bags with books that we may want to read while on the flight and our laptop computer with files that we may want to work on while waiting for the connecting flight. For people who are frequently on the move, remembering to download all the files onto the laptop that may be needed on the next trip easily can become a cumbersome task. Furthermore, on returning from the trip, the files have to be uploaded to a computer that is backed up regularly.

A more serious problem may arise if a file taken on a trip, say, a Word document, is shared among various people. If the file is shared only for reading, then there is no problem. However, if the shared file is modifiable, then the following situation may arise: Consider that Alice is sharing a file with Bob. If Alice or Bob (but not both) are permitted to modify the file, then while Alice is on the trip, Bob will not have access to the most up-to-date version of the shared file if Alice has modified it. Conversely, Alice will have access only to the stale version of the shared document while she is on the trip if Bob also has modification rights to the file and has changed it while she is on her trip. In the case where Alice has read-only permission to the file, on returning from her trip, there is no need for her to upload the file to her system. In the other case, if she has the modification rights to the file, the modified version on her laptop has to be uploaded to the shared storage so that Bob can have access to the most recent version of the file in the future.

Now suppose that both Alice and Bob have modification rights to the shared document. This may lead to two different versions of the file, Alice's shared document and Bob's shared document, if both Alice and Bob modify the file while Alice is on her trip. These modifications may be inconsistent with each other. For example, Alice may have deleted the sentence, "Bob is a very hardworking person" from the shared document, and Bob may have modified the same sentence to, "Bob not only is very hardworking but also is a very sincere person." Such incompatible versions of the same file are difficult to merge into a single version in the absence of any guidelines or user assistance. For example, if Alice is Bob's boss, a rule to reconcile inconsistent change may be, "In case of conflict, Alice's changes stay, and Bob's changes are deleted." If no such rule exists or is not applicable, then user intervention is necessary to resolve such conflicting changes. Fortunately, in the real world, most of the time people resolve among themselves who is the "owner" of the file. However, in the cyberspace, processes may be sharing files, making it unclear who is the owner of the document at any particular time.

3.1 Challenges

The various characteristics of wireless networks and mobile computing environments pose new challenges for distributed data management. Mobile wireless networks are predominantly of two types: *architecture-based* and *architecture-less*. In both types of networks, the wireless links may be of low bandwidth (in comparison with wired links) and subject to frequent disconnections, leading to *weakly connected* mobile clients. Consequently, mobile clients often can be disconnected from their data servers. These disconnections can be either *voluntary* (e.g., when the user disables the wireless network interface card [NIC] to conserve battery power) or *involuntary* (e.g. when the user moves to an area where wireless service is not available). Irrespective of the cause of the disconnection, a user would like to have some connection transparency—in the sense that he would like to have access to data vital to the application with which he is working. Weak connectivity of mobile clients in a mobile wireless computing platform creates a new challenge for data management: *how to ensure high data availability in mobile computing environments where frequent disconnections may occur because the clients and server may be weakly connected.* To allow access to critical data even during periods of disconnections, the distributed file system Coda uses data caching to improve availability at the expense of transparency.

A second characteristic of mobile computing environments is the severe constraints on the availability of resources (such as battery power) at the mobile node. A typical node in such environments has limited power and processing resources. This characteristic leads to another challenge for data management in mobile computing environments, specifically, *how to minimize resource consumption (e.g., energy and bandwidth) for data management while ensuring a desirable level of data consistency.*

The third characteristic of some wireless networks is the presence of *asymmetric communication links.* For example, in architecture-based wireless networks, the downstream (base station to mobile nodes) communication link capacity is usually much higher than the upstream (mobile to base station) capacity. To make matters worse, mobile nodes may have to compete with several other mobile nodes to get access to an upstream channel (using some medium access techniques, such as ALOHA). In some cases, the mobile node may even not have the capability to perform uplink communication. Competing with other mobile nodes for an uplink channel and requesting a data item from the server are expensive in terms of battery power consumption because a mobile node may have to keep its network interface powered up from the time it initiates the request to the time the response arrives from the server. *How can the asymmetric nature of wireless connectivity be exploited to ensure low data access latency and resource consumption?* This is a new challenge posed by this characteristic feature of wireless networks.

The fourth characteristic of mobile computing environments is that data may be *location- and time (context)–dependent.* For example, a mobile user may query various databases periodically to retrieve both location-dependent and time-dependent information. A traveler visiting a city may want to know the list of restaurants in her vicinity and may go over that list each day of her stay to find a new place to try. Similarly, a mobile salesman may need to know the most up-to-date price list while making a business deal with a customer. In the case of the traveler, the list of restaurants does not change as long as the location of the traveler does not change. In the case of the salesman, the price list does not change as long as prices of items on the list do not change. Caching and prefetching can be an effective technique in both these cases to reduce the impact of low-bandwidth and intermittent wireless links in a mobile environment. However, novel cache invalidation and prefetching techniques are needed to maintain location-dependent and time-dependent information.

Context-dependent data impose another problem for cache management algorithms—the decision to cache or replace a data item now also depends on the context (e.g., location) of the mobile node, in addition to the temporal or spatial locality in the reference pattern. *How do you enhance existing cache management techniques for context-dependent data?* This is another challenge that needs to be addressed for mobile computing environments.

An *architectureless wireless network* or a *mobile ad hoc network* (MANET) consists solely of mobile computing devices within mutual wireless communication range. These types of networks are different from architecture-based wireless networks in the sense that there are no dedicated network infrastructure devices in a MANET. Because of the limited radio propagation range of wireless devices, the route from one device to another may require multiple hops. In some scenarios, to communicate with the outside world, a few devices that have network connections with outside base stations (or satellites) can serve as the gateways for the ad hoc network. We would refer to such ad hoc networks as *weakly connected ad hoc networks* because they do have some infrastructure support but not as much as the mobile nodes in infrastructure-based networks. Such weakly connected ad hoc networks may be used for emergency hospitals for disaster relief in remote areas and military operations in remote areas.

In general, the MANET environment also has the two features of wireless computing environments: *weak connectivity* and *resource constraints.* Consequently, data availability and bandwidth/energy efficiency still need to be addressed in MANETs. However, data management schemes developed for an architecture-based wireless network cannot be used directly to solve the data management problems in a MANET for various reasons. First, gateways of ad hoc networks are unreliable mobile computing devices, whereas base stations are reliable dedicated networking devices. Gateways communicate with local hosts using low-bandwidth, unreliable links, such as radio frequency wireless links, or possibly with remote hosts through high-latency, unreliable links, such as satellite channels. On the other hand, base stations communicate with remote hosts through high-speed wired networks. Second, MANETs are inherently peer-to-peer (P2P) networks. In contrast, architecture-based wireless networks mostly follow the client-server (CS) paradigm. The inherent P2P structure requires rethinking data management approaches to improve data management performance in MANET environments. For example, cooperative caching has been used on the Internet to provide more cache space and faster speeds. Specifically, the NLANR caching hierarchy consists of many backbone caches, and those caches can obtain data from each other using Hyper-Text Transfer Protocol (HTTP) and the Internet Caching Protocol (ICP) (NLANR, 2002). However, these cooperative caching schemes do not address the special concern of MANET environments, namely, weak connectivity and severe resource constraints, and thus are being adapted for MANETs.

3.2 Data Dissemination

The most prevalent use of a mobile computer is for accessing information on remote data servers, such as Web servers and file servers. Usually, remote data are accessed by sending a request (or a query) to the remote server. In response to the query, the data server sends the requested data. This access mode is known as *pull mode* or *on-demand mode*. However, in pull mode, the mobile node has to explicitly send a query. This requires competing for wireless access to send the query on the uplink channel and then waiting for the response. The entire process consumes the mobile node's precious battery power. Further, wireless bandwidth is consumed—which in some cases may be scarce. The problem is further exacerbated if multiple users in a wireless cell request the same data.

Another way in which data can be delivered to a mobile node is by pushing the data to it. The data server periodically broadcasts the data (along with some indexing information) on a broadcast (or multicast) channel. The indexing information is used by a mobile client to determine when the data in which it is interested will be available on the broadcast channel. As shown in Fig. 3.2, the mobile node can use this information to conserve its battery power.

There are several advantages to this pushing (or publishing) mode of data dissemination. First, broadcasting the frequently requested data

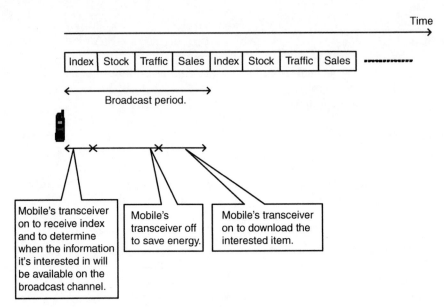

Figure 3.2 Broadcast channel access scheme for conserving mobile's energy.

items, called *hot data items,* conserves bandwidth because it eliminates repetitive on-demand data transfers for the same data item to different mobile nodes. This makes the push mode highly scalable because the same amount of bandwidth is consumed irrespective of how many mobile nodes simultaneously need a particular data item. Second, this mode of data transfer also conserves the mobile node's energy by eliminating the energy-consuming uplink transmission from a mobile node to the data server. Obviously, not all the data items can be provided on the broadcast channel, and hence both modes need to be supported in a general mobile computing environment. As illustrated in Fig. 3.3, a cell's wireless bandwidth is divided into three logical channels:

1. *Uplink request channel*—shared by all the mobile clients to send queries for data to the server

2. *On-demand downlink channel*—for the server to send on-demand request data items to the requesting mobile client

3. *Broadcast downlink channel*—to periodically broadcast the hottest data items

From the implementation perspective, it is convenient to partition the available wireless bandwidth into two physical channels: on-demand and broadcast. The physical on-demand channel is accessed

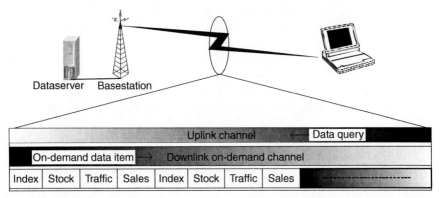

Figure 3.3 Channel allocation to support both push and pull mode of data dissemination.

using a distributed medium access technique such as ALOHA. Both the uplink request channel and the on-demand channel are mapped to the same physical on-demand channel. The broadcast channel is slotted such that a data item (page) can fit in each slot. The data server constructs a broadcast schedule (which page to broadcast when), and the mobile node can tune in at the beginning of the slot in which the data item in which it is interested is going to be broadcast to download that data item.

3.2.1 Bandwidth allocation for publishing

Partitioning the available wireless bandwidth among on-demand and broadcast channels is an interesting problem. Consider that the available wireless bandwidth that needs to be partitioned between bandwidth for an on-demand channel B_o and bandwidth for a broadcast channel B_b is B (that is, $B = B_o + B_b$). The data server has n data items $D_1, D_2, \ldots, D_n$; each of size S, and each data query is of size R. Further, assume that D_1 is the most popular data item, with popularity ratio p_1; D_2 is the next most popular data item, with popularity ratio p_2; and so on. The popularity ratio (or access probability) is the (long-term) ratio of the number of requests for a data item over the total number of requests by all the mobile nodes (the value of each p_i, $1 \leq i \leq n$, is between 0 and 1). For simplicity, assume that each mobile node generates requests at an average rate of r.

Let us try to compute the average access time T over all data items. Obviously, $T = T_b + T_o$, where T_o is average access time to access an on-demand item, and T_b is average time to access a data item from the broadcast channel. The average time to service an on-demand request is $(S + R)/B_o$. If all the data items are provided only on-demand, the average request rate for all the on-demand items will be $M \times r$, where M is the number of mobile nodes in the wireless cell. We know from queuing theory that the average service time (which includes queuing delay) would increase rapidly as the query generation rate ($M \times r$) approaches the service rate [$B_o/(S + R)$]. Thus it is easy to see that as the number of mobile nodes increases, the average query generation rate increases, and when it does so beyond some point, the average service time exceeds the acceptable server time threshold. This shows that allocating all the bandwidth to the on-demand channel has poor scalability.

At the other extreme, all the data items can be provided only on the broadcast channel. That is, all the bandwidth is allocated to the broadcast channel. If all the data items are published on the broadcast channel with the same frequency (ignoring their popularity ratio), then a mobile node would have to wait, on average, for $n/2$ data items before it gets the data items from the broadcast channel. Thus the average access time for a data item would be $(n/2) \times (S/B_b)$. Note that this is independent of number of mobile nodes in the cell. However, the average access time will increase as the number of data items to broadcast increases. Even if the data server employs a more intelligent schedule in which hotter data items are broadcast more frequently, the average access time increases with n. This is so because increasing the broadcast frequency of some items necessarily decreases the frequency of others (i.e., increasing their access time).

Consider the simple case of broadcasting two data items D_1 and D_2. Assume that D_1 is much more popular than D_2 (that is, $p_1 \gg p_2$). One may be tempted to broadcast D_1 all the time, but that would cause the access time of D_2 to be infinite (i.e., D_2 is never available). In fact, it can be determined that the minimum average access time is achieved when the frequency f_1 of broadcasting D_1 and the frequency f_2 of broadcasting D_2 are in same ratio as the square root of p_1 and the square root of p_2. Specifically, $f_1 = \sqrt{p_1}/(\sqrt{p_1} + \sqrt{p_2})$, and $f_2 = \sqrt{p_2}/(\sqrt{p_1} + \sqrt{p_2})$. For example, assume that $p_1 = 0.9$ and $p_2 = 0.1$ (i.e., data item D_1 is nine times more popular than D_2). For this case, the minimum average access time is achieved when $f_1 = 0.75$ and $f_2 = 0.25$; i.e., D_1 should be broadcasted only three times more often than D_2, even though it is nine times more popular than D_2. This is a surprising result indeed. In general, if there are n data items with popularity ratio $p_1, \ldots, p_n$, they should be broadcast with frequencies $f_1, \ldots, f_n$, where $f_i = \sqrt{p_i}/Q$, and $Q = \sqrt{p_1} + \sqrt{p_2} + \cdots + \sqrt{p_n}$, in order to achieve minimum average access latency: $p_1 t_1 + p_2 t_2 + \cdots + p_n t_n$, where $t_1, \ldots, t_n$ are average access latencies of $D_1, \ldots, D_n$, respectively (Imieliński and Viswanathan, 1994).

As we have mentioned, the publishing mode helps to save the energy of the mobile node, although at the cost of increasing access latency. Hence the goal is to put as many hot items on the broadcast channel as

For i = N down to 1 do:

Begin

1. Assign D_1, ... , D_i to the broadcast channel

2. Assign D_{i+1}, ... , D_N to the on-demand channel

3. Determine the optimal value of B_b and B_o to minimize the average access time T

   as follows:

   a. Compute To by modeling on-demand channel as M/M/1 (or M/D/1) queue

   b. Compute Tb by using the optimal broadcast frequencies f1, ... , fi.

   c. Compute optimal value of Bb which minimizes the function T = To + Tb.

4. if T <= L then break

End

Figure 3.4   Imieliński and Viswanathan Algorithm for determining optimal bandwidth allocation between on-demand and broadcast channel.

possible under the constraint that the average access latency is below a certain threshold L. The algorithm by Imieliński and Viswanathan (1994), which is shown in Fig. 3.4, can be used to determine what data items to put on the broadcast channel, as well as the appropriate bandwidth allocation for it.

3.2.2 Broadcast disk scheduling

Broadcast scheduling deals with determining how often to publish a certain data item—once the decision regarding which data items to publish has been made. A novel way to view a broadcast channel is to regard it as an extension to the memory hierarchy of the mobile node—a sort of memory disk in the air. The broadcast channel itself can be structured as multiple virtual disks, each spinning at different rates. The hottest set of data items is allocated to the fastest-spinning disk, the next hottest data item to next-fastest disk, and so on. For example, consider the example illustrated in Fig. 3.5. There are nine data items, with item D_1 assigned to disk 1, items D_2 through D_5 assigned to disk 2, and items D_6 through D_9 assigned to disk 3. As can be seen from the broadcast schedule, the items on disk 1 appear four times as frequently as those on disk 3, and the items on disk 2 appear two times as frequently as those on disk 3. In effect, disk 1 is spinning fastest, followed by disk 2 and then disk 3.

Figure 3.6 gives the AAFZ algorithm developed by Acharaya, Alonso, Franklin, and Zodnik for deriving the schedule for the broadcast channel

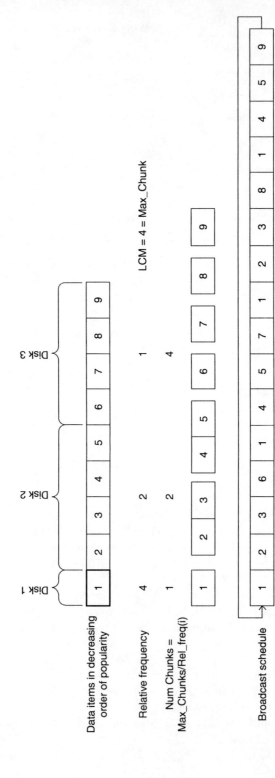

Figure 3.5 Example derivation of broadcast schedule for broadcast disks.

1. Order the data items from hottest to coldest

2. Partition the list into multiple ranges, called disks. Each disk consists of data items which nearly same popularity ratio. Let the number of disks chosen be num_disks.

3. Choose the relative frequency of broadcast for each disk.

4. Cluster the items in each disk into smaller units called chunks: num_chunk(i) = max_chunks/rel_freq(i), where max_chunks is the least common multiple of relative frequencies.

5. Create broadcast schedule as follows:

    a. For i = 0 to mak_chunks – 1

        i. For j = 1 to n

            1. k = i mod num_chunks(j)

            2. Broadcast chunk Cj,k

        ii. Endfor

    b. Endfor

Figure 3.6 Acharya, Alonso, Franklin, Zodnik algorithm for scheduling broadcast disks.

that achieves a selected relative spinning of the disks. It takes as input the number of disks and the assignment of data items to the disk, as well as the relative spinning frequencies of the disks. For example, the chosen relative frequencies for the three disks in Fig. 3.5 are rel_freq(disk 1) = 4, rel_freq(disk 2) = 2, and rel_freq(disk 3) = 1. The figure also illustrates the intermediate computation performed by the algorithm to determine the broadcast schedule.

Another way to minimize latency and energy consumption for accessing data items is to cache them locally. We next discuss issues related to the caching of data at mobile computers and look at some of the techniques designed for cache management in mobile computing.

3.3 Mobile Data Caching

Caching is a very important performance enhancing technique in the computing world. A typical computer system consists of several cache memories. A *cache* is a small, fast memory for holding frequently used data. In fact, the different types of memories in a computer system, such as disk, main memory, and cache, can be viewed as a part of a *memory hierarchy* in which the closer a memory unit is to the processor, the smaller is its capacity and *access time. How do we reduce memory read/write latency in the presence of a memory hierarchy?* This is the main problem for cache management in such systems. To achieve this, the system needs to (transparently) decide what data to cache (to copy from a farther memory unit to a closer memory unit) so that the cache (memory unit closer to the processor) can satisfy as many memory requests as possible—effectively achieving average access time close to that of the fast memory unit at the effective price of the large memory unit. Cache management techniques try to predict which data items are most likely to be used in the future and either copy them to a memory unit closer to the processor when they are accessed for the first time (i.e., on a *cache miss*) or *prefetch* them well in advance so that they are available in the cache memory when they are needed. Many cache replacement algorithms, such as least-recently used (LRU) and prefetching algorithms, have been developed to solve this problem (Silberschatz and Galvin, 1998).

3.3.1 Caching in traditional distributed systems

Data caching has been used traditionally in distributed (as well as nondistributed) systems to improve performance (data access latency) by exploiting temporal and spatial locality of the reference pattern. For such distributed systems, it is assumed that a high-bandwidth, low-latency network connects the clients and servers and that network failures are infrequent. In cases where access latency is high and/or bandwidth is not sufficient, data prefetching in conjunction with data caching can be used to hide communication latency and to cope with limited bandwidth. Both data caching and prefetching are performed transparently so that the applications do not have to do anything different when either is used. The only difference from the user's (application's) perspective is that the results are obtained faster (or in case of applications such as video streaming, the quality of the output is better). Thus proper use of caching/prefetching in such cases can only improve the situation and cause no harm.

Further, in distributed systems and network computing environments, cache management schemes need to handle two different scenarios. In the first scenario, the data in the *shared memory* or the *server* can be *read or written* by different processors or *clients* concurrently. In the second scenario, the data are *read-only* for the clients. Further, many times the data may be replicated onto multiple servers for improving fault tolerance and availability. An example of the former scenario is a distributed file system, whereas an instance of the latter scenario is seen often on the World Wide Web (WWW). In these environments, when a client accesses a cached data item, the server or another client could have just modified that data item. Thus, in distributed and network computing environments, the most crucial problem for caching is *how to maintain data consistency among the clients and the servers.* This problem is more difficult to solve than the problem of ensuring that the copy of the data in the higher-level memory is updated, such as through write-through or write-back schemes (Silberschatz and Galvin, 1998), when the cached copy of the data in the lower-level memory has been updated. The complexity arises from the various failures that may occur—server failure, network failure, and client failure. To maintain cache consistency in distributed or network computing environments, different approaches have been developed. These approaches include polling every time, adaptive time to live (TTL) (Gwertzman and Seltzer, 1996), server invalidation (Cao and Liu, 1998), and leases-based invalidation (Cao and Liu, 1998; Yin et al., 1998).

In addition to caching for improving data access performance, caching is used to improve data availability by trading it off against consistency. Before understanding this trade-off, we next discuss the traditional consistency maintenance scheme.

3.3.2 Cache consistency maintenance

Cache consistency maintenance is required to ensure that whenever a data item of value is fetched from the cache, it satisfies certain currency requirements. Several cache consistency models are supported in distributed systems. Under a strong cache consistency model, the cache consistency maintenance protocol guarantees that the value of a data item x provided to the application at any time is the most recent value of x. This is to say, if the data item x were accessed at time t, then the value of x would be the value of the last write to x that occurred before time t. A *strong cache consistency* requirement implies that whenever data item x is accessed, checks are performed in consultation with remote servers to determine whether the cached value of x is the most current. In implementations of strong cache consistency protocols such as that employed by Network File System (NFS), if the remote server is inaccessible because either the network connecting the client and the server is down or the server itself is down, then the cache access to x just waits until the server becomes accessible. In some implementations, the client machine even blocks any further cache access, practically freezing the client machine. Maintaining such strict consistency requirements can make mobile computing environments unusable when disconnections become frequent.

Several cache consistency maintenance schemes have been designed for different distributed applications, such as the WWW, distributed file systems, and CS databases. These techniques can be classified based on whether they detect or avoid access to stale data. These techniques provide different degrees of consistency guarantees with varying costs and availability in the presence of disconnections.

The classic techniques for maintaining cache coherency, such as *callbacks* and *validity checks*, were designed for computing environments consisting of workstations connected via reliable wired links. These techniques are not suitable for mobile computing environments consisting of battery-powered laptops connected via wireless links. For example, *callback breaks* (invalidation messages) lost while a mobile client is disconnected from the network may result in stale cached items in the client's cache (Fig. 3.6). Thus, in the absence of any other mechanisms, a client may have to invalidate the entire cache on each disconnection. Refetching valid data from the server is wasteful of both wireless bandwidth and battery power. The same is true when validation checks are used for every data item before using it. To address these drawbacks of traditional techniques, new methods have been developed for maintaining cache consistency in mobile environments that take into account the constraints and features of such environments.

3.3.3 Performance and architectural issues

It is challenging to design caching strategies for mobile environments because an efficient strategy should take into consideration various issues such as

1. Data access pattern
2. Data update rate
3. Communication/access cost
4. Mobility pattern of the client
5. Connectivity characteristics (disconnection frequency, available bandwidth)
6. Data currency requirements of the user (user expectations)
7. Context dependence of the information

In mobile computing environments, the cache management schemes need to address the following problems:

- How to reduce client-side latency
- How to maintain cache consistency between various caches and the servers

- How to ensure high data availability in the presence of frequent disconnections
- How to achieve high energy/bandwidth efficiency
- How to determine the cost of a cache miss and how to incorporate this cost in the cache management scheme
- How to manage location-dependent data in the cache
- How to enable cooperation between multiple peer caches

The first two problems are not new to mobile computing environments. They have been studied extensively in distributed systems and wired network computing environments. The subsequent four problems stem from the features of mobile computing environments. And the last problem is one of taking advantage of the P2P paradigm of ad hoc networks. In the next section we present several cache management schemes that have been developed to address these problems specific to mobile computing environments. Before looking at these management schemes, let us look at some of the cache organization issues. These include

1. Where do we cache?
2. How many levels of caching do we use (in the case of hierarchical caching architectures)?
3. What do we cache (when do we cache a data item and for how long)?
4. How do we invalidate cached items?
5. Who is responsible for invalidations? What is the granularity at which the invalidation is done?
6. What data currency guarantees can the system provide to users?
7. What are the costs involved? How do we charge users?
8. What is the effect on query delay (response time) and system throughput (query completion rate)?

For example, in a mobile environment, data can be cached at the server, at a proxy, or at a client. Furthermore, the data at each cache depend on the frequency, pattern, and cost of access. In the case of a server, data cached depend on the aggregate access pattern and the cost of retrieving requested data from the data store (i.e., I/O cost). In the case of the proxy and client, the caching strategy has to take into account the access pattern, communication cost, and update rate of the data being cached. In reality, data can be cached at several places. In order to maintain the consistency of the cached data, different strategies are employed based on the communication cost and design goals. If the goal is to minimize uplink queries by a mobile client (saving battery power of the mobile node) and to minimize contention on narrow-bandwidth wireless links (saving both bandwidth and energy), then the server can broadcast invalidation reports periodically (Barbara and Imieliński, 1994). The invalidation report can be organized in different ways and at various data granularity levels (Jing et al., 1997; Mummert and Satyanarayanan, 1994). We describe these and other major techniques to address the preceding issues and characteristics of mobile environments in the next section.

3.4 Mobile Cache Maintenance Schemes

Data caching is especially important in mobile computing environments for improving data availability and access latencies particularly because these computing environments are characterized by narrow-bandwidth wireless links and frequent voluntary and involuntary disconnections from the static network. However, these very features of mobile environments, coupled with the need to support seamless mobility, make maintaining a consistent cache at the mobile client a challenging task. There are several reasons for this. First, the underlying cache maintenance protocol should not overburden wireless resources and the mobile device. Second, these protocols should be energy-efficient, tolerant of disconnections, and adaptive to varying the QoS provided by the wireless network.

3.4.1 A taxonomy of cache maintenance schemes

Before going into details of some important mobile cache maintenance scheme, let us first understand the main characteristics of these schemes through the following classification.

As we saw earlier, there are two different cache consistency requirements: *strong cache consistency* and *weak cache consistency*. Strong cache consistency specifies that the cached data always be up-to-date. Polling every time and invalidating data on modification are two approaches to achieve strong cache consistency. On the other hand, weak cache consistency allows some degree of data inconsistency. TTL-based consistency strategies are used when it is sufficient to guarantee weak consistency for a data item.

In general, there are three basic strategies for maintaining cache consistency: polling every time, TTL-based, and invalidation-based (Cao and Liu, 1998). With *polling-every-time* and *TTL-based* caching strategies, the client initiates the consistency verification; i.e., the client is responsible for verifying the data consistency before using them. In a TTL-based caching strategy, every cached data item is assigned a TTL value, which can be estimated based on the data item's update history. For example, the adaptive TTL approach in Cate (1992) estimates TTL based on the age of a data item. When the user request arrives for a data item x, if data item x's residence time has exceeded its TTL value, the client sends a message to the server to ask if x has changed. Based on the server's response, the client may get a new copy of x from the server (if data item x has changed since the last time the client cached x) or just use the cached copy to answer the user's request (if data item x has not been modified since the last time the client received a copy of x). For the polling-every-time approach, every time a data item is requested, the clients need to poll the server to verify if the cached data item has changed. The polling-every-time caching strategy can be thought of as a special type of TTL-based scheme, with the TTL field equal to zero for every data item. Polling-every-time and TTL-based approaches are used in many existing Web caches.

On the other hand, with *invalidation-based strategies*, the server initiates the cache consistency verification. Invalidation-based cache strategies are further classified into *stateless* and *stateful* approaches (Barbara and Imieliński, 1994). In a *stateless* approach, the server does not maintain information about the cache contents of the clients; i.e., the server does not know what data are cached or how long they have been cached by a particular client. This type of cache maintenance strategy is discussed in many papers, including those by Barbara and Imieliński (1994, 1995), Cao and Liu (1998), Jing and colleagues (1997), and Tan, Cai, and Ooi (2001). Stateless approaches can be further categorized into synchronous and asynchronous approaches. In *asynchronous* approaches, invalidation reports are sent out on data modification. In *synchronous* approaches, the server sends out invalidation reports periodically. For a detailed comparison and evaluation of different approaches in this category, interested readers should refer to Tan, Cai, and Ooi (2001).

In *stateful* approaches, a server keeps track of the cache contents of its clients. Although stateful approaches also can be categorized into synchronous and asynchronous approaches, there are hardly any schemes in the *stateful synchronous* category. One example of a *stateful asynchronous* approach has been the Asynchronous Stateful (AS) scheme proposed by Kahol et al., 2001. In this approach, a home location cache (HLC) is maintained for every client which is used to record the data items cached by that client and their last modification times. Based on this information, a client's HLC can generate asynchronously invalidation reports specific to that client.

The preceding taxonomy is not restricted to mobile computing environments. Here, however, we are going to concentrate on the cache consistency strategies specified for mobile computing environments. In these environments, the main challenge for consistency maintenance strategies is caused by weak connectivity and the resource constraints of mobile computing environments.

3.4.2 Cache maintenance for push-based information dissemination

Traditional CS information systems use pull-based communication schemes for information access, in which clients initiate data transfers by requesting data from a server. Such pull-based techniques are not suitable for architecture-based wireless networks because they require substantial upstream communications. To make use of the downstream communication capacity, *push-based* information system architectures have been developed, where data are pushed from the servers to clients (Imieliński, Viswanathan, and Badrinath, 1994a, 1994b; Acharya et al., 1995; Acharya, Franklin, Zdonik, 1997). The idea is that the server periodically broadcasts frequently accessed data (hot data items) on a broadcast channel. The mobile node can tune into the broadcast channel at the start of the broadcast, determine when the data items in which it is interested will be available on the channel by reading the index information, and then go to sleep until the time when the data item is on the channel.

Such push-based architectures present a new challenge for cache management. Traditionally, caches are used to store the most frequently used data, but in push-based environments, the cost of obtaining the data also should be considered. For example, consider the following scenario from Acharya and colleagues (1995): Assume that data item x is accessed 1 percent of the time at a client C and is also broadcasted 1 percent of the time and that another data item y is accessed 0.5 percent of the time at C but is broadcast only 0.1 percent of the time. This implies that the time period between two occurrence of data item y on the broadcast channel ty is 10 times the broadcast period tx of data item x. If we choose to cache x instead of y, then the client will experience longer delays when a cache miss happens for y. This will adversely affect the average data access delay. Intuitively, the penalty for not caching a data item on the average data access latency is proportional to the product of access frequency and broadcast period. For this example, the product of access frequency and broadcast period for data item y is higher than that for data item x.

Therefore, new cache management algorithms have been developed for pushed-based information systems that take into account the cost of a cache miss. Note that in traditional cache systems all cache misses are assumed to have the same cost. This is not so in push-based information access architectures. In general, an important metric to evaluate these cache management algorithms is the achievable *hit ratio*, the fraction of total data requests satisfied from the cache. This metric depends not only on cache management algorithms but also on the cache size and the particular request pattern. In mobile computing environments, the hit ratio should not be the only metric to evaluate cache management algorithms (Satyanarayanan, 1996) because its underlying assumption is that all cache misses have equal cost. This assumption does not necessarily hold in weakly connected environments, where cache miss cost also depends on data size and timing. New metrics representing different cache costs in mobile computing environments therefore are needed. Let us consider one such metric called PIX.

Acharya and colleagues (1995) proposed a cost-based page-replacement algorithm using PIX. Suppose that the access probability of data item d is P and the broadcast frequency is X, then the PIX value of d is P/X. For the preceding example, the algorithm replaces item x with y because y has a higher PIX value. Note that, in general, the broadcast pattern of the server also should be considered for the client's cache management algorithms.

3.4.3 Broadcasting invalidation reports

In this section we describe cache consistency management schemes based on broadcasting invalidation reports. This technique is for wireless cellular environments. We begin describing the *broadcasting timestamp* (BT) scheme from Barbara and Imieliński (1995). As illustrated in Fig. 3.7, a data server periodically broadcasts *invalidation reports* that consist of all the invalidations in a *window* of the last w time units. An *invalidation* consists of a pair of the form (id, ts), where id is the identifier of a data item that was modified at most w time units ago, and ts is the timestamp denoting the time when it was modified, that is, $t - w \leq ts < t$, where t is the time when the invalidation report was sent by the server. On receipt of an invalidation report with timestamp t, denoted as IR(t), within w time units from last invalidation report, a mobile client m performs the following operation to validate its cache: For each data item id in m's cache with (id, ts) found in report IR(t), if the timestamp of id in m's cache is less than ts, then the cached entry for id is stale, and it is deleted from m's cache; otherwise, the timestamp of the cached entry of id is set to t.

The window size w decides how long a client can sleep (be disconnected from the network) and still be able to validate the cached entry of a data item. A mobile client that sleeps longer than w time units has to discard all the items in its cache (or revalidate them before use). All the clients in the wireless cell experience a delay for any access to a cached entry until the arrival of the next invalidation report. On receipt of an IR, a mobile client validates the cached entry.

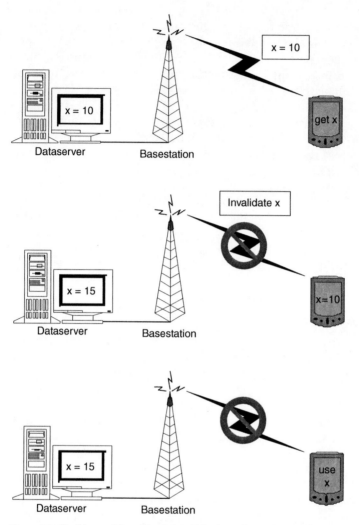

Figure 3.7 Problem with callback invalidation scheme in mobile computing environment.

There are several works based on the idea of periodically broadcasting invalidation reports by Barbara and Imieliński (1994). Jing and colleagues (1997) have proposed a scheme to adjust the size of the invalidation report to optimize the use of wireless bandwidth while retaining the effectiveness of cache invalidation. Liu and Maguire (1996) have proposed a two-level caching scheme based on mobility agents that takes into account the mobility pattern of the user to restrict broadcasting of the reports in the neighborhood of the user's current location. Hu and Lee (1998) have proposed broadcasting invalidation report methods that take into account the update and query rates/patterns and client disconnection time to optimize the uplink query cost. These invalidation report schemes have the following characteristics:

1. They assume a stateless server and generally do not address the issue of mobility directly. An exception to this is the work by Liu and Maguire, 1996.

2. The entire cache is invalid if the client is disconnected for a period larger than the period of the broadcast—making scalability of such schemes questionable for large databases.

3.4.4 Disconnected operation

In order to increase the utility of mobile computers, it's highly desirable that users are able to continue working on their tasks irrespective of the state of connectivity. This, however, is problematic when the application data is shared by several other clients. Let us consider Coda's approach to this problem of permitting disconnected operation on shared data.

Disconnected computing (operation) is possible whenever some information is better than no information at all. In other words, whenever data availability is more important than data consistency. On the other hand, when obtaining current data is important, then disconnected operation (or modification of data) should not be permitted. In distributed systems design, a trade-off exists between availability and consistency. The Coda file system provides support for disconnected operations on shared files in UNIX-like environments (Kistler and Satyanarayanan, 1992). The goal is to support file operations even while the user is disconnected from the network. Caching at clients is used along with server replication to improve availability of the data. Coda also uses two mechanisms for cache coherency. While the client is reachable from at least one server, a callback mechanism is used. When disconnection occurs, access to possibly stale data is permitted at a client for the sake of improving availability. On reconnection, only those modifications at the client are committed which do not cause any conflict. A balance between the speed of validating the cache (after a disconnection) and the accuracy of invalidations is achieved by maintaining version timestamps on volumes (a subtree in the file system hierarchy). However, validating the entire cache on every reconnection may put an unnecessary burden on the clients. Furthermore, since Coda is a distributed file system, it assumes a stateful server, which may not be appropriate for other applications, such as Web caching. Coda clients hoard files while they are connected to the data server. Hoarding is a cache maintenance scheme to facilitate disconnected operations. The main issues in hoarding are

1. What data items (files) do we hoard?

2. When and how often do we perform hoarding?

3. How do we deal with cache misses?

4. How do we reconcile the cached version of the data item with the version at the server?

Coda uses a prioritized scheme to determine what to hoard. It uses user-assigned priorities on data items (files) along with the data access pattern information. Periodically, a *hoard walk* is performed on the cache to ensure that no uncached object has a higher priority than any cached object. Hoard walking is a way to ensure that the cache is in *equilibrium* in the sense that it meets user expectations about availability when disconnected from the network.

3.4.5 Asynchronous stateful (AS) scheme

In this section we describe a caching scheme called the *asynchronous stateful* (AS) scheme (Kahol et al., 2001).

AS uses *asynchronous* invalidation reports (callbacks) to maintain cache consistency, as opposed to a broadcasting timestamp scheme, which sends invalidation reports periodically. Further, invalidation reports are sent to a mobile host only when some data change. The AS scheme is illustrated in Fig. 3.8. For each mobile client (host), a *home location cache* (HLC) is maintained by its *home agent* (HA) to assist with handling disconnections. The HA is similar in concept to the HA in Mobile IP (see Chap. 2) because all the messages between the mobile client and the data server pass through the HA. The difference is that the HA here is used to assist in cache consistency maintenance, whereas the HA in Mobile IP is used for location management and routing. Further, the HA in AS architecture can be maintained at any trusted static host in the network and can be moved closer to the mobile host for improving performance.

AS assumes the following computing scenario: The application program runs on the mobile host as a client process and communicates with the data server through messages; i.e., the client sends an *uplink request* (query) for the data it needs to the data server, and the server responds by sending the requested data on the *downlink*. In order to minimize the number of uplink requests, the client caches some data in its local memory. The objective of this scheme is to minimize the overhead for mobile hosts to validate their cache on reconnection, to allow stateless servers, and to minimize the bandwidth requirement. These

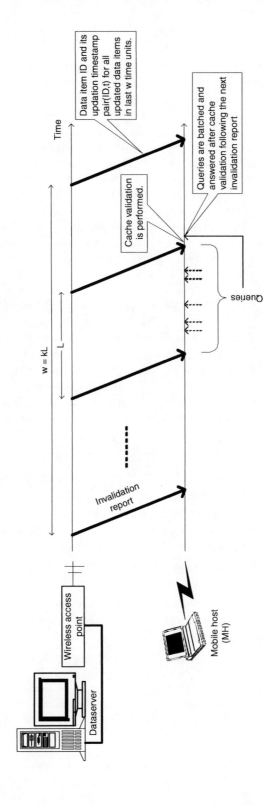

Figure 3.8 Broadcasting timestamp scheme.

objectives are achieved by buffering the invalidation messages in a mobile host's HLC by its HA. The HLC can be viewed as a proxy in the CS proxy architecture for adaptive computing we saw in Chap. 1. Here the HLC helps in adapting to a client's arbitrary disconnection pattern.

A mobile host's HLC has an entry for each data item cached by the host in which it maintains only the timestamp when that data item was last invalidated. At the cost of this extra overhead of maintaining an HLC, a mobile host can continue to use its cache even after prolonged periods of disconnection from the network. This is a considerable savings over the broadcasting timestamp scheme we saw previously because in the broadcasting timestamp scheme, the cache has to be discarded if the disconnection period exceeds the invalidation window duration w.

In the AS scheme, a mobile host is considered to be in two modes: *awake* or *asleep*. When a mobile host is awake—connected to the network and hence the data server—it can receive invalidation messages. Thus this state includes both active and dozing CPU modes. A mobile host can be disconnected from the network either voluntarily or involuntarily. From the perspective of the mobile host's cache maintenance, it is irrelevant whether the invalidation was lost due to voluntary disconnection (e.g., switching off the laptop) or involuntary disconnection (e.g., wireless link failure, handoff, etc.). Thus a disconnected client is considered to be in sleep mode. The term *wakeup* is used to indicate the reestablishment of connection between the mobile host and the data server.

The AS scheme makes the following assumptions:

- Whenever the data server updates any data item, an invalidation message is sent out to all HAs via the wired network (IP Multicast can be used for this purpose). The HAs forward this message to the relevant mobile hosts; thus, when a mobile host is roaming, it gets the invalidation message if it is not disconnected.

- No message is lost due to communication failure or otherwise in the wired network.

- A mobile host can detect whether or not it is connected to the network. Whenever the mobile host is disconnected, it suspends answering any query from the application.

- A mobile host informs its HA before it stores (or updates) any data item in its local cache.

The AS scheme works as follows. The HA keeps track of what data have been locally cached at its mobile hosts (cache state information of the mobile host). In general, the HLC for mobile host m is a list of records $(x, T,$ invalid flag) for each data item x locally cached at m, where x is the identifier of a data item, and T is the timestamp of the last invalidation of x. The timestamp is the same as that provided by the server in its invalidation message. The invalid flag (in the HLC record for the specific data item) is set to TRUE for data items for which an invalidation has been sent to the mobile host but no acknowledgment has yet been received. The cache timestamp that is sent by the mobile host with its query message serves as an implicit acknowledgement for the invalidation messages.

Each mobile host maintains a local cache of data items that it accesses frequently. Before answering any queries from the application, it checks if the requested data are in a consistent state. Callbacks from an HA are used to achieve this goal. When an HA receives an invalidation message from a data server, the HA determines the number of mobile hosts that are using the data by consulting their HLCs and sends an invalidation report to each of them. When a mobile host receives the invalidation message, it discards that data item from its local cache. When a mobile host receives (from the application) a query for a data item, it satisfies the query from its local cache if the data item is present in the cache, which saves latency, bandwidth, and battery power; otherwise, an uplink request to the HA for the data item is required. The HA asks the server for the data item on behalf of the mobile host. On receiving the data item, the HA adds an entry to the mobile host's HLC for the requested data item and forwards the data item to the mobile host.

A mobile host alternates between active modes and sleep modes. In the sleep mode, a mobile client is unable to receive any invalidation messages sent to it by its HA, and it suspends processing of any queries from the applications. The following timestamp-based scheme is used by which the HA can decide which invalidations it needs to retransmit to the mobile host. Each client maintains a timestamp for its cache called the *cache timestamp*. The cache timestamp is the timestamp of the last message received by the mobile host from its HA. The client includes the cache timestamp in all its communications with the HA. The HA uses the cache timestamp to discard invalidations that it no longer needs to keep and to decide which invalidations it needs to resend to the client. On receiving a message with timestamp t, the HA discards any invalidation messages with timestamp less than or equal to t from the mobile host's HLC. Further, it sends an invalidation report consisting of all the invalidation messages with timestamps greater than t in the mobile host's HLC to the mobile host.

When a mobile host wakes up after a sleep, it sends a *probe* message to its HA with its cache timestamp. The probe message is piggybacked on the first query after waking up to avoid unnecessary probing. In response to this probe message, the HA sends an invalidation report. In this way, a mobile host can determine which data items changed while it was disconnected. A mobile host defers answering all queries that it receives after waking up until it has received the invalidation report from its HA. In this scheme, the time at which the mobile host got disconnected is not needed. Simply by maintaining a cache timestamp, both wireless link failures and voluntary disconnections are handled. Even if the mobile host wakes up and then immediately goes back to sleep before receiving the invalidation report, consistency of the cache is not compromised because it would use the same value of its cache timestamp in its probe message after waking up and hence get the correct information in the invalidation report. Thus AS can handle arbitrary sleep patterns of the mobile host.

Consider the example scenario shown in Fig. 3.9. Initially, the cache timestamp of the mobile host is t_0, and mobile host's cache has two data items with IDs x and z. When the HLC receives an invalidation message notifying it that x has changed at the server at time t_1, it adds the invalidation message to mobile host's HLC and also forwards the invalidation message to the mobile host with data item ID and timestamp, i.e., $(x; t_1)$. On receiving the invalidation message from the HLC, the mobile host updates its cache timestamp to t_1 and deletes data item x from its cache. Later, when the mobile host wants to access y, it sends a data request with $(y; t_1)$ to the HLC. In response to the data request, the HLC fetches and forwards the data item associated with y to the mobile host and adds $(y; t_2)$ to the mobile host's HLC, where t_2 is the timestamp of the last update provided by the data server. The mobile host updates its timestamp to t_2 and adds y to its cache. Now suppose that the mobile host gets disconnected from the network, and the invalidation message for y is lost because of this disconnection. When the mobile host wakes up, it ignores any invalidation messages it receives (until the first query) because later, for its first query after waking up, it sends a probe message (invalidation check message) to the HLC. The HLC uses the timestamp in this probe message to determine the invalidations missed by the mobile host and sends an invalidation report with all the invalidations missed by the mobile host. In this case, the HLC determines from the mobile host's cache timestamp t_2 that the mobile host has missed invalidations for y and z, so it resends them to the mobile host.

The AS scheme ensures, in absence of any loss of invalidation reports in wired networks, that the data returned to a mobile client is at most t seconds old, where t is the maximum latency of forwarding an invalidation report from the server to the client via its HA. This is different from the weak-consistency schemes, e.g., the Coda scheme, where a disconnected client is allowed both read and write access to its cached data. When the client gets reconnected, its cached data are reintegrated with the server data. The user needs to resolve any conflicts arising from independent modification of data by multiple disconnected clients. Coda's design is suitable for its goals, i.e., providing UNIX-like file

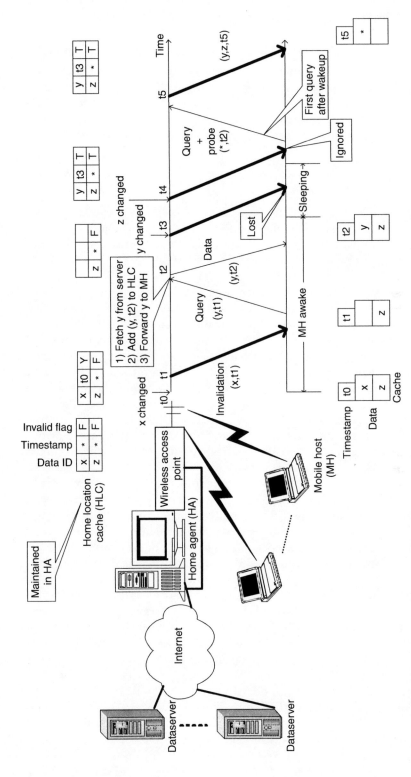

Figure 3.9 AS scheme for cache consistency maintenance.

system semantics in an environment where write-write conflicts are rare (e.g., college campus environments) and access to stale data does not lead to dire consequences or is detectable by the application. In contrast, the AS scheme is designed for applications that require strict data currency guarantees and in which access to stale data is undesirable. Such applications include access to critical data such as bank account information and air traffic information.

3.4.6   To cache or not to cache?

Consider the problem of deciding whether to cache a particular data item x (e.g., a Web page) at a mobile node m. If x is never modified, then the decision is obvious—it is always beneficial to maintain a copy of x at m as long as resources are available to do so. However if x is a modifiable data item, then the decision to cache x at m is not straightforward. The problem arises because the future access and modification pattern (who will modify x when) of the data item is not known a priori. Static allocation schemes such as "always cache data item at m" (SA-always) and "never cache data item at m" (SA-never) are obviously not optimal. SA-always is costly when the data item is modified frequently at the server. Whenever the data item is modified at the server, the mobile node has to be contacted to update its cached copy. This uses precious wireless bandwidth and the mobile node's battery energy. In case the mobile node does not access the modified data item subsequently, the resources consumed to update the copy at the mobile node go to waste. However, if the mobile node vigorously accesses the updated version of the data item, then SA-always conserves resources because now the mobile node does not have to contact the server multiple times to ensure that it has the most recent copy of data item x. Similarly, SA-never is costly when the data item is modified infrequently at the server. In fact, the frequency of modification should be considered in relation to the access rate of data item x by m. If the cache management scheme can somehow foresee the access patterns, then it can determine dynamically which scheme to use in the future—SA-never or SA-always. Prediction of the future access pattern is facilitated by temporal locality in the access pattern. Dynamic optimization algorithms are known as *online algorithms* in computer science parlance. Online algorithm performance is compared with that of *offline algorithms* (algorithms that have complete knowledge of the future in making the optimization decision) in terms of competitive ratio. Informally, an online algorithm is α competitive if the quality of the optimization decision is always within a factor of α of the optimal offline algorithm.

Sistla, Wolfson, and Huang (1998) have developed an online algorithm called a *sliding-window dynamic data allocation scheme* to determine when to cache a data item at a mobile client. Intuitively, the scheme decides to choose between SA-always and SA-never based on the access pattern in the recent past. That is, the data item is allocated/deallocated at the mobile node based on the access pattern history. This history is maintained as a window of size k (a parameter of the allocation scheme), and it consists of "relevant" access operations—a read operation at the mobile node (rm) and a write operation at the server (ws). The "irrelevant" operations (a write operation at the mobile node [wm] and a read operation at the server [rs]) are not considered because the "cost" of these operations remains the same whether or not the mobile node has a copy of the data item—in the case of wm, the write always needs to be propagated to the server (an assumption of this scheme), and in the case of rs, the server always has the up-to-date copy as a consequence of the previous assumption. Let us assume a cost model in which each message costs a unit. Under this message-passing cost model, the cost of relevant operations are as follows:

- Each rm costs 1 unit if the mobile node does not have the copy of the data item; otherwise, its cost is 0 units.

- Each ws costs 1 unit if the mobile node has the copy of the data item (because the write has to be propagated to the mobile); otherwise, its cost is 0 unit.

A schedule ψ is a sequence of relevant operations. For example, $\psi =$ (ws, rm, rm, ws) is a schedule with two writes by the server separated by two reads by the mobile node. The cost of the schedule ψ under allocation scheme A, denoted as cost(ψ, A), is the sum of the cost of each operation in ψ under allocation scheme A. For example, cost(ψ, SA-always) is $1 + 0 + 0 + 1 = 2$, and cost(ψ, SA-never) is also $2 = 0 + 1 + 1 + 0$.

Now consider a dynamic allocation scheme in which the data item is allocated at the mobile node after the first ws operation and deallocated after second rm operation. Assuming that the allocation and deallocation of data items at the mobile cost nothing, the cost of the schedule is 0. In reality, there is no free lunch. When the data item is allocated at the mobile node, some state information may have to be sent to the mobile node by the server. Also, when the data item is deallocated at the mobile node, some state information may have to be sent to the server by the mobile node.

In the sliding-window scheme, the state information is the window itself. When the data item is allocated at the mobile node, the mobile node is in charge of maintaining the window because it sees all the relevant operations. However, when the data item is not cached at the mobile node, the window is maintained at the server. Under the message-passing cost model, each of these operations costs 1 unit. Continuing the preceding example, the cost of ψ still would be 2 when we take into account the allocation and deallocation costs. The dynamic scheme will incur a much lower cost for schedules that have a long sequence of ws operations followed by a long sequence of rm operations. For example, a schedule with m ws operations followed by n rm operations will have a cost of m and n for SA-always and SA-never, respectively. For the dynamic scheme, however, the cost of this schedule would be just 1 unit because the data item can be allocated at the mobile node after the last write operation at the server, resulting in zero cost for all the subsequent reads at the mobile node. In fact, the dynamic scheme is the optimal offline scheme.

The sliding-window scheme with window size k [SW(k)] maintains the last k relevant operations. After the window is full, each subsequent relevant operation is added to the window, and the oldest operation is deleted from the window. The SW(k) makes the deallocation/allocation after each relevant operation has been added to the window based on the following rules:

- *Case "data item is not cached at the mobile node."* If the window has more rm operations than ws operations (i.e., number of rm operations is greater than $k/2$), then allocate the data item at the mobile node.

- *Case "data item is cached at the mobile node."* If the window has more ws operations than rm operations, then deallocate the data item from the mobile node.

These algorithms are shown to be competitive with respect to the optimal offline algorithm. In this work, it is implicitly assumed that the mobile client is always connected to the network. This is necessary to maintain the history information, which is maintained dynamically either at the mobile client or at the server. Furthermore, some additional issues involving implementation, such as caching multiple data items and finite cache sizes (cache replacement policy), have to be considered in order to use such data allocation schemes in practice.

3.5 Mobile Web Caching

Mobile Web caching mechanisms employ a TTL-based cache consistency maintenance scheme. In the TTL-based cache consistency strategy, the clients are responsible for polling the server to verify that the cached data is up-to-date. WebExpress (Housel, Samaras, Lindquist, 1998) and Mowgli (Liljeberg et al., 1996) are good examples of TTL-based cache consistency strategies adapted to wireless networks. Here we discuss how these approaches help to solve the problem of frequent disconnections and narrow bandwidth.

3.5.1 Handling disconnections

The caching system in Mowgli supports the disconnected mode in various ways. First, Mowgli chooses to validate documents only when explicitly requested to do so by the user. Second, Mowgli suffixes each hypertext link in a document with an indicator that tells the user if the referred document would have to be fetched from a server. In addition, Mowgli offers the user access to cached documents directly through a cache inventory. In this way, the user can stay safely within the bounds of the cache. Third, the cache maintained by the Mowgli agent is persistent, which means that it is stored on disk and retained over multiple browsing sessions.

3.5.2 Achieving energy and bandwidth efficiency

WebExpress uses cyclic redundancy codes (CRC) to help reduce unnecessary transmission in maintaining cache coherency. When a requested object in the client cache has exceeded the coherency interval defined by the client, the CRC of this object is transmitted to the server to determine the difference between the fresh copy and the cached copy. A new copy will be fetched only when the difference exceeds a specified threshold.

WebExpress also employs the *differencing technique*. To update the cached data item, the entire data item is not refetched from the server; instead, WebExpress updates the cached data item based on its difference from the fresh copy. For dynamic Web page requests, a common base object is cached on both the client and the server sides. The difference between these versions is calculated and transmitted to the client in response to a new request.

Some techniques for allowing the client to further conserve energy during the wait period of *on-demand* or *pull-based* information access have been developed recently. For example, see Krashinsky and Balakrishnan (2002).

Protocol optimization.   In WebExpress, each client connects to the server with a single Transmission Control Protocol/Internet Protocol (TCP/IP) connection. All requests and responses are multiplexed over this connection to avoid repeated connection establishment overhead. In addition, when the client establishes a connection with the server, it sends its capabilities on only the first request, and the server maintains this state information to avoid multiple transmissions of the same information. In Mowgli, there are two levels of protocol optimization. At the transport level, TCP over the wireless part of the network is replaced by MTCP, which is a lightweight protocol, has minimal packet headers, involves as few round trips over the wireless link as possible, and also provides improved fault tolerance. At the application level, HTTP is replaced by the binary-encoded protocol MHTTP, which supports the predictive upload of documents and document objects.

3.6 Summary

In this chapter we saw various techniques that have been developed to minimize the overhead of accessing data by a mobile client. The broadcast nature of wireless communication is exploited to provide scalable and energy-efficient publish-mode access to data. This mode is also useful in a communication environment with asymmetric channel capacity. Caching of frequently accessed data locally at a mobile node can further reduce the overheard. However, a problem arises when the mobile node gets disconnected and the application needs to modify cached data. Hoarding is a scheme that tries to ensure availability of data during disconnection. Data objects that are modified when the mobile node is disconnected from the network have to be reintegrated with the primary copy at the server on reconnection. Several interesting algorithmic problems associated with doing so, such as optimal bandwidth allocation to uplink and downlink channels, computation of an optimal broadcast schedule, and designing an optimal cache replacement scheme, are active areas of research.

3.7 References

Acharya, S., M. Franklin, and S. Zdonik, "Balancing Push and Pull for Data Broadcast," in *Proceedings of the ACM SIGMOD, Tuscon AZ, ACM Press, New York, NY.* 1997, p. 183.

Acharya, S., R. Alonso, M. Franklin, and S. Zdonik, "Broadcast Disks: Data Management for Asymmetric Communication Environments," in *Proceedings of the ACM SIGMOD International Conference on the Management of Data, San Jose CA, ACM Press New York, NY.* 1995, p. 199.

Barbara, D., and T. Imieliński, "Sleepers and Workaholics: Caching Strategies for Mobile Environments," in *Proceedings of the ACM SIGMOD Conference on Management of Data, Minneapolis Minnesota, ACM Press New York NY.* 1994, p. 1.

Barbara, D., and T. Imieiński, "Sleepers and Workaholics: Caching Strategies for Mobile Environments (Extended Version). *International Journal on Very Large Data Bases* 4(4):567–602, 1995.

Cao, P., and C. Liu, "Maintaining Strong Cache Consistency in the World Wide Web," *IEEE Transcations on Computers* 47(4):445–457, 1998.

Cate, V., "Alex—A Global File System," in *Proceedings of 1992 USENIX File System Workshop, Ann Arbor MI, USENIX Berkeley CA.* 1992, p. 1.

Housel, B. C., Samaras, G., and Lindquist, D. B., "WebExpress: A Client/Intercept Based System for Optimizing Web Browsing in a Wireless Environment," *Mobile Networks and Applications* 3:419–431, 1998.

Hu, Q., and D. K. Lee, "Cache Algorithms Based on Adaptive Invalidation Reports for Mobile Environments," *Cluster Computing* 1:39, 1998.

Imieliński, T., S. Viswanathan, and B. R. Badrinath, "Power Efficient Filtering of Data on Air," in *Proceedings of the 4th international conference on extending database technology, Cambridge, UK, Springer-Verlag NY, Inc. New York NY.* 1994a, p. 245.

Imieliński, T., S. Viswanathan, and B. R. Badrinath, "Energy Efficient Indexing on Air," in *Proceedings of 1994 ACM SIGMOD International Conference on Management of Data, Minneapolis MN, ACM Press.* 1994b, p. 25.

Imieliński, T., and S. Viswanathan, "Adaptive Wireless Information Systems," in *Proceedings of SIGDBS (Special Interest Group in DataBase Systems) Conference Tokyo, Japan ACM press*, 1994, p. 19.

Kahol, A., S. Khurana, S. Gupta, and P. Srimani, "A Strategy to Manage Cache Consistency in a Disconnected Distributed Environment," *IEEE Transactions on Parallel and Distributed Systems* 12(7):686–700, 2001.

Kistler, J., and M. Satyanarayanan, "Disconnected Operation in the Coda File System," *ACM Transactions on Computer Systems* 10(1):3, 1992.

Krashinsky, R., and H. Balakrishnan, "Minimizing Energy for Wireless Web Access with Bounded Slowdown," in *Proceedings of MOBICOM 2002. Atlanta GA*, 2002, p. 119.

Liu, G. Y., and G. Q. Maguire, Jr., "A Mobility-Aware Dynamic Database Caching Scheme for Wireless Mobile Computing and Communications," *Distributed and Parallel Databases* 4:271–288, 1996.

Liljeberg, M., H. Helin, M. Kojo, and K. Raatikainen, "Enhanced Services for World-Wide Web in a Mobile WAN Environment," University of Helsinki, Department of Computer Science, Series of Publications C, No. C-1996-28, 1996.

Mummert, L., and M. Satyanarayanan, "Large Granularity Cache Coherence for Intermittent Connectivity," Proceedings of the 1994 Summer USENIX Conference, Boston June, USENIX Berkeley CA, 1994, p. 279.

National Laboratory for Applied Network Research (NLANR), "A Distributed Testbed for National Information Provisioning"; available at *http://ircache.nlanr.net/*.2002

Satyanarayanan, M., "Fundamental Challenges in Mobile Computing," in *Proceedings of the Fifteenth Annual ACM Symposium on Principles of Distributed Computing.* Philadelphia, PA, ACM Press 1996, p. 1.

Silberschatz, A., and P. Galvin, *Operating System Concepts.* Reading, MA: Addison-Wesley, 1998.

Sistla, A. P., O. Wolfson, and Y. Huang, "Minimization of Communication Cost Through Caching in Mobile Environments," *IEEE Transactions on Parallel and Distributed Systems* 9(4):378–389, 1998.

Yin, J., L. Alvisi, M. Dahlin, et al., "Using Leases to Support Server-Driven Consistency in Large-Scale Systems," in *Proceedings of the 18th International Conference on Distributed Computing Systems.* Amsterdam, the Netherlands, IEEE Computer Science Press, 1998, p. 285.

Context-Aware Computing

The success of mobile computing is contingent on how gracefully the system adapts to changes in the environment. In Chap. 1 we introduced two approaches to developing adaptive mobile systems: the *application-transparent* approach and the *application-aware* approach. The application-transparent approach uses system software, such as the underlying operating system and networking software, to adapt to changes in operating conditions, such as a substantial reduction in available bandwidth or a drop in the level of battery power, in an application-independent manner. In contrast to this approach, application-aware adaptation uses collaboration between the system software and the application software to adapt to changes in the availability of computing and communication resources.

A step further in the direction of application-aware adaptation is the *context-aware computing paradigm.* In context-aware computing, the application adapts not only to changes in the availability of computing and communication resources but also to the presence of contextual information, such as who is in the vicinity, the time of day, where the system currently resides, the current emotional state of the user, the action the user is performing, and the intention with which that action is being performed. A context-aware (or context-sensitive) application requires contextual information that must be gathered from various sources, such as sensors that are embedded in the environment, devices that are worn by end users, repositories of historical data tracking use of the application, and information contained in user profiles.

Let's consider an illustrative scenario that demonstrates context-aware computing. For example, suppose that you are a tourist visiting Bombay (Mumbai), India, for the first time. You have your personal digital assistant (PDA) with wireless service from a service provider that covers Bombay. Being an adventurous tourist, you have not made hotel reservations in advance. Fortunately, you have a context-based Web information service (WISE) available to you from your wireless service provider. The first thing you do is ask WISE for suggestions for hotels. WISE takes into account information such as your itinerary, preferences (such as price range), and current location and provides you with a list of hotels that are nearby. You then select one of the hotels and invoke an online hotel reservation form. Most of the details in the form are filled in automatically based on your preferences, which the system reads from your user profile. For example, it knows that you prefer to pay the hotel bill using your credit card, which earns you frequent flier miles, and that you favor staying in nonsmoking rooms.

This chapter will discuss how such context-aware applications can be developed. In later chapters of this book we discuss various sensing technologies and their applications, along with communication protocols employed in sensor networks formed by these sensing devices. In this chapter we assume the existence of such an infrastructure and concentrate on issues related to design and development of context-aware applications such as (1) how contextual information can be provided to the application and (2) how the application can be developed to react to the changing contextual information.

4.1 Ubiquitous or Pervasive Computing

Pioneering work in context-aware computing was started in the early 1990s at Xerox PARC Laboratory and Olivetti Research, Ltd. (now part of AT&T Laboratories Cambridge), under the vision of *ubiquitous computing.* (Since the mid-1990s, ubiquitous computing also has been known as *pervasive computing.*) Marc Weiser, in his seminal paper entitled, "The Computers of 21st Century" (Weiser, 1991), envisioned that in accordance with Moore's law, future computing environments would consist of very cheap (disposable) interconnected specialized computers all around us, some embedded in our surroundings and others worn by us (Fig. 4.1). However, if the usage model of ubiquitous computing systems follows the trend of the usage model of mainframe and personal computers, where a substantial effort is required on the part of users to accomplish any computing or communication tasks, then we would be constantly distracted by these numerous devices. The aim of ubiquitous computing is to design computing infrastructures in such a manner that they integrate seamlessly with the environment and become almost invisible. This is analogous to the "profound technologies," such as the electric motor (a typical automobile has more than 25 motors) and writing technology (our environment is equipped with whiteboards, notepads, Post-It notes for various needs without being a distraction), that "weave

Figure 4.1 Ubiquitous computing vision.

themselves into the fabric of everyday life until they are indistinguishable from it" (Weiser, 1991).

To meet the goals of ubiquitous computing systems, Weiser and Brown (1996) suggest designing *calm* computing technology, which, rather than always being at the center of our attention, empowers our peripheral attention and has the agility to move fluidly between the periphery and the center of our attention. Researchers in Project Aura at Carnegie Mellon University (http://www-2.cs.cmu.edu/~aura) have found that to minimize human distraction, ubiquitous computing systems have to be *proactive* in anticipating the future demands of the user and *adaptive* (*self-tunable*) in order to be able to respond better to future user demands. The ability to sense and process context is fundamental to making a system proactive and self-tunable.

In this chapter the emphasis will be on context-aware mobile computing. Understandably, this form of computing is essential to meet the goals of ubiquitous computing. Motion is an integral part of our daily life, and any ubiquitous system that does not reasonably support mobility of computing devices will have difficulty in becoming "invisible" to the user. Essentially, this form of computing is broader than mobile computing because it concerns not just *mobility of computers* but, more important, *mobility of people*.

4.2   What Is a Context? Various Definitions and Types of Contexts

As human beings, we are adept in our understanding and use of context in our daily activities. We routinely use contextual information, such as who is in our vicinity or where we are, to modulate our responses to or interactions with other people. As we mature, we learn which contextual information is important given a particular situation. For example, when we are with our close friends, we rarely worry about how we speak: "Like, you know what I mean." However, in a formal situation, we are careful in what we say and how we say it.

To use contextual information in adapting applications, there should exist a means to capture, store, and process context. However, what exactly is context? Let us look at an English definition of context. *Merriam-Webster's Collegiate Dictionary* defines *context* as "(1) the parts of a discourse that surround a word or passage and can throw light on its meaning; (2) the interrelated conditions in which something exists or occurs."

The word *context* has its origin in the Latin verb *contexere*, meaning "to weave together." However, such dictionary definitions of the word *context* are too general to be of much use in developing context-aware applications. These definitions do not help in determining what contextual information an application should try to acquire or how it should use the contextual information it has. Attempts to provide a computer-friendly definition of context has led to numerous definitions, falling under two broad categories: *enumeration-based*, in which context is defined in terms of its various categorizations, and *role-based*, in which context is defined in terms of its role in context-aware computing.

4.2.1   Enumeration-based

An example of a definition of context in terms of various categories is Chen and Kotz's (2000) refinement of Schilit's definition of context: *Context* consists of the following categories:

1. *Computing context* includes network connectivity, communication costs, communication bandwidth, and local resources, such as printers, displays, and workstations.

2. *User context* includes user profiles, location, and people in the vicinity of the user.

3. *Physical context* includes lighting and noise levels, traffic conditions, and temperature.

4. *Temporal context* includes time of day, week, month, and season of the year.

5. *Context history* is the recording of computing, user, and physical context across a time span.

As we develop and use increasingly context-aware applications, a universally acceptable definition of context may evolve. For now, we can try to identify some essential types of context using these lines from a poem by Rudyard Kipling to guide us: "I keep six honest serving men. They taught me all I knew. Their names are What and Why and When and How and Where and Who." The following five W's of context can form the core of different context types used by an application (Abowd and Mynatt, 2000):

- *Who (social context)*. This consists of information such as user identification and identification of people near the user. A context-aware system can use the identification of the person who is using the system to determine how to respond based on the user's preference. For example, WISE used the user's ID to get information about the preferences. Programs can use an end-user's information to implicitly perform certain actions.

- *What (functional context)*. This consists of information about what tasks the user is performing.

- *Where (location context)*. This consists of information about where the system is currently located. This information can be raw location information, such as the latitude and longitude of the user, or it can be obtained at a higher level, such as the number of the room in which the system is currently operating. Location context is the most prolifically used type of context for developing context-aware applications.

- *When (temporal context)*. This is the same as the temporal context defined earlier.

- *Why (motivating context)*. This specifies why the user is performing a certain task. This is one of the most difficult types of contextual information to determine.

In addition to these different kinds of contextual information, one also can consider information such as a user's emotional state (*emotional context*) and information about the environment, such as room temperature and illumination level (*environmental context*) (Satyanarayanan, 2002).

Context can be categorized further into low-level context and high-level context (Chen and Kotz, 2000). *Low-level context* information can be sensed directly using sensors or through simple processing, e.g., by accessing a database, e.g., room temperature or devices that are in the vicinity of a user. *High-level context* information may involve the amalgamation of low-level context information as well as sophisticated processing, such as machine vision or artificial intelligence (AI) techniques. For example, low-level context information such as the current location of the user and the current time can be combined from the user's calendar to obtain information about a user's current social situation, inferring that the user "is in a meeting,", "attending a lecture," or "waiting at the airport."

4.2.2 Role-based

Perhaps a more useful way to discern what context is and how to use it is to look at context in terms of how it can be used by mobile applications. Chen and Kotz (2002) define context as "the set of environmental states and settings that either determines an application's behavior or in which an application event occurs and is interesting to the user."

Based on this definition, two types of context can be identified. *Active context* is the contextual information used by the application to adapt its behavior, whereas *passive context* is the contextual information that is not critical for application adaptation but is provided to the user to enhance his or her understanding of the situation (Chen and Kotz, 2002).

The task of building a context-aware application that adapts its response to so many different types of contexts is not only challenging but also quite daunting. Figure 4.2 illustrates the various types of contexts we have discussed. In the rest of this chapter we will discuss various existing software tools that aid in the development of context-aware applications. First, however, we will study different kinds of context-aware applications that have been developed by various researchers in industry and academia around the world.

4.3 Context-Aware Computing and Applications

Context-aware computing devices and applications respond to changes in the environment in an intelligent manner to enhance the computing environment for the user (Pascoe, 1997). Context-aware applications tend to be mobile applications for obvious reasons: (1) The user's context fluctuates most frequently when a user is mobile, and (2) the need for context-aware behavior is greatest in a mobile environment. At a minimum, context-aware applications should be proactive in acquiring contextual information and adapt their response based on the acquired information. The response itself can be proactive (automatically initiated

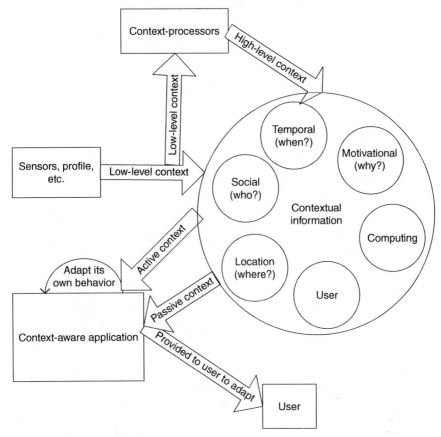

Figure 4.2 Various types of contexts.

by the system or application) or reactive (in response to the user's request). A context-aware application can perform various tasks, which may include providing a context-aware user interface, presenting contextual and noncontextual information to the user, context-sensitive information services, and either proactive (such as automatic reconfiguration) or reactive (such as context-sensitive querying) context-aware adaptation of behavior.

4.3.1 Core capabilities for context awareness

Applications should posses some of the following capabilities in order to be characterized as context-aware (Pascoe, 1998):

- *Contextual sensing* refers to the detection of various environmental states and how they are presented to the user. A basic context-aware application is one that presents the contextual information it obtains in a user-friendly form. For example, the location information obtained from a location sensor such as global positioning satellite (GPS) device can be presented to the user via a map display annotated with a "you are here" marker.

- *Contextual adaptation* is the capability of the system to adapt its behavior by using contextual information.

- *Contextual resource discovery* is the capability by which a system can discover available resources, which it can use to better adapt to the user's needs.

- *Contextual augmentation* refers to having the capability to associate contextual information with some digital data.

4.3.2   Types of context-aware applications

In order to get a better understanding of context-aware applications, let us now consider several different types of useful context-aware applications. In the past decade or so, various context-aware applications have been developed, and many methods for classifying context-aware applications have been proposed (Schilit, Adams, and Want, 1994; Dey, Salber, and Abowd, 2002). According to Schilit, Adams, and West (1994), context-aware applications can be classified along three different dimensions: function or service type, initiating agent, and adaptation (Table 4.1).

Function or service type.   Context-aware applications have been developed to perform various tasks. However, they can be broadly classified into applications whose primary task is related to providing information or actuating some command.

Table 4.1   Types of Context-Aware Applications (Schilit, 1994)

Application type	Function/service type	Initiation	Adaptation
Contextual selection	Information	Manual	Information
Information presentation	Information	Manual	User interface
Contextual information	Information	Manual	Information
Automatic contextual reconfiguration	Any	Automatic	Information or system
Contextual command	Command	Manual	Command
Context-triggered actions	Command	Automatic	None
Contextual tagging	Information	Automatic	Information

Initiation.   The context-aware application either can be initiated explicitly by the user (manual) or can be invoked implicitly by the application.

Adaptation.   Adaptation performed by a context-aware application consists of various types: information, system, user interface, and command (behavior).

Contextual selection refers to the selection and presentation of physical or virtual objects based on the user's context. The most important type of contextual selection in mobile computing is *proximity selection* (Schilit, Adams, and Want, 1994), where the selection of objects is based primarily on the location context of the user, e.g., (1) selection of resources and devices such as printers or speakers in the vicinity of the user, (2) selection of places such as gas stations and restaurants closest to the current location of the user, and (3) selection or addressing of objects with which the user is currently interacting, such as the group of people in the room to whom the user wants to send information. Under the umbrella of ubiquitous computing, many other types of contextual selections are possible. For example, one can perform selection based on the social context of the user. An example might be a user who wants to get information about the items in his family members' "wish list" when doing online Christmas shopping at the last minute. Another type of contextual selection is the selection of objects based on the temporal context of the user. As an example, consider a scenario in which you have made a large number of trips during the last summer and you would like to obtain a list of all the interesting people you met, along with their contact information.

Contextual selection applications need to employ various user-interface techniques to assist the user in the selection of objects in the physical or virtual world based on the user's context. An important issue is how to present the information to the user so that the relevance of the information is implicit in the presentation. In proximate selection, applications may employ various visual effects to convey the relative ordering of the objects to the user. For example, different sized fonts can be used to display a spatial relationship between an object and the user (the closer the object, the larger is the font size).

In order to minimize computation and communication resources, a contextual selection application may fine-tune the granularity and accuracy of information presented to the user. Suppose that it is almost lunchtime, and you are driving around in an unfamiliar city looking for a restaurant. Under such a scenario, the application may update the information regarding closer restaurants more frequently than that of distant restaurants.

Contextual information applications modulate their responses based on the context of the user. In contrast, *contextual selection applications* provide information about the context itself. These applications can be viewed as parameterizing user's queries with contextual information. The main contextual information that is used is the location of the user.

Similar to contextual information applications, *contextual commands* modulate their behavior based on the current context of the user. For example, a print command may, by default, print to the printer nearest to the user.

Automatic contextual reconfiguration applications automatically adapt the system configuration in response to a change in context. For example, the application may configure itself to use the display device available to the user.

Context-triggered actions are simple IF-THEN rules used to specify how context-aware software should respond automatically to contextual changes. The condition in the IF-THEN statement is based on the contextual predicates. There is an action associated with each IF-THEN rule that is performed when the associated condition becomes true.

4.3.3 Developing context-aware applications

In general, the following steps are used for developing context-aware applications:

1. Identifying relevant context

2. Specifying context-aware behaviors

3. Integrating with mechanisms for acquisition of contextual information

The first step is application-dependent, and the third step is platform-dependent. Thus we will not discuss them any further. Since the techniques to achieve the second step can be used in several applications, let's look at two different approaches to specifying context-aware behavior: context-triggered actions and Stick-E notes.

Context-triggered actions

Watchdog and contextual reminder for active badges.   The Watchdog program was designed for a UNIX environment that is coupled with an Active Badge Location System. The Active Badge is an electronic tag that periodically broadcasts a unique identifier for the purpose of determining the location of the wearer. Events such as "arriving," "departing," and "settled in" are generated by the Active Badge Location System, which gives information about the mobility of the wearer. If the system does not hear from a certain badge for certain duration, a "missing" event is generated. The badge incorporates a single finger button that, when pressed twice, generates an "attention" event.

Watchdog monitors Active Badge activity and executes relevant UNIX shell commands as required. The user specifies the context-triggered actions in a configuration file when the Watchdog program is first started. The configuration file contains descriptions of Active Badge events and actions to perform in the following format:

```
badge location event-type action
```

The `badge` and `location` are strings that match the badge wearer's ID and the last sighting location, respectively. The `event-type` is a badge event type: `arriving`, `departing`, `settled-in`, `missing`, or `attention`.

Whenever an event of the type `event-type` is generated by the badge at the specified `location`, Watchdog invokes the `action` with a set of UNIX environment variables as parameters, which include the badge owner, owner's office, sighting location, and name of the nearest host. For example,

```
john any attention "emacs -display $NEARESTHOST:0.0"
```

specifies that an Emacs window should start at a nearby host whenever the attention signal is received from tag john.

Contextual Reminders is an application for ParcTabs. Contextual Reminders provide a more expressive way of specifying reminders. The following example shows a set of predicates that makes use of date, time, location, and proximity to another person to trigger a reminder.

```
after April 15
between 10 and 12 noon
in room 35-2200
with {User Adams}
with {Type Display} having {Features Color}
```

This is similar to the Stick-E Note system we describe next.

Stick-E note. Stick-E Note is a technology that has been developed to facilitate the creation of context-aware applications by nonprogrammers (Brown, 1995; Pascoe, 1997). It is motivated by the paradigm of the Post-It Note, those yellow sticky notes used to put down reminders at prominent spots in one's environment. The Stick-E Note is designed with the assumption that a user is moving around with a personal digital assistant (PDA). The PDA has wireless connectivity to a communication network and is equipped with various sensors. such as a GPS transceiver.

Each Stick-E Note consists of two parts:

1. *Context.* A context that the Stick-E Note is attached to, which can consist of a location, the identity of nearby users, and a time (where, who, and when).

2. *Content.* The content that the note represents. This could be information, actions, and interfaces.

A Stick-E Note is an Standard Generalized Markup Language (SGML) document with a context section and a body. The following is an example Stick-E Note that will display a reminder on one's PDA to pick up a library book when the person is in the vicinity of a particular library:

```
<note>
<at> "Noble Engineering Library"
<body>
Pick up the book from interloan library section.
```

The <at> tag identifies the location, in this case the library where the book is being held for the user. The <body> tag specifies the message to be displayed on the PDA when the user is at the Noble Engineering Library. This Stick-E Note only uses the location, or *where*, context. A point to remember here is that the note assumes an infrastructure for translating low-level contextual information (the current location coordinates obtained from GPS) and high-level contextual information (the user is in the vicinity of the Noble Engineering Library). This note can be modified to include optional information, such as a frequency for triggering (displaying) the note, as illustrated below:

```
<note>
<at> "Noble Engineering Library"
<optional>
<triggering-frequency> once
<body>
Pick up the book from interloan library section.
```

This note will be triggered only once and will avoid the problem of the note being triggered repeatedly whenever the user happens to be in the vicinity of the library.

4.4 Middleware Support

Context-aware applications need support for the acquisition and delivery of contextual data. Several methods have been proposed to accomplish this. Some are general ways to handle any form of contextual data. However, since location information is the most widely used contextual information, many approaches have been developed for providing location context. There are, however, some issues common to any infrastructure that supports the acquisition and delivery of contextual information.

The main complication in developing context-aware applications stems from the very nature of contextual information:

1. It is acquired from various heterogeneous and distributed sources:
   a. *Hardware and software sensors*—obtained from various sensors, such as motion detectors, noise and temperature sensors, and location systems.
   b. *System recorded input*—such as user-system interaction history. Context information history is essential for applications such as context-based retrieval.
   c. *Other applications:*
      i. *User's personal computing space*—such as those obtained from schedules, calendars, address books, contact lists, and to-do lists (Satyanarayanan, 2002).
      ii. *Distributed computing environment*—such as those obtained from applications running in the vicinity of these devices, e.g., services provided by the infrastructure of a shopping mall or freeway system.

2. The same type of contextual information may have to be obtained from different sources at different times. For example, a mobile device's location information may be acquired from GPS receivers if the mobile device is outside or from an indoor positioning system if the mobile is inside a building.

3. The low-level contextual information obtained directly from sensors must be abstracted to be useful to the application. As an example, GPS receivers provide location information in terms of latitudes and longitudes. However, a tour-guide application may need location information in a form such as "user is on Lemon Street going west" or "user is near Goldwater Center."

4. Context awareness is most relevant when the environment is highly dynamic, such as when a user is mobile. Changes in the context must be detected in real time and conveyed to the applications as soon as possible so that the applications, particularly those using context-triggered commands and contextual automatic reconfiguration, have ample time to adapt their behaviors.

4.4.1 Contextual services

As we have seen, in order for an application to be responsive to a rich set of contextual information, it needs to interact with various distributed and heterogeneous sources. The task of developing context-aware applications can be facilitated greatly by a middleware infrastructure that can provide the following services (Dey, Sabler, and Abowd, 2001):

1. *Context subscription and delivery service.* A service to which applications can subscribe that delivers context events back to the applications.

2. *Context query service.* A service for applications to query current context.

3. *Context transformation service.* A service that transforms low-level contextual information into high-level contextual information. More generally, it transforms contextual information in one form into contextual information in a form suitable for an application.

4. *Context synthesis service.* A service that fuses various *types* of contextual information.

5. *Discovery and management service.* A service that helps in the discovery of available services and manages all the sensors and software components used by these services. Further, it can provide a "white page" service to enable applications to locate a particular service and a "yellow page" service to enable applications to obtain the set of available services having certain attributes.

4.4.2 Actuator service

As opposed to contextual services that help an application acquire contextual input from sensors, an actuator is a service that helps an application perform a context-dependent output function.

4.4.3 An example: context toolkit

Several general-purpose approaches have been designed to facilitate development of context-aware applications by freeing them from the low-level and tedious task of context acquisition. Hopefully, this will accelerate the pace at which context-aware applications are developed in the future. Currently, there exists no commercially available product. All systems that exist to date have been developed in academic projects. In fact, some of the systems we have described are still in development and may currently have additional features beyond those described here. In this section we describe Context Toolkit's approach to providing support for developing context-aware applications.

Context toolkit.   The Context Toolkit, created at Georgia Tech, is developed around the paradigm of widgets, similar to GUI widgets used for developing graphic user interfaces (GUIs). It is an object-oriented framework that allows retrieval of contextual information through polling and callbacks. The toolkit provides several software components that a software designer can use for context acquisition:

1. *Context widgets.* A context widget is a software component that provides context acquisition and delivery service. A widget serves as an

interface between sensors and applications. It provides an abstraction layer allowing access to heterogeneous and distributed sensors in a uniform manner. Every context widget has a state and a behavior. The state is a set of attributes, and the behavior is a set of callback functions that are triggered by context changes. As opposed to GUI widgets, context widgets are persistent entities that can be shared by multiple applications. Since historical information is useful for predicting the future actions or intentions of the user, context widgets can store all the contextual information they gather automatically and make the history available to the applications.

2. *Context interpreters.* A context interpreter provides a context transformation service to the application. For example, a tourist guide application can use an interpreter to convert GPS data to street names. A context interpreter also can be used to determine the high-level context, such as a social or motivating context. For example, a context interpreter can be used to interpret context from all the widgets in a conference room to determine that a meeting is in progress.

3. *Context aggregators.* Aggregators merge context data from various sources (context widgets and interpreters) to represent context that is associated with an entity. An entity can be a person, room, another software system, or a hardware device, among many possibilities.

4. *Discoverer.* A discoverer provides discovery and management services for sources of contextual information. It maintains a registry of the capabilities of widgets that exist in the framework. An application can use a discoverer to find a particular component with a specific name. It also can use a discoverer to find a set of components that match a specific set of capabilities.

4.4.4 Providing location context

In Chap. 2 we saw that tracking the location information (point of attachment to the network) of a mobile host is essential for delivering messages to it. In the realm of mobile and ubiquitous computing, the location of the mobile user (e.g., the room in which the user is located or the resources located near a user) can be exploited by various context-aware applications to adapt their behavior and to provide various location-sensitive services to the user. For example, knowledge of the room a user is in can be used by a teleporting application to automatically transfer a user's desktop session to the computer nearest to her.

A location information system (LIS) provides location information to the application. A LIS locates objects representing either a person or resources inside areas (Spreitzer and Theimer, 1993). We are mostly interested in issues related to providing location information about people. A LIS can obtain location information about the user from various sources:

1. Indoor locating systems such as infrared-based active badges or ultrasound-based bats (Harter et al., 1999)

2. Wireless nanocell communication activity

3. Outdoor locating systems such as GPS

4. Device input activity from various computers

5. Motion sensors and cameras

6. Explicit information from the user.

Several issues relate to providing the location information of the user. First and foremost is the issue of privacy. Many people feel uncomfortable if their whereabouts are tracked continuously and are freely accessible to anyone. Thus it is important to provide user control over location information. Another important issue is the accuracy of the location information. There are two aspects to this: spatial resolution and temporal resolution of the locating system. Spatial resolution depends on the accuracy of the underlying sensing technology. Temporal resolution depends on such factors as how frequently LIS updates the location information of each user and how sensitive the sensing technology is (e.g., how small a change in someone's location can the sensing technology detect). The accuracy of the location information has implications on the kinds of applications that can be implemented. For example, consider a technology such as the Active Badge, which provides room-level resolution. This enables teleporting applications such as migrating a user's desktop session from the user's office computer to a computer in the conference room. However, supporting applications, which require finer-grained location information, such as flicking a window from one's laptop computer to that of a neighbor sitting next to the user, would be difficult (Spreitzer and Theimer, 1993).

4.5 Summary

Context-aware computing is a new and rapidly evolving field. Already we are seeing availability of context-aware applications. For example, the Google search engine has started using location information to provide location-dependent results to search queries (see *http://local. google.com/lochp*). We are seeing the emergence of proximity-based services, such as serendipity (created by a team of students at MIT; see *http://edition.cnn.com/2004/TECH/ptech/03/19/mobile.dating. reut/*), which match profiles of people stored on their cell phones to socially connect two strangers with similar profiles who happen to be close proximity of each other (about 10 m for Bluetooth-enabled cell phones). "Bluejacking" (see *http://www.bluejackq.com/*) is a trick one can pull on an unsuspecting stranger in close proximity who also happens to have a Bluetooth-enabled phone. By using your Bluetooth-enabled phone to first locate another Bluetooth-enabled phone in your proximity, you can then send a surprise text message, such as "I like your pink sweater ☺"

These are simple context-aware computing applications, but they illustrate the power to change the ways in which we interact and use computing and communication devices. This chapter has described some fundamental issues related to context-aware computing, such as what is a context and how can context-aware behavior can be specified, We also looked at some of the middleware support and services that are being developed to help in the development of more sophisticated context-aware applications.

4.6 References

Abowd, G. D., and E. D. Mynatt, "Charting Past, Present, and Future Research in Ubiquitous Computing," *ACM Transactions on Computer-Human Interaction* 7(1):29, 2000.

Brown, P. J., "The Stick-E Document: A Framework for Creating Context-Aware Applications," *Electronic Publishing* 8(2–3):259, 1995.

Chen, G., and D. Kotz, "A Survey of Context-Aware Computing Research," Darthmouth Computer Science Technical Report TR2000-381, Hanover, NH, 2000.

Dey, A. K., "Understanding and Using Context," *Personal Ubi Comp* 5(1):4, 2001 (*http://link.springer.de/link/service/journals/00779/tocs/t1005001.htm*).

Dey, A. K., D. Salber, and G. D. Abowd, "A Conceptual Framework and a Toolkit for Supporting the Rapid Prototyping of Context-Aware Applications," *Human-Computer Interaction (HCI) Journal* 16(2–4):97, 2001 (*http://www.cc.gatech.edu/fce/ctk/pubs/HCIJ16.pdf*).

Harter, A., A. Hopper, P. Steggles, et al., "The Anatomy of a Context Aware Application," in *Proceedings of the Fifth Annual ACM/IEEE International Conference on Mobile Computing and Networking (MOBICOM '99)*. 1999, p. 59.

Pascoe, J., "The Stick-E Note Architecture: Extending the Interface Beyond the User," in *Proceedings International Conference on Intelligent User Interfaces*. Orlando, Florida, pp 261–264, 1997.

Pascoe, J., "Adding Generic Contextual Capabilities to Wearable Computing," in *Proceedings of the Second International Symposium on Wearable Computers (ISWC pp 92–99, Pittsburgh, Oct.* 1998.

Satyanarayanan, M., "Challenges in Implementing a Context-Aware System," *IEEE Pervasive Computing* 1(3):2, 2002.

Satyanarayanan, M., "Pervasive Computing: Vision and Challenges," *IEEE Personal Communications* 8(4):10, 2001.

Schilit, W. N., N. I. Adams, and R. Want, in *Proceedings of the Workshop on Mobile Computing Systems and Applications*. New York: IEEE, 1994, p. 85 (*ftp://ftp.parc.xerox.com/pub/schilit/wmc-94-schilit.ps*).

Spreitzer, M., and M. Theimer, "Providing Location Information in a Ubiquitous Computing Envrionment," in *Proceedings of the Fourteenth ACM Symposium on Operating Systems, Ashville, NC, Dec. 1993.* 1993, p. 270.

Weiser, M., "The Computer of 21st Century," *Scientific American,* September 1991. 265(3): 94–104.

Weiser, M., and J. S. Brown, "Designing Calm Technology," *PowerGrid Journal* 1(1): XX (No page number available since it appeared in an online journal), 1996 (*http://powergrid.electiciti.com/1.01*).

Introduction to Mobile Middleware

5.1 What Is Mobile Middleware?

Middleware is software that supports mediation between other software components, fostering interoperability between those components across heterogeneous platforms and varying resource levels. For example, middleware can serve as plumbing, allowing applications that do not normally support disconnected operation to do so through clever use of data hoarding (see Chap. 1). Ideally, beyond bridging heterogeneous systems, middleware should be transparent, robust, efficient, secure, and based on open standards. All of these requirements support the goal of making software development in association with middleware easier than "from scratch." This chapter and the next two chapters address three major types of middleware for mobile computing— middleware to support application adaptation (which was covered briefly in Chap. 1), mobile agent systems, and service discovery frameworks. The concept of context-aware computing (see Chap. 4) can also be encapsulated in middleware frameworks.

Chapter 6 covers adaptation middleware and agents, whereas Chap. 7 covers service discovery. Adaptation is a general concept and is arguably mandatory—it allows applications to offer reasonable performance to users across disparate environments. Mobile agents are a special type of adaptive middleware, which extend the reach of data servers and help mobile devices conserve energy. Service discovery frameworks allow mobile devices to change configuration quickly and easily, depending on available services. The emphasis is on fundamental design issues, including choices of communication strategies, trade-offs between application-level and operating-system-level support, and security. Current and historical systems are surveyed to learn which decisions were made in the design and why—this type of understanding is viewed throughout as more important than learning all the nuances of a particular system.

The remaining sections in this chapter provide a relatively nontechnical overview of adaptation, mobile agents, and service discovery frameworks. These are representative of the kinds of middleware used to support interesting mobile applications.

5.2 Adaptation

Many enthusiastic mobile users—their belts hanging low with personal digital assistant (PDA), cellular phone, and portable FAX machine—proclaim (enthusiastically, of course) that the next mobile technology, whether it be a new wireless networking interface or a faster mobile processor, finally will narrow the gap between mobile devices and traditional wired devices. The gap never really narrows. The hunger for computing resources is insatiable, and new, hungrier applications always appear. The new desktop computer shames the newest PDA. And 10-Gb Ethernet makes third-generation (3G) wireless seem not so fast after all. Therefore, mobile applications must fight to make the computing experience of their users tolerable—to make that gap seem not quite so wide. To that end, mobile applications must adapt their behaviors and expectations to conserve scarce resources and to adjust quality of service (QoS)—essentially, a guarantee of performance—to sustainable levels.

For example, a mobile user is likely to be more tolerant of a lower-quality audio stream than a "high-quality" stream that constantly stutters and pops owing to inadequate bandwidth. Similarly, a sequence of still frames probably is more desirable than a highly "pixelated" video stream that never resolves a clear image. On the other hand, a mobile application must not maintain low QoS in the face of abundant resources.

How should adaptation be supported? At one extreme, each application could fend for itself, trying to monitor resources and adapt appropriately. At the other extreme, applications might be blissfully ignorant, depending entirely on their host operating system to perform adaptation. Mobile middleware allows an application to take a middle ground (or for a legacy application, to be retrofitted to take the middle ground whether it knows about it or not), participating in the adaptation process. This makes writing individual applications far less tedious because an application developer need not reinvent the world every time he opens Emacs. Adaptation middleware monitors resources, drawing on policies determined by the user or by application developers to assist applications in modifying application expectations to match available resources. In adapting, a mobile application typically does not change its *core* behavior—a video player remains a video player—but instead takes its user down a slightly different road, increasing or decreasing the fidelity of the data stream delivered to the user.

5.3 Agents

Mobile agents put the action where the data are, allowing programs to move autonomously about a network in order to access remote resources. In a mobile agent system, programs migrate directly to servers, gain access to data or computational resources, and potentially migrate again, eventually returning to their "home base" to deliver results. By allowing a program to migrate directly to a server to gain access to data, several benefits are realized. The first is that disconnected operation is easily supported—a mobile user can dispatch an agent, disconnect (potentially because network resources will be unavailable for a time, as the user migrates away from an 802.11 network, for example), and then reconnect later to allow the agent to return home. Another benefit is that agents can gain access to large amounts of data to solve a problem, even if a mobile user's network resources are scant. For example, a mobile agent might migrate to a server containing thousands of PDF documents, sift through the documents to build a bibliography, and then return home with only the bibliography in tow. A third benefit is that mobile agent systems allow the functionality of servers to be expanded *dynamically*. A skeptic of mobile agents might argue that any task performed by agents could also be performed using traditional client-server methods, and he would be correct. By inserting new code into a server, for example, the "creation of a bibliography" problem (above) could be solved. However, such enhancements require that a server's code be modified every time a new enhancement is desired, and this makes administration of the server much more difficult. With agents, existing computing infrastructure can be used to solve problems not envisioned by the creators of that infrastructure.

5.4 Service Discovery

Service discovery middleware extends the client-server (CS) paradigm to include dynamic discovery of services and more dynamic interaction between clients and services. This type of middleware directly supports the extended client server model, which was discussed in Chap. 1. Service discovery is appropriate for both traditional wired networks and wireless networks but is particularly exciting for mobile computing environments because it allows peripheral-poor mobile devices to discover needed services on demand. For example, imagine a mobile user with a service discovery–enabled PDA, sitting in a coffee shop, sipping a cup of coffee. The user is not familiar with the city he or she is visiting but decides to see a movie. His or her PDA dynamically discovers the Bluetooth access point attached to the ceiling of the shop, using the Bluetooth Service Discovery Protocol. He or she then uses a search engine on the Web to locate a nearby movie theater and determine the show times. The site has a map showing the location of the theater, so the user issues a print command to get a hard copy of the map. The PDA dynamically discovers a Jini printer, determines that the cost is appropriate ($0.10 per page), and prints the document. A minute later the server delivers a coffee refill and the printout.

With service discovery middleware, developers can quickly develop highly dynamic CS systems that are self-healing and support "plug and play" for individual components. The concepts embodied in service discovery architectures are not completely new; however, service discovery frameworks standardize the environments in which to deploy highly dynamic, self-healing CS architectures.

6

Middleware for Application Development: Adaptation and Agents

Application development for mobile computers is a difficult task—on their own, applications are faced with a myriad of challenges: limited power and processing speed, varying levels of network connectivity, completely disconnected operation, and discovery of needed services. The goal of mobile middleware is to provide abstractions that reduce development effort, to offer programming paradigms that make developing powerful mobile applications easier, and to foster interoperability between applications. *Service discovery*, or the art of dynamically discovering and advertising services, is the subject of the next chapter. This chapter examines two other important types of middleware for mobile computing—*adaptation* and *agents*. The first, adaptation, was first discussed in Chap. 1. We revisit this topic in this chapter. Recall that adaptation helps applications to deal intelligently with limited or fluctuating resource levels. The second type of middleware, mobile agents, provides a powerful and flexible paradigm for access to remote data and services.

6.1 Adaptation

Mobile computers must execute user- and system-level applications subject to a variety of resource constraints that generally can be ignored in modern desktop environments. The most important of these constraints are power, volatile and nonvolatile memory, and network bandwidth, although other physical limitations such as screen resolution are also important. In order to provide users with a reasonable computing environment, which approaches the best that currently available resources will allow, applications and/or system software must adapt to limited or fluctuating resource levels. For example, given a sudden severe constraint on available bandwidth, a mobile audio application might stop delivering a high-bit-rate audio stream and substitute a lower-quality stream. The user is likely to object less to the lower-quality delivery than to the significant dropouts and stuttering if the application attempted to continue delivering the high-quality stream. Similarly, a video application might adjust dynamically to fluctuations in bandwidth, switching from high-quality, high-frame-rate color video to black-and-white video to color still images to black-and-white still images as appropriate. A third example is a mobile videogame application adjusting to decreased battery levels by modifying resolution or disabling three-dimensional (3D) features to conserve power.

6.1.1   The spectrum of adaptation

At one end of the spectrum, adaptation may be entirely the responsibility of the mobile computer's operating system (OS); that is, the software for handling adaptation essentially is tucked under the OS hood, invisible to applications. At the other end, adaptation may be entirely the responsibility of individual applications; that is, each application must address all the issues of detecting and dealing with varying resource levels. Between these extremes, a number of *application-aware strategies* are possible, where the OS and individual application each share some of the burden of adaptation. While applications are involved in adaptation decisions, the middleware and/or OS provides support for resource monitoring and other low-level adaptation functions. The spectrum of adaptation is depicted in Fig. 6.1. In this part of the chapter, we are concerned primarily with middleware for adaptation, that is, software interfaces that allow applications to take part in the adaptation process. Pure system-level adaptation strategies, those which take place in a mobile-aware file system such as Coda (e.g., caching and hoarding), are covered elsewhere in this book.

6.1.2   Resource monitoring

All adaptation strategies must measure available resources so that adaptation policies can be carried out. For some types of resources—cash, for example—monitoring is not so difficult. The user simply sets limits and appropriate accounts. For others, more elaborate approaches are required. The Advanced Configuration and Power Interface (ACPI) provides developers with a standardized interface to power-level information on modern devices equipped with "smart" batteries. Accurately

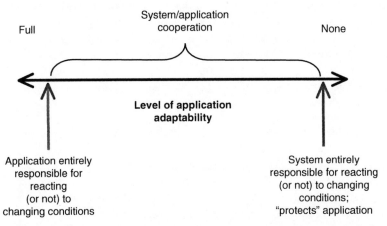

Figure 6.1 At one end of the spectrum of adaptation, applications are entirely responsible for reacting to changing resource levels. At the other end of the spectrum, the operating system reacts to changing resource levels without the interaction of individual applications.

measuring network bandwidth over multihop networks is more difficult. Some approaches are described in Lai and Baker (1999) for the interested reader. Whatever methods are used to measure resource levels have a direct impact on the effectiveness of the entire adaptation process because accurate measurement of resource levels is critical to making proper adaptation decisions.

6.1.3 Characterizing adaptation strategies

The Odyssey project (Noble et al., 1997; Noble, 2000) at Carnegie Mellon University was one of the first application-aware middleware systems, and it serves as a good model for understanding application-aware adaptation. In describing the Odyssey system, Satyanarayanan proposed several measures that are useful for classifying the goodness of an adaptation strategy. We describe these—*fidelity, agility*, and *concurrency*—below.

Fidelity measures the degree to which a data item available to an application matches a reference copy. The reference copy for a data item is considered the exemplar, the ideal for that data item—essentially, the version of the data that a mobile computer would prefer given no resource constraints. Fidelity spans many dimensions, including perceived quality and consistency. For example, a server might store a 30-frame-per-second (fps), 24-bit color depth video at 1600×1200 resolution in its original form as shot by a digital video camera. This reference copy of the video is considered to have 100 percent fidelity. Owing to resource constraints such as limited network bandwidth, a mobile host may have to settle for a version of this video that is substantially reduced in quality (assigned a lower fidelity measure, perhaps 50 percent) or even for a sequence of individual black-and-white still frames (with a fidelity measure of 1 percent). If the video file on the server is replaced periodically with a newer version and a mobile host experiences complete disconnection, then an older, cached version of the video may be supplied to an application by adaptation middleware. Even if this cached version is of the same visual quality as the current, up-to-date copy, its fidelity may be considered lower because it is not the most recent copy (i.e., it is *stale*).

While some data-dependent dimensions of fidelity, such as the frame rate of a video or the recording quality of audio, are easily characterized, others, such as the extent to which a database table is out of date or a video is not the most current version available, do not map easily to a 0 to 100 percent fidelity scale. In cases where there is no obvious mapping, a user's needs must be taken into account carefully when assigning fidelity levels. More problematic is the fact that fidelity levels are in general type-dependent—there are as many different types of fidelity-related adaptations as there are types of data streams; for example, image compression schemes are quite different from audio compression schemes. Generally, an adaptation strategy should provide the highest fidelity possible given current and projected resource levels. Current adaptation middleware tends to concentrate on the present. Factoring projected resource levels into the equation is an area for future research.

Agility measures an adaptation middleware's responsiveness to changes in resource levels. For example, a highly agile system will determine quickly and accurately that network bandwidth has increased substantially or that a fresh battery has been inserted. *An adaptation middleware's agility directly limits the range of fidelity levels that can be accommodated.* This is best illustrated with several examples, which show the importance of both speed and accuracy. For example, if the middleware is very slow to respond to a large increase in network bandwidth over a moderate time frame (perhaps induced by a user resting in an area with 802.11 WLAN connectivity), then chances to perform opportunistic caching, where a large amount of data are transferred and hoarded in response to high bandwidth, may be lost. Similarly, an adaptation middleware should notice that power levels have dropped substantially before critical levels are reached. Otherwise, a user enjoying a high-quality (and power-expensive) audio stream may be left with nothing, rather than a lower-quality audio stream that is sustainable.

Agility, however, is not simply a measure of the speed with which resource levels are measured; accuracy is also extremely important. For example, consider an 802.11a wireless network, which is much more sensitive to line-of-sight issues than 802.11b or 802.11g networks. A momentary upward spike in available bandwidth, caused by a mobile host connected to an 802.11a network momentarily having perfect line of sight with an access point, should not necessarily result in adjustments to fidelity level. If such highly transient bandwidth increases result in a substantial increase in fidelity level of a streaming video, for example, many frames may be dropped when bandwidth suddenly returns to a lower level.

The last measure for adaptation middleware that we will discuss is *concurrency*. Although the last generation of PDAs (such as the original Pilot by Palm, Inc.) used single-threaded operating systems, capable of executing only one application at a time; newer PDAs, running newer versions of Palm OS, variants of Microsoft Windows, and Linux, run full-featured multitasking OSs. Thus it is reasonable to expect that even the least powerful of mobile devices, not to mention laptops that run desktop operating systems, will execute many concurrent applications, all of which compete for limited resources such as power and network bandwidth. This expectation has a very important implication for adaptation: Handling adaptation at the left end of the spectrum (as depicted in Fig. 6.1), where individual applications assume full responsibility for adapting to resource levels, is probably not a good idea. To make intelligent decisions, each application would need to monitor available resources, be aware of the resource requirements of all other applications, and know about the adaptation decisions being made by the other applications. Thus some system-level support for resource monitoring, where the OS can maintain the "big picture" about available resources needs and resource levels, is important.

6.1.4 An application-aware adaptation architecture: Odyssey

In this section we examine the Odyssey architecture in greater detail. In the spectrum of adaptation, Odyssey sits in the middle—applications are *assisted* by the Odyssey middleware in making decisions concerning fidelity levels. Odyssey provides a good model for understanding the issues in application-aware adaptation because the high-level architecture is clean, and the components for supporting adaptation are clearly delineated. The Odyssey architecture consists of several high-level components: the *interceptor*, which resides in the OS kernel, the *viceroy*, and one or more *wardens*. These are depicted in Fig. 6.2. The version of Odyssey described in Nobel and colleagues (1997) runs under NetBSD; more recent versions also support Linux and FreeBSD. To minimize changes to the OS kernel, Odyssey is implemented using the Virtual File System (VFS) interface, which is described in great detail for kernel hacker types in Bovet and Cesati (2002). Applications interact with

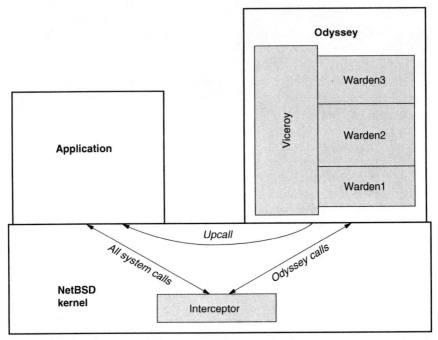

Figure 6.2 The Odyssey architecture consists of a type-independent viceroy and a number of type-specific wardens. Applications register windows of acceptable resource levels for particular types of data streams and receive notifications is when current resource levels fall outside the windows.

Odyssey using (mostly) file system calls, and the interceptor, which resides in the kernel, performs redirection of Odyssey-specific system calls to the other Odyssey components.

The basic Odyssey model is for an application to choose a fidelity level for each data type that will be delivered—e.g., 320 × 240 color video at 15 fps. The application then computes resource needs for delivery of each stream and registers these needs with Odyssey in the form of a "window" specifying minimum and maximum need. The viceroy monitors available resources and generates a callback to the application when available resources fall outside registered resource-level window. The application then chooses a new fidelity level, computes resource needs, and registers these needs, as before. Thus applications are responsible for deciding fidelity levels *and* computing resource requirements—the primary contribution that Odyssey makes is to monitor resources and to notify applications when available resources fall outside constraints set by the application. Before describing a sample Odyssey application, the wardens and viceroy are discussed in detail below.

Wardens. A *warden* is a type-specific component responsible for handling all adaptation-related operations for a particular sort of data stream (e.g., a source of digital images, audio, or video). Wardens sit between an application and a data source, handling caching and arranging for delivery of data of appropriate fidelity levels to the application. A warden must be written for each type of data source. An application typically must be partially rewritten (or an appropriate proxy installed) to accept data through a warden rather than through a direct connection to a data source, such as a streaming video server.

Viceroy. In Odyssey, the viceroy is a type-independent component that is responsible for global resource control. All the wardens are statically compiled with the viceroy. The viceroy monitors resource levels (e.g., available network bandwidth) and initiates callbacks to an application when current resource levels fall outside a range registered by the application. The types of resources to be monitored by the viceroy in Odyssey include network bandwidth, cache, battery power, and CPU, although the initial implementations of the Odyssey architecture did not support all these resource types.

6.1.5 A sample Odyssey application

We now turn to one of the sample applications discussed in Nobel and colleagues (1997): the *xanim* video player. The xanim video player was modified to use Odyssey to adapt to varying network conditions, with three fidelity levels available—two levels of JPEG compression and black-and-white frames. The JPEG compression frames are labeled 99 and 50 percent fidelity, whereas the black-and-white content is labeled 1 percent fidelity. Integration of xanim with Odyssey is illustrated in Fig. 6.3. A "video warden" prefetches frames from a video server with the appropriate fidelity and supplies the application with metadata for the video being played and with individual frames of the video.

The performance of the modified xanim application was tested using simulated bandwidths of 140 kB/s for "high" bandwidth and 40 kB/s for "low" bandwidth. A number of strategies were used to vary bandwidth: *step up*, which holds bandwidth at the low level for 30 seconds, followed by an abrupt increase to high bandwidth for 30 seconds; *step down*, which reverses the bandwidth levels of step up but maintains the same time periods; *impulse up*, which maintains a low bandwidth over a 60-second period with a single 2-second spike of high bandwidth in the middle; and *impulse down*, which maintains high bandwidth for 60 seconds with a single 2-second spike of low bandwidth in the middle. Both high and low bandwidth levels are able to support black-and-white video and the lower-quality (50 percent fidelity) JPEG video. Only the high bandwidth level

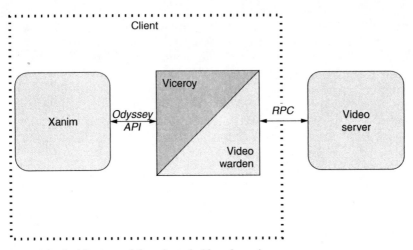

Figure 6.3 Architecture of the adapted video player in .y.

is sufficient for the 99 percent fidelity JPEG frames to be delivered without substantial numbers of dropped frames.

In the tests, Odyssey maintained average fidelities of 73, 76, 50, and 98 percent for step up, step down, impulse up, and impulse down, respectively, all with less than 5 percent dropped frames. In contrast, trying to maintain the 99 percent fidelity rate by transferring high-quality video at all times, ignoring available network bandwidth, resulted in losses of 28 percent of the frames for step up and step down and 58 percent of the frames for impulse up. Several other adapted applications are discussed in the Odyssey publications.

6.1.6 More adaptation middleware

Puppeteer.   For applications with well-defined, published interfaces, it is possible to provide adaptation support without modifying the applications directly. The Puppeteer architecture allows component-based applications with published interfaces to be adapted to environments with poor network bandwidth (a typical situation for mobile hosts) without modifying the application (de Lara, Wallach, and Zwaenepoel, 2001). This is accomplished by outfitting applications and data servers with custom proxies that support the adaptation process. A typical application adaptation under Puppeteer is a retrofit of Microsoft PowerPoint to support incremental loading of slides from a large presentation or support for progressive JPEG format to speed image loading. Both these adaptations presumably would enhance a user's experience when handling a large PowerPoint presentation over a slow network link.

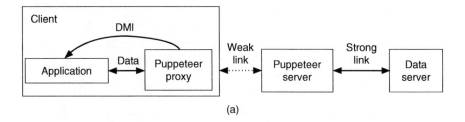

(a)

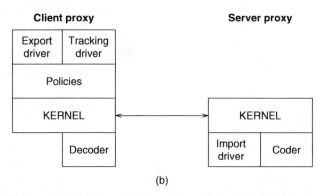

(b)

Figure 6.4 (a) Illustrates the overall Puppeteer architecture, where client applications interact with data servers through proxies. DMI is the Data Manipulation Interface of the applications, which allows Puppeteer to view and modify data acted on by the application. The relationship between client-side and server-side proxies is illustrated in (b).

The Puppeteer architecture is depicted in Fig. 6.4. The Puppeteer provides a kernel that executes on both the client and server side proxies, supporting a document type called the Puppeteer Intermediate Format (PIF), a hierarchical, format-neutral format. The kernel also handles all communication between client and server sides. To adapt a document, the server and client side proxies communicate to establish a high-level PIF skeleton of the document. Adaptation policies control which portions of the document will be transferred and which fidelities will be chosen for the transmitted portions. For example, for a Microsoft PowerPoint document, selected slides may transferred, with images rendered at a lower fidelity than in the original presentation. The *import driver* and *export driver* parse native document format to PIF and PIF to native document format, respectively. Transcoders in Puppeteer perform transformations on data items to support the adaptation policies. For example, a Puppeteer transcoder may reduce the quality of JPEG images or support downloading only a subset of a document's data. A typical Puppeteer-adapted application operates as follows:

- When the user opens a document, the Puppeteer kernel instantiates an appropriate import driver on the server side.

- The import driver parses the native document format and creates a PIF format document. The skeleton of the PIF is transmitted by the kernel to the client-side proxy.

- On the client side, policies available to the client-side proxy result in requests to transfer selected portions of the PIF (at selected fidelities) from the server side. These items are rendered by the export driver into native format and supplied to the application through its well-known interface.

- At this point, the user regains control of the application. If specified by the policy, additional portions of the requested document can be transferred by Puppeteer in the background and supplied to the application as they arrive.

Coordinating adaptation for multiple mobile applications.   Efstatiou and colleagues (2002a, 2002b) argue that what's missing from most current adaptation middleware architectures is coordination among adaptive applications. Odyssey and Puppeteer, for example, support sets of independently adapting applications but do not currently assist multiple applications in coordinating their adaptation strategies. When multiple applications are competing for shared resources, individual applications may make decisions that are suboptimal. At least three issues are introduced when multiple applications attempt to adapt to limited resources—*conflicting adaptation, suboptimal system operation,* and *suboptimal user experience.*

Several sample scenarios illustrate these concerns. First, consider a situation where a number of applications executing on a mobile host with limited power periodically write data to disk. This would occur, for example, if two or more applications with automatic backup features were executing. Imagine that the mobile host maintains a powered-down state for its hard drives to conserve energy. Then, each time one of the automatic backup facilities executes, a hard disk on the system must be spun up. If the various applications perform automatic backups at uncoordinated times, then the disk likely will spin up quite frequently, wasting a significant amount of energy. If the applications coordinated to perform automatic backups, on the other hand, then disk writes could be performed "in bulk," maximizing the amount of time that the disk could remain powered down. This example illustrates suboptimal system operation despite adaptation.

Another issue when multiple applications adapt independently is conflicting adaptation. Imagine that one application is adapting to varying power, whereas another application is adapting to varying network bandwidth. When the battery level in the mobile device becomes a concern, then the power-conscious application might throttle its use of the network interface. This, in turn, makes more bandwidth available, which might trigger the bandwidth-conscious application to raise fidelity levels for a data stream, defeating the other application's attempt to save energy.

A third issue is that in the face of limited resources, a user's needs can be exceedingly difficult to predict. Thus some user participation in the adaptation process probably is necessary. To see this, imagine that a user is enjoying a high-bandwidth audio stream (Miles Davis, *Kind of Blue?*) while downloading a presentation she needs to review in 1 hour. With abundant bandwidth, both applications can be well served. However, if available bandwidth decreases sharply (because an 802.11 access point has gone down, for example, and the mobile host has fallen back to a 3G connection), should a lower-quality stream be chosen and the presentation download delayed because Miles is chewing up a few tens of kilobits per second? Or should the fun stop completely and the work take precedence?

Efstatiou and colleagues propose using an adaptation policy language based on the event calculus (Kowalsky, 1986) to specify global adaptation policies. The requirements for their architecture are that a set of extensible adaptation attributes be sharable among applications, that the architecture be able to centrally control adaptation behavior, and that flexible, system-wide adaptation policies, depending on a variety of issues, be expressible in a policy language. Their architecture also allows human interaction in the adaptation process both to provide feedback to the user and to engage the user in resolving conflicts (e.g., Miles Davis meets downloading PowerPoint). Applications are required to register with the system, providing a set of adaptation policies and modes of adaptation supported by the application. In addition, the application must expose a set of state variables that define the current state of the application. Each application generates events when its state variables change in meaningful ways so that the adaptation architecture can determine if adaptive actions need to be taken; for example, when a certain application is minimized, a global adaptation policy may cause that application to minimize its use of system resources. A *registry* in the architecture stores information about each application, and an *adaptation controller* monitors the state of the system, determining when adaptation is necessary and which applications should adapt. Another policy language–driven architecture advocating user involvement is described in Keeney and Cahill (2003).

6.2 Mobile Agents

We now turn to another type of mobile middleware, mobile agent systems. Almost all computer users have used mobile code, whether they realize it or not—modern browsers support Javascript, Java applets, and other executable content, and simply viewing Web pages results in execution of the associated mobile code. Applets and their brethren are mostly *static*, in that code travels from one or more servers to a client and is executed on the client. For security reasons, the mobile code often is prevented from touching nonlocal resources. Mobile agents are a significant step forward in sophistication, supporting the migration of not only code but also state. Unlike applets, whose code typically travels (at an abstract level at least) one "hop" from server to client, mobile agents move freely about a network, making autonomous decisions on where to travel next. Mobile agents have a mission and move about the network extracting data and communicating with other agents in order to meet the mission goals.

Like adaptation middleware, mobile agent systems (e.g., Cabri, Leonardi, and Zambonelli, 2000; Gray, 1996, 1997; Gray et al., 1998, 2000; Bradshaw et al., 1999; Lange and Oshima, 1998; Peine and Stoplmann, 1997; Wong, Paciorek, and Moore, 1999; Wong et al., 1997) support execution of mobile applications in resource-limited environments, but mobile agent systems go far beyond allowing local applications to respond to fluctuating resource levels. A mobile agent system is a dynamic client-server (CS) architecture that supports the migration of mobile applications (agents) to and from remote servers. An agent can migrate whenever it chooses either because it has accomplished its task completely or because it needs to travel to another location to obtain additional data. An alternative to migration that an agent might exercise is to create one or more new agents dynamically and allow these to migrate. The main idea behind mobile agents is to get mobile code as close to the action as possible—mobile agents migrate to remote machines to perform computations and then return home with the goods.

For example, if a mobile user needs to search a set of databases, a traditional CS approach may perform remote procedure calls against the database servers. On the other hand, a mobile agents approach would dispatch one or more applications (agents) either directly to the database servers or to machines close to the servers. The agents then perform queries against the database servers, sifting the results to formulate a suitable solution to the mobile user's problem. Finally, the mobile agents return home and deliver the results.

The advantages of this approach are obvious. First, if bandwidth available to the mobile user is limited and the database queries are complicated, then performing a series of remote queries against the servers might be prohibitively expensive. Since the agents can execute a number of queries much closer to the database servers in order to extract the desired information, a substantial amount of bandwidth might be saved (of course, transmission of the agent code must be taken into account). Second, continuous network connectivity is not required.

The mobile user might connect to the network, dispatch the agent, and then disconnect. When the mobile user connects to the network again later, the agent is able to return home and present its results. Finally, the agents are not only closer to the action, but they also can be executed on much more powerful computers, potentially speeding up the mining of the desired information.

Of course, there are substantial difficulties in designing and implementing mobile agent systems. After briefly discussing the motivations for mobile agent systems in the next section, those challenges will consume the rest of this chapter.

6.2.1   Why mobile agents? And why not?

We first discuss the advantages of mobile agents at a conversational level, and then we look at the technical advantages and disadvantages in detail. First, a wide variety of applications can be supported by mobile agent systems, covering electronic commerce (sending an agent shopping), network resource management (an agent might traverse the network, checking versions of installed applications and initiating upgrades where necessary), and information retrieval (an agent might be dispatched to learn everything it can about Thelonious Monk).

An interesting observation made by Gray and colleagues (2000) is worth keeping in mind when thinking about agent-based applications: While *particular* applications may not make a strong case for deployment of mobile agent technology, *sets* of applications may make such a case. To see this point, consider the database query example discussed in the preceding section. Rather than using mobile agents, a custom application could be deployed (statically) on the database servers. This application accepts jobs (expressing the type of information required) from a mobile user, performs a sequence of appropriate queries, and then returns the results. Since most of the processing is done off the mobile host, the resource savings would be comparable to a mobile agents solution.

Similarly, little computational power on the mobile host is required because much of the processing can be offloaded onto the machine hosting the custom application. However, what if a slightly different application is desired by a mobile user? Then the server configuration must be changed. Like service discovery protocols, covered in Chap. 7, mobile agent systems foster creation of powerful, *personalized* mobile applications based on common frameworks. While individual mobile applications can be written entirely without the use of agent technologies, the amount of effort to support a changing *set* of customized applications may be substantially higher than if mobile agents were used.

Mobile agent systems provide the following set of technical advantages (Milojicic, Douglis, and Wheel, 1998):

- *The limitations of a single client computer are reduced.* Rather than being constrained by resource limitations such as local processor power, storage space, and particularly network bandwidth, applications can send agents "into the field" to gather data and perform computations using the resources of larger, well-connected servers.

- *The ability to customize applications easily is greatly improved.* Unlike traditional CS applications, servers in an agent system merely provide an execution *environment* for agents rather than running customized server applications. Agents can be freely customized (within the bounds of security restrictions imposed by servers) as the user's needs evolve.

- *Flexible, disconnected operation is supported.* Once dispatched, a mobile agent is largely independent of its "home" computer. It can perform tasks in the field and return home when connectivity to the home computer is restored. Survivability is enhanced in this way, especially when the home computer is a resource-constrained device such as a PDA. With a traditional CS architecture, loss of power on a PDA might result in an abnormal termination of a user's application.

Despite these advantages, mobile agent architectures have several significant disadvantages or, if that is too strong a word, disincentives. One is that neither a killer application nor a pressing need to deploy mobile agent technology has been identified. Despite their sexiness, mobile agents do not provide solutions to problems that are otherwise unsolvable; rather, they simply seem to provide a good framework in which to solve certain problems. In reflections on the Tacoma project (Milojicic, Douglis, and Wheel, 1998), Johansen, Schneider, and van Renesse note that while agents potentially reduce bandwidth and tolerate intermittent communication well, bandwidth is becoming ever more plentiful, and communication is becoming more reliable. As wireless networking improves and mobile devices become more powerful and more prevalent, will mobile agents technologies become less relevant? Further, while a number of systems exist, they are largely living in research laboratories. For mobile agent systems to meet even some of their potential, widespread deployment of agent environments is required so that agents may travel freely about the Internet.

A related problem is a lack of standardization. Most mobile agent systems are not interoperable. Some effort has gone into interoperability for agent systems, but currently, there seem to be no substantial market pressures forcing the formation of a single (or even several) standards for mobile agent systems. The Mobile Agents System Interoperability Facility (MASIF; Milojicic et al., 1999) is one early attempt at fostering agent interoperability for Java-based agent systems.

All the disadvantages just discussed are surmountable with a little technical effort—apply a good dose of marketing, and most disappear. There is a killer disadvantage, however, and that is *security*. Even applets and client-side scripting languages (such as Javascript), which make only a single hop, scare security-conscious users to death, and many users turn off Java, Javascript, and related technologies in their Web browsers. Such users maintain this security-conscious stance even when interacting with Web sites in which they place significant trust because the potential for serious damage is high should the sandbox leak. Security for mobile agent systems is far more problematic than simple mobile code systems such as Java applets because agents move autonomously.

There are at least two broad areas of concern. First, agents must be prevented from performing either unintentional or malicious damage as they travel about the network. Could an agent have been tampered with at its previous stop? Is it carrying a malicious logic payload? Does it contain contraband that might be deposited on a machine? Will the agent use local resources to launch a denial-of-service attack against another machine? Essentially, if agents are to be allowed to get "close to the action," then the "action" must be convinced (and *not* just with some marketing) that the agents will not destroy important data or abuse resources. Second, the agents themselves must be protected from tampering by malicious servers. For example, an agent carrying credit card information to make purchases on behalf of its owner should be able to control access to the credit card number. Similarly, an agent equipped with a proprietary data-mining algorithm should be able to resist reverse engineering attacks as it traverses the network.

6.2.2 Agent architectures

To illustrate the basic components of mobile agent architectures, a high-level view of Telescript (White in Milojicic, Douglic, and Wheel, 1998) works well. Telescript was one of the first mobile agent systems, and while it is no longer under development, many subsequent systems borrowed ideas from Telescript. There are a number of important components in the Telescript architecture: *agents, places, travel, meetings, connections, authorities*, and *permits*. These are depicted in Fig. 6.5. Each of these components is described in detail below.

Places. In a mobile agent system, a network is composed of a set of places—each place is a location in the network where agents may visit. Each place is hosted by a server (or perhaps a user's personal computer) and provides appropriate infrastructure to support a mobile agent migrating to and from that location. Servers in a network that do not

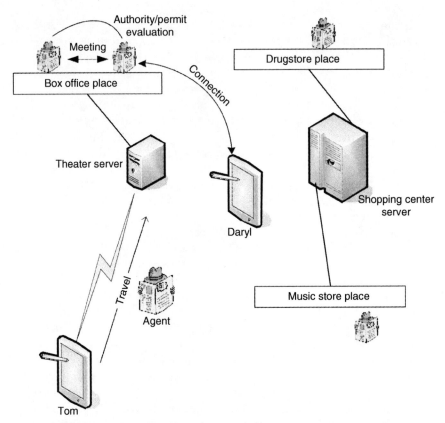

Figure 6.5 The major Telescript components are illustrated above. Tom has just dispatched an agent which has not yet arrived at the theater server. When Tom's agent arrives, it will interact with the static agent in the box office place to arrange for theater tickets. Daryl previously dispatched an agent to purchase tickets and has a connection with her agent in the box office place, so she can actively negotiate prices. Daryl's agent and the box office agent have identified each other through their respective authorities and permits associated with Daryl's agent have been evaluated to see what actions are permitted. The static agents in the drugstore and music store places, which both reside on a shopping center server, are currently idle. To interact with the drugstore or music store agents, Daryl or Tom's agents will have to travel to the drugstore and music store places.

offer a "place" generally will not be visitable by agents. Places offer agents a resting spot in which they can access resources local to that place through a stationary agent that "lives" there, interacting with other agents currently visiting that place.

Travel.   Travel allows agents to move closer to or to colocate with needed resources. For example, an agent dispatched by a user to obtain tickets to a jazz concert and reservations at one of several restaurants (depending on availability) might travel from its home place to the place hosted by the jazz club's box office before traveling to the places hosted by the restaurants. The primary difference between mobile code strategies such as Java applets and agents is that agents travel with at least part of their state intact—after travel, an agent can continue the computation it was engaged in at the instant that travel was initiated. Migration is studied in further detail below in the section entitled, "Migration Strategies."

Meetings.   Meetings are local interactions between two or more agents in the same *place*. In Telescript, this means that the agents can invoke each other's procedures. The agent in search of jazz tickets and a restaurant reservation (discussed under "Travel" above) would engage in meetings with appropriate agents at the ticket office and at the restaurant's reservation office to perform its duty.

Connections.   Connections allow agents at different places to communicate and allow agents to communicate with human users or other applications over a network. An agent in search of jazz tickets, for example, might contact the human who dispatched it to indicate that an additional show has been added, although the desired show was sold out (e.g., "Is the 11 P.M. show OK?"). Connections in Telescript require an agent to identify the name and location of the remote agent, along with some other information, such as required quality of service. This remote communication method, which tightly binds two communicating agents (since both name and location are required for communication), is the most restrictive of the mechanisms discussed in further detail below in the section entitled, "Communication Strategies."

Authorities.   An agent's or place's authority is the person or organization (in the real world) that it represents. In Telescript, agents may not withhold their authority; that is, anonymous operation is not allowed—the primary justification for this limitation is to deter malicious agent activity. When an agent wishes to migrate to another location, the destination can check the authority to determine if migration will be permitted. Similarly, an agent may examine the authority of a potential destination to determine if it wishes to migrate there. Implementation of authorities in an untrusted network is nontrivial and requires strong cryptographic methods because an agent's authority must be unforgeable.

Permits.   Permits determine what agents and places can do—they are sets of capabilities. In general, these capabilities may have virtually any form, but in Telescript they come in two flavors. The first type of capability determines whether an agent or place may execute certain types of instructions, such as instructions that create new agents. The second type of capability places resource limits on agents, such as a maximum number of bytes of network traffic that may be generated or a maximum lifetime in seconds. If an agent attempts to exceed the limitations imposed by its permits, it is destroyed. The actions permitted an agent are those which are allowed by both its internal permits and the place(s) it visits.

Other issues. A number of details must be taken into account when designing an architecture to support mobile agents, but one of the fundamental issues is the choice of language for implementation of the agents (which might differ from the language used to implement the agent *architecture*). To support migration of agents, all computers to which an agent may migrate must share a common execution language. While it is possible to restrict agents to a particular computer architecture and OS (e.g., Intel 80×86 running Linux 2.4), clearly, agent systems that can operate in heterogeneous computer environments are the most powerful. Compiled languages such as C and C++ are problematic because agent executables must be available for every binary architecture on which agents will execute. Currently, interpreted languages such as Java, TCL, and Scheme are the most popular choices because many problems with code mobility are alleviated by interpreted languages. In cases where traditionally compiled languages such as C++ are used for implementation of agents, a portable, interpreted byte code typically is emitted by a custom compiler to enable portability (e.g., see Gray et al., 2000). Java is particularly popular for mobile agents because Java has native support for multithreading, object serialization (which allows the state of arbitrary objects to be captured and transmitted), and remote procedure calls.

Other factors, aside from the implementation language for agents, include migration strategies, communication, and security. Migration and communication strategies are discussed in detail below. A thorough treatment of security in agent systems is beyond the scope of this chapter.

6.2.3 Migration strategies

To support the migration of agents, it must be possible to either capture the state of an agent or to spawn an additional process that captures the state of the agent. This process state must then be transmitted to the remote machine to which the agent (or its child, in the case of spawning an additional process) will migrate. This is equivalent to process checkpointing, where the state of a process, including the stack, heap, program code, static variables, etc., is captured and stored for a later resuscitation of the process. Process checkpointing is a very difficult problem that has been studied in the operating systems and distributed systems communities for a number of years, primarily to support fault tolerance and load balancing (Jul et al., 1988; Douglis and Ousterhout, 1991; Plank, 1995). In general, commodity operating systems do not provide adequate support for checkpointing of processes, and add-on solutions [e.g., in the form of libraries such as libckpt (Plant, 1995)] are nonportable and impose significant restrictions, such as an inability to reconstitute network connections transparently. A number of research operating systems have been designed that better support process migration, but since none of these is viable commercially (even in the slightest sense), they are not currently appropriate platforms.

Checkpointing processes executing inside a virtual machine, such as Java processes, are a bit easier, but currently most of these solutions (Richard and Tu, 1998; Sakamoto, Sekiguchi, and Yonezawa, 2000; Truyen et al., 2000) also impose limitations, such as restrictions on the use of callbacks, network connections, and file activity. The virtual machine itself can be checkpointed, but then the issues of portability discussed earlier reemerge, and network connections and file access will still pose problems. So where is this going? The punch line is that if commodity operating systems are to be targeted by agent systems—and for wide-scale deployment, this must be the case—then completely capturing the state of general processes to support migration is rife with problems.

One solution is to impose strong restrictions on the programming model used for mobile agents. Essentially, this entails capturing only the essential internal state of an agent, i.e., sufficient information about its current execution state to continue the computation on reconstitution, combined with a local cleanup policy. This means that an agent might perform a local cleanup, including tearing down communication connections and closing local files, before requesting that the agent middleware perform a migration operation. For example, in Aglets (Lange and Oshima, 1998), which is a Java-based mobile agents system, agents are notified at the beginning of a migration operation. It is the responsibility of an individual agent, on receiving such a notification, to save any significant state in local variables so that the agent can be properly "reconstituted" at the new location. Such a state may include the names of communication peers, loop indices, etc. Agent migration in Aglets begins with an agent initiating a migration (its own or that of another agent) by invoking dispatch(). A callback, onDispatch(), will be triggered subsequently, notifying the agent that it must save its state. After the migration, the agent's onArrival() callback will be invoked so that the agent can complete its state restoration.

6.2.4 Communication strategies

Communication among agents in a mobile agent system can take many forms, including the use of traditional CS techniques, remote procedure call, remote method invocation (e.g., using Java's RMI), mailboxes, meeting places, and coordination languages. Each of these communication strategies has advantages and disadvantages, some of which are exacerbated in mobile agent systems. One consideration is the degree of temporal and spatial locality exhibited by a communication scheme (Cabri, Leonardi, and Zambonelli, 2000).

Temporal locality means that communication among two or more agents must take place at the same physical time, like a traditional telephone conversation. Interagent communication mechanisms exhibiting temporal locality are limiting in a mobile agent's architecture because all agents participating in a communication must have network connectivity at the time the communication occurs. If an agent is in transit, then attempts to communicate with that agent typically will fail.

Spatial locality means that the participants must be able to name each other for communication to be possible—in other words, unique names must be associated with agents, and their names must be sufficient for determining their current location. Some of the possible communication mechanisms for interagent communication are discussed below.

Traditional CS communication.   Advantages of traditional CS mechanisms such as sockets-based communication, Remote Method Invocation (RMI) in Java, and CORBA include a familiar programming model for software developers and compatibility with existing applications. Significant drawbacks include strong temporal and spatial locality—for communication to be possible, agents must be able to name their communication peers and initiate communication when their peers are also connected. RMI and other communication mechanisms built on the Transmission Control Protocol/Internet Protocol (TCP/IP) also require stable network connectivity; otherwise, timeouts and subsequent connection reestablishments will diminish performance significantly. Examples of agents systems that use traditional CS mechanisms are D'Agents (Gray et al., 1998) and Aglets (Lange and Oshima, 1998). In Aglets, an agent first must obtain another agent's proxy object (of type AgletProxy) before communication can take place. This proxy allows the holder to transmit arbitrary messages to the target and to request that the target agent perform operations such as migration and cloning (which creates an identical agent). To obtain a proxy object for a target agent, an agent typically must provide both the name of the target agent and its current location. If either agent moves, then the proxy must be reacquired.

Meeting places.   Meeting places are specific places where agents can congregate in order to exchange messages and typically are defined statically, avoiding problems with spatial locality but not temporal locality. In Ara (Peine and Stolpmann, 1997), meeting places are called *service* *points* and provide a mechanism for agents to perform local communication. Messages are directed to a service point rather than to a specific agent, eliminating the need to know the names of colocated agents.

Tuple spaces. Linda-like tuple spaces are also appropriate for interagent communication. Linda provides global repositories for tuples (essentially lists of values), and processes communicate and coordinate by inserting tuples into the tuple space, reading tuples that have been placed into the tuple space, and removing tuples from the tuple space. Tuple spaces eliminate temporal and spatial bindings between communicating processes because communication is anonymous and asynchronous.

Retrieval of objects from a tuple space is based on the content or type of data, so choosing objects of interest is easier than if objects were required to be explicitly named. One agent architecture that provides a Linda-like communication paradigm is Mobile Agent Reactive Spaces (MARS; Cabri, Leonardi, and Zambonelli, 2000). MARS extends Java's JavaSpaces concept, which provides read(), write(), take(), readAll(), and takeAll() methods to access objects in a JavaSpace. The extensions include introduction of reactions, which allow programmable operations to be executed automatically on access to certain objects in an object space. This allows, for example, a local service to be started and new objects introduced into the object space, all based on a single access to a "trigger" object by an agent. While distributed implementations of the Linda model exist, MARS simply implements a set of independent object spaces, one per node. Agents executing on a particular node may communicate through the object space, but agents executing on different nodes cannot use the object spaces to communicate directly. MARS is intended as a communication substrate for other mobile agent systems rather than as an independent mobile agent system.

6.3 Summary

This chapter introduced two types of middleware, adaptation middleware and mobile agent systems. Adaptation middleware assists applications in providing the best quality of service possible to users, given the widely fluctuating resource levels that may exist in mobile environments. Mobile agents provide an alternative to static client/server systems for designing interesting mobile applications that access remote data and computational services. Rather than issuing remote procedure calls against distant services, mobile agents migrate code closer to the action to reduce communication and computational requirements for mobile hosts. When a mobile agent has completed its tasks, it can then return home to present the results to the user (or to another application). Both of these types of middleware are complementary to service discovery frameworks, which are the subject of the next chapter.

6.4 References

"Advanced Configuration and Power Interface," *http://www.acpi.info/*.

Ahuja, S., N. Carriero, and D. Gelernter, "Linda and Friends," *IEEE Computer* 19(8):26, 1986.

Bharat, K. A., and L. Cardelli, "Migratory Applications," in *Proceedings of the Eighth Annual ACM Symposium on User Interface Software and Technology,* November 1995.

Bovet, D., and M. Cesati, *Understanding the Linux Kernel,* 2d ed. O'Reilly, 2002.

Bradshaw, J. M., M. Greaves, H. Holmback, W. Jansen, T. Karygiannis, B. Silverman, N. Suri, and A. Wong, "Agents for the Masses?" IEEE Intelligent Systems, March–April 1999.

Cabri, G., L. Leonardi, and F. Zambonelli, "MARS: A programmable coordination architecture for mobile agents," *IEEE International Computing* 4(4), 2000.

de Lara, E., D. S. Wallach, and W. Zwaenepoel, "Puppeteer: Component-Based Adaptation for Mobile Computing," in *Proceedings of the 3rd USENIX Symposium on Internet Technologies and Systems* (USITS-01), Berkeley, CA, March 2001.

Douglis, F., and Ousterhout J., "Transparent Process Migration: Design Alternatives and the Sprite Implementation," *Software: Practice and Experience* 21(8), August 1991.

Efstratiou, C., A. Friday, N. Davies, and K. Cheverst, "A Platform Supporting Coordinated Adaptation in Mobile Systems," in *Proceedings of the 4th IEEE Workshop on Mobile Computing Systems and Applications (WMCSA'02).* Callicoon, NY: U.S., IEEE Computer Society, June 2002, pp. 128–137.

Efstratiou, C., A. Friday, N. Davies, and K. Cheverst, "Utilising the Event Calculus for Policy Driven Adaptation in Mobile Systems," in Lobo, J., Michael, B. J., and Duray N. (eds): *Proceedings of the 3rd International Workshop on Policies for Distributed Systems and Networks (POLICY 2002).* Monterey, CA., IEEE Computer Society, June 2002, pp. 13–24.

Fuggetta, A., G. P. Picco, and G. Vigna, "Understanding Code Mobility," *IEEE Transactions on Software Engineering* 24(5):May 1998.

Gray, R. S., "Agent Tcl: An Extensible and Secure Mobile-Agent System," in *Proceedings of the Fourth Annual Tcl/Tk Workshop (TCL '96),* Monterey, CA, July 1996.

Gray, R. S., "Agent Tcl: An Extensible and Secure Mobile-Agent System," PhD thesis, Dept. of Computer Science, Dartmouth College, June 1997.

Gray, R. S., D. Kotz, G. Cybenko, and D. Rus, "D'Agents: Security in a Multiple-Language, Mobile-Agent System," *Mobile Agents and Security,*1419:154–187, 1998.

Gray, R. S., D. Kotz, G. Cybenko, and D. Rus, "Mobile Agents: Motivations and State-of-the-Art Systems," Dartmouth Computer Science Department Technical Report TR2000-365, 2000.

Hjalmtysson, H., and R. S. Gray, "Dynamic C++ Classes: A Lightweight Mechanism to Update Code in a Running Program," Proceedings of the 1998 USENIX Technical Conference, 1998.

Joseph, A. D., J. A. Tauber, and M. Frans Kaashoek, "Mobile Computing with the Rover Toolkit," *IEEE Transactions on Computers: Special Issue on Mobile Computing,* 46, March 1997.

Jul, E., H. Levy, N. Hutchinson, and A. Black, "Fine-Grained Mobility in the Emerald System," *ACM Transactions on Computer Systems* 6 1, February 1988.

Keeney, J., and V. Cahill, "Chisel: A Policy-Driven, Context-Aware, Dynamic Adaptation Framework," in *Proceedings of the Fourth IEEE International Workshop on Policies for Distributed Systems and Networks (POLICY 2003),* 2003.

Kowalsky, R. "A Logic-Based Calculus of Events," *New Generation Computing* 4:67, 1986.

Lai, K., and M. Baker, "Measuring Bandwidth," in *Proceedings of IEEE Infocom'99,* March 1999.

Lange, D., and M. Oshima: *Programming and Deploying Java Mobile Agents with Aglets*, Reading, MA, Addison Wesley, 1998.

Litzkow, M., and M. Solomon, "Supporting Checkpointing and Process Migration Outside the Unix Kernel," in *Proceedings of the 1992 Winter USENIX Technical Conference*, 1992.

Milojicic, D., M. Breugst, I. Busse, et al, "MASIF: The OMG Mobile Agent System Interoperability Facility," in *Proceedings of the International Workshop on Mobile Agents (MA'98)*, Stuttgart, September 1998.

Milojicic, D., F. Douglis, and R. Wheel (eds), *Mobility: Processes, Computers and Agents*. ACM Press, 1999.

Nahrstedt, K., D. Xu, D. Wichadukul, and B. Li, "QoS-Aware Middleware for Ubiquitous and Heterogeneous Environments," *IEEE Communications* 39(11): 140–148, November 2001.

Noble, B., "System Support for Mobile: Adaptive Applications," *IEEE Personal Computing Systems* 7(1):44, 2000.

Noble, B., M. Satyanarayanan, D. Narayanan, J. E. Tilton, J. Flinn, and K. R. Walker, "Agile Application-Aware Adaptation for Mobility," in *Proceedings of the 16th ACM Symposium on Operating Systems Principles*, Saint-Malo, France, 5–8 Oct. 1997, pp. 276–287.

Peine, H. "Security Concepts and Implementations for the Ara Mobile Agent System," in *Proceedings of the Seventh IEEE Workshop on Enabling Technologies: Infrastructure for the Collaborative Enterprises*, Stanford University, June 1998.

Peine H., and T. Stolpmann: "The Architecture of the Ara Platform for Mobile Agents," in *Proceedings of the First International Workshop on Mobile Agents (MA '97)*, Vol. 1219 of *Lecture Notes in Computer Science*. Berlin, Springer, 1997.

Plank, J. "Libckpt: Transparent Checkpointing under Unix," in *Proceedings of the Usenix Winter 1995 Technical Conference*, New Orleans, January 1995.

Ranganathan, M., A. Acharya, S. Sharma, and J. Saltz, "Network-Aware Mobile Programs," in *Proceedings of the 1997 USENIX Technical Conference*, 1997, pp. 91–104.

Richard, G. G., III, and S. Tu, "On Patterns for Practical Fault Tolerant Software in Java," in *Proceedings of the 17th IEEE Symposium on Reliable Distributed Systems*, 1998, pp. 144–150.

Sakamoto, T., T. Sekiguchi, and A. Yonezawa, "Bytecode Transformation for Portable Thread Migration in Java," in *Proceedings of the Joint Symposium on Agent Systems and Applications/Mobile Agents (ASA/MA)*, September 2000, pp. 16–28.

Truyen, E., B. Robben, B. Vanhaute, T. Coninx, W. Joosen, and P. Verbaeten, "Portable Support for Transparent Thread Migration in Java," in *Proceedings of the Joint Symposium on Agent Systems and Applications / Mobile Agents (ASA/MA)*. September 2000, pp. 29–43.

Walsh, T., N. Paciorek, and D. Wong, "Security and Reliability in Concordia," in *Proceedings of the Thirty-First Annual Hawaii International Conference on System Sciences*, vol. 7. January 1998.

Wong, D., N. Paciorek, and D. Moore, "Java-Based Mobile Agents," *Communications of the ACM* 42(3):XX, March 1999.

Wong, D., N. Pariorek, T. Walsh, et al. "Concordia: An Infrastructure for Collaborating Mobile Agents," in *Proceedings of the First International Workshop on Mobile Agents (MA '97)*, Vol. 1219 of *Lecture Notes in Computer Science*. Berlin, Springer, 1997.

Service Discovery Middleware: Finding Needed Services

Mobility introduces interesting challenges for the delivery of services to clients because mobile devices typically are more resource-poor than their wired counterparts and know much less about their current environment. While a desktop computer typically has ready access to many peripheral devices such as printers, scanners, and tape backup, mobile clients are generally not bound to particular infrastructure, trading those bonds for increased freedom. In turn, mobile clients depend more on dynamic interaction with their environment, discovering services as needed.

Service discovery frameworks (e.g., Arnold et al., 1999; Guttman et al., 1999; Salutation, 1999; UPnP, 2003; Czerwinski et al., 1999) make networked services significantly less tedious to deploy and configure and can be used to build rich mobile computing environments. In a service discovery–enabled network, for example, a printer becomes usable (and discoverable) as soon as it is plugged in. This reduces configuration hassles and saves valuable systems administration time because the printer adjusts to its surroundings with little additional help. Similarly, users have a more hassle-free experience: Service discovery-enabled clients (e.g., a word processor) can find and use the printer immediately without forcing the user to search manually for the printer, identify its type, and then download and install device drivers. If the printer is removed—say, in order to upgrade it to a model with more capacity—and replaced with another, the new printer will integrate into the network just as easily. And a mobile client roaming away from the printer (perhaps leaving the office, headed for some afternoon work in a coffee shop) will disassociate itself from that printer and find another appropriate one when necessary. Used in this way, service discovery extends the local "plug and play" technology that (usually) works in Windows environments conceptually to the network and between different platforms. This naturally makes device mobility less painful—moving a device from home to an office and then to a friend's home requires no reconfiguration if the same service discovery framework is available in all the locations.

Services that are more interesting than the usual examples—printing, scanning, etc.— can also be enabled by service discovery technologies. A service discovery-enabled key chain in a user's purse could turn on lights, transfer desktop settings, or adjust stereo systems as the user moves about. The same device also might suck up electronic business cards automatically when the user attends a meeting or make a copy of a diagram scribbled on a service discovery-enabled whiteboard. Remote file storage services can be deployed to extend the limited storage capacity of small mobile computers such as personal digital assistants (PDAs). These ideas are not revolutionary—the late Marc Weiser wrote about these issues many years ago—but service discovery technologies *standardize* the software environment for creating such environments, making both implementation and compatibility more straightforward.

Broadly, a *service discovery framework* is a collection of protocols for developing highly dynamic client-server (CS) applications that standardizes a number of common mechanisms for interaction between clients and services. Central questions in the design of CS systems include:

- What types of services are available?
- Where are the services?
- How can clients contact services?
- What protocols do client and service use?
- How can we make CS systems self-healing?
- How can we effectively manage system load or enhance fault tolerance by adding redundant components?
- How safe is it for a client to use a service or for a service to interact with a particular client?

Service discovery frameworks provide a context to answer questions of this sort in a standard fashion. The service discovery frameworks that have been proposed to date, for example, Jini (Arnold et al., 1999), Universal Plug and Play (UPnP, 2003), Service Location Protocol (Guttman et al., 1999), and the Salutation Architecture (Salutation, 1999) have a lot of common ground. The basic components and operations necessary to address the preceding questions are generally well understood. All service discovery frameworks support the concepts of *client* and *service*, defined in the typical sense: Clients need things, and services provide them. For example, a service discovery–enabled LCD projector might provide wireless projection services to a laptop client during a paper presentation. Similarly, a service discovery-enabled telescope might provide a user with a high-resolution image of Saturn. Service discovery frameworks differ in their concrete design particulars but are remarkably similar at a high level.

The most basic interactions between clients and services are service advertisement and service discovery. *Service advertisement* allows services to announce their presence when they enter the network and to announce their departure from the network. The advertisement typically includes necessary contact information and descriptive attributes or information that allows these descriptive attributes to be discovered. From the client point of view, *service discovery* allows clients to discover dynamically services present either in their local network environment or on a larger scale (e.g., in the global Internet). In some cases, services are sought directly; in others, one or more service catalogs are discovered, and these catalogs are queried for needed services. The discovery attempt typically includes information about the type of services needed, including the standardized service type name(s) and service characteristics. These characteristics might identify the specific (geographic) location of a service, device capabilities (e.g., duplex capability for a printer) in the case of hardware services, and accounting information, such as the cost to use the service.

Whether services are sought directly or a catalog is consulted, a client needs very little information about its environment—it can locate services (or service catalogs) dynamically with little or no static configuration. Similarly, service characteristics such as the protocols necessary for communication can be determined dynamically. Service discovery frameworks also standardize the operation of service catalogs, garbage collection facilities, security, and the development of protocols for communication between clients and services.

The primary advantage provided to both developers and end users by service discovery protocol suites is standardization—none of the CS interactions (such as discovery and advertisement) are particularly magical and could be developed in an ad hoc fashion by any competent programmer. However, standardization brings "out of the box" compatibility to a diverse set of clients and services. Of course, an implementation of a service discovery framework necessarily provides concrete implementations of discovery, advertisement, and eventing, potentially saving a developer a significant amount of effort over developing sophisticated CS systems from scratch. This chapter examines the common components of service discovery frameworks in detail, exploring a range of possible design strategies and drawing specific design choices from the range of currently deployed frameworks. The chapter is not intended as a primer for application *development* under service discovery—this would require substantially more space, and there are several good books that cover development (e.g., Richard, 2002; Newmarch, 2000; Jeronimo and Weast, 2003).

7.1 Common Ground

In this section we explore the common ground among service discovery protocol suites in more detail. Looking at the common characteristics of service discovery technologies gives us a good idea of what they provide for end users and programmers. Subsequent sections in this chapter examine design strategies for most of the mechanisms, discussing some of the approaches taken by current frameworks. Most service discovery frameworks address a large subset of the following concepts:

- *Standardization of services.* To support dynamic discovery of services, service types must be standardized. In the standardization process, the essence of a service is defined; this includes the operations that the service supports, the protocols it uses, and descriptive attributes that provide additional information about the service.

- *Discovery of services.* Needed services may be discovered on demand with minimal prior knowledge of the network. This is the point of service discovery. Typically, clients can search for services by type ("digital camera") or by descriptive attributes ("manufactured by Cameras, Inc."), or both. The *richness* of provided search facilities varies considerably among current service discovery offerings. More powerful search facilities allow clients to fine-tune their discovery requests more carefully, whereas lightweight facilities are appropriate for a wider range of devices, including devices with severe resource constraints such as cellular phones.

- *Service "subtyping."* Clients occasionally may be interested in a very specific type of service—e.g., a high-resolution color laser printer might be required to print a digital photograph. In other cases, only basic black-and-white printing services are required (e.g., to print a shopping list). Service subtyping allows a client to specify a needed service type with as much (or as little) detail as necessary. Service subtyping allows the bare essence of a service type to be standardized and more specific instances of a service type to inherit and expand on this essence.

- *Service insertion and advertisement.* Service advertisement allows the dynamic insertion into and removal of services from a network, providing an extension of "plug and play" technologies into a networked environment. Services slip into a network with a minimum of manual configuration and advertise their availability either directly to clients or to servers maintaining catalogs of services. Conversely, services leaving a network in an orderly fashion (as opposed to crashing) can advertise their departure. A primary difference between service discovery technologies and relatively static information services such as the Domain Name Service (DNS; Mockapetris, 1987) or Dynamic Host Configuration Protocol (DHCP; Droms, 1997) is that service discovery technologies allow highly dynamic updates—services appearing or disappearing result in immediate reconfiguration of the network. In contrast, DNS and DHCP rely on static files or databases that are configured by systems administrators with higher levels of authority than typical users.

- *Service browsing.* Browsing allows clients to explore the space of currently available services without a priori knowledge of the network environment and without any specific service types in mind. Service browsing is to "window shopping" as service discovery is to a focused attempt to buy a specific item. Information obtained through service browsing may be presented to the user in a graphic user interface. A user then chooses to interact with whichever services are sufficiently interesting.

- *Service catalogs.* Although some discovery frameworks operate entirely in a peer-to-peer fashion, putting clients directly in touch with services from the very start, some support catalogs that maintain listings of available services. When service catalogs are implemented, services perform advertisement against one or more catalogs rather than interacting with clients directly. Similarly, clients query catalogs for needed services rather than searching the network for services. There are some substantial advantages to using service catalogs, including greater flexibility in deploying services beyond the local network segment and a reduction in multicast traffic. These advantages are covered in later sections.

- *Eventing.* Eventing allows asynchronous notification of interesting conditions (e.g., a needed service becoming available or an important change in the state of a service, such as a printer running out of paper or a long computation being completed). An eventing mechanism replaces polling, providing more timely notifications of important events, making software development more straightforward, and reducing the burden on network resources.

- *Garbage collection.* Garbage collection facilities remove outdated information from the network, including advertisements associated with defunct services and subscriptions to eventing services. Without garbage collection, performance could suffer significantly, as clients try to contact nonexistent services or services continue to perform operations (such as eventing) on behalf of crashed clients. Garbage collection is critical for the proper operation of service catalogs as well, which would overflow with outdated information in the absence of such a facility.

- *Scoping.* Service discovery frameworks typically provide a mechanism for controlling the scope of both service discovery and service advertisement. Scoping is addressed in two different ways. The first controls the extent of multicast communication, often by using administratively scoped multicast (Meyer, 1998). The purpose of this sort of scoping is to provide administrative control over the multicast radius for advertisement and discovery in order to limit traffic on networks whose (service) inhabitants are unlikely to be useful to distant clients and services. For example, dynamic discovery of a printer in China is unlikely to be useful to a person sitting in a coffee shop in San Francisco. The other sort of scoping associates names with services to create service groups. This type of scoping is simpler and does not actually reduce the radius of multicast-based communication. For example, assigning the name "UNO CS DEPT" to certain services allows clients to specify during discovery attempts that they are interested in services within the control of the Department of Computer Science at the University of New Orleans. The messages in the discovery attempt actually may travel far beyond the bounds of the Department of Computer Science but will be ignored.

7.2 Services

Services provide benefits to clients, such as file storage, printing, faxing, and access to high-performance computing facilities in the same sense as the "server" in traditional CS environments. The dynamicity added to the CS paradigm by service discovery introduces some new concerns, such as globally unique identifiers for services so that individual service instances can be tracked, how services are located, and methods for standardization. These are addressed in this section.

7.2.1 Universally unique identifiers

It is useful to be able to identify services uniquely, particularly in large networks where many instances of the same service type may be present. Assigning *universally unique identifiers* (UUIDs) to services has several benefits: It allows clients to search for a specific service by its identifier (clients can be statically configured to seek specific service instances in this manner), and it allows clients and service catalogs to determine if two service instances (perhaps discovered at different times) are in fact the same service. Since service discovery protocols are targeted at a wide variety of network types, from small home networks with only a few nodes and little infrastructure to large corporate networks, it is important that universally unique identifiers can be created without relying on a centralized controller.

At first, it might seem difficult to generate universally unique identifiers without global infrastructure, but in fact, most computers with network interfaces already contain at least one unique identifier—an IEEE 802.3 Media Access Control (MAC) address. MAC addresses are 6 byte quantities with 3 bytes allocated to a vendor ID and 3 bytes to a vendor-specific serial number. For example, MAC addresses beginning with bytes 0x0000A0 are allocated to Sanyo, Inc. MAC addresses alone are insufficient for universally unique service IDs for a number of reasons. The first is that some services may be hosted on devices with no assigned MAC address. A second reason is that MAC addresses can be configured in software, potentially resulting in duplicates. Another reason is that a device with a single MAC address may host many services, each of which requires a universally unique ID. Finally, it is not generally desirable to tie a service to a particular machine. If the hardware fails or the service is migrated to higher-performance hardware, it is nice if the service can make the transition with its identity intact.

We describe Jini's method for creating universally unique identifiers because it is typical. Jini's method is similar to that described in ISO-11578, with an additional mechanism to allow UUIDs to be created without reference to an MAC address, and is adequate for a broad class of applications. Note that even when a MAC address is available, for security reasons, it may be desirable to generate UUIDs without reference to this address (Goland et al., 1999). In addition, some high-level languages, including Java, cannot generate UUIDs based on MAC addresses programmatically without using native methods (such as JNI—the Java Native Interface, which allows interaction between Java code and traditionally compiled code, such C or C++).

UUIDs in Jini are 128 bits long and are created using a combination of random numbers, a measure of the current time, and possibly a MAC address. The most significant 64 bits of the identifier are composed of a 32-bit *time_low* field, a 16-bit *time_mid* field, a 4-bit *version* number, and a 12-bit *time_hi* field. The least significant 64 bits of the identifier are composed of a 4-bit *variant* field, a 12-bit *clock_seq* field, and a 48-bit *node* field. The variant is always 2. The *version* field can contain either 1 or 4. If the version field contains 1, then the *node* field is set to a 48-bit MAC address, the *clock_seq* field is set to a random number, and the three time fields are set to a 60-bit measure of elapsed time (in 100-ns increments) from midnight, October 15, 1582. If the version field is 4, then the other fields (except for the variant) are set to a random number.

If a MAC address is not used to generate a UUID, then RFC 2518 contains some words of wisdom on choosing the 48-bit random number. First, to avoid conflict with "real" MAC addresses, the high-order bit should be set to 1. This is the unicast/multicast bit and will never be set in MAC addresses assigned to network interfaces. The other 47 bits should be created using a random-number generator of cryptographic quality. A UUID for a service instance generally should be created once and then stored permanently. Depending on the nature of the service, the UUID may be stored in ROM, in nonvolatile RAM, or on disk.

7.2.2 Standardization

An essential component in service discovery frameworks is a standardization process for new service types. For clients to discover needed services (in an abstract sense, printers, scanners, high-performance compute services), it must be possible for the client to specify service types in a standard way. This is more than a simple naming problem—beyond the initial discovery of services of a specific type, a client must know how to interact with the service—how to make the service "do its thing." This requires the standardization process to capture the "essence" of a service type, which in turn requires answers to the following questions:

- *What does it mean to be a service of type X?* In other words, what do clients want from an instance of X?

- *What operations are appropriate for services of type X?* For a basic type (e.g., a printer), these operations should be the essential ones. Using service subtyping, operations more appropriate for specialized instances of a service type (e.g., a color printer with duplex capabilities) can be introduced.

- *What protocols does an instance of type X use?*

- *What descriptive attributes are required to adequately describe the characteristics and capabilities of a service?* For a printer, such attributes would reveal whether the printer is capable of duplex printing, color, a speed rating (e.g., in pages per minute), and the like. In addition, the attributes might reveal the e-mail address of a human being responsible for proper operation of the printer, the manufacturer of the device, and a URL for documentation associated with this brand of printer.

There are currently two schools of thought for standardization of services: using textual descriptions, which are independent of the language in which a service will be implemented, and describing services in terms of an interface (in the object-oriented sense) for a particular programming language. We discuss each of these design choices in detail in the following sections.

7.2.3 Textual descriptions

Most of the proposed service discovery frameworks use textual descriptions for standardizing services. The Service Location Protocol (SLP), Universal Plug and Play (UPnP), Salutation, and the Bluetooth Service Discovery Protocol (SDP), among the current commercial protocol suites, all use programming language–independent textual descriptions to describe services. Ninja, which is primarily a research prototype emphasizing advanced security features for service discovery, does as well.

There are two problems in constructing a textual service description. The name of the service type must be standardized so that clients have a mechanism for specifying needed services. In addition, the names of standard attributes must be defined so that capabilities and characteristics of service instances can be determined either during the initial discovery attempt or during a postdiscovery service interrogation. The second issue is choosing the particular *protocol* used between a client and a service instance. Some service discovery frameworks, such as SLP, use an attribute to specify an external protocol that is used for client/service communication; SLP is not concerned with the *definition* of this protocol. For example, the specification for a printer in SLP might indicate that the Line Printer Daemon Protocol (LPD) is to be used (McLaughlin, 1987). Another document—in this case, RFC 1179— precisely defines the protocol. Other frameworks, such as UPnP, define the client/service protocol in conjunction with the definition of the service and attribute names.

To illustrate service standardization using a textual approach, we examine the specification of a fictitious blender device in UPnP and then an echo service in SLP. These are provided to whet your appetite. You are strongly encouraged to consult the specifications for additional details.

UPnP uses XML description documents to specify services. A description document for our blender service type might look like this:

```
<?xml version="1.0"?>
<root xmlns="urn:schemas-upnp-org:device-1-0">
  <specVersion>
    <major>1</major>
    <minor>0</minor>
  </specVersion>
  <URLBase>http://10.0.0.13:5431</URLBase>
  <device>
    <deviceType>urn:schemas-upnp-org:device:blender:1</deviceType>
    <friendlyName>UPnP Blender</friendlyName>
```

```
<manufacturer>University of New Orleans
                Dept. of Computer Science
</manufacturer>
<manufacturerURL>http://www.cs.uno.edu/</manufacturerURL>
<modelDescription>UPnP-compatible blender with Accublend Whirring
</modelDescription>
<modelName>Plug-N-Blend Deluxe</modelName>
<modelNumber>UBlend9873A</modelNumber>
<modelURL>http://www.upnpblend.com/</modelURL>
<serialNumber>999954321</serialNumber>
<UDN>uuid:Upnp-Blender-1_0-1234567890001</UDN>
<UPC>123456789</UPC>
<serviceList>
  <service>
    <serviceType>
      urn:schemas-upnp-org:service:PowerSwitch:1
    </serviceType>
    <serviceId>urn:upnp-org:serviceId:PowerSwitch1</serviceId>
    <controlURL>/upnp/control/power1</controlURL>
    <eventSubURL>/upnp/event/power1</eventSubURL>
    <SCPDURL>/blenderpowerSCPD.xml</SCPDURL>
  </service>
  <service>
    <serviceType>
      urn:schemas-upnp-org:service:SpeedControl:1
    </serviceType>
    <serviceId>
      urn:upnp-org:serviceId:SpeedControl1
    </serviceId>
    <controlURL>/upnp/control/speed1</controlURL>
    <eventSubURL>/upnp/event/speed1</eventSubURL>
    <SCPDURL>/blenderspeedSCPD.xml</SCPDURL>
  </service>
  <service>
    <serviceType>
      urn:schemas-upnp-org:service:Bowl:1
    </serviceType>
    <serviceId>urn:upnp-org:serviceId:Bowl1</serviceId>
    <controlURL>/upnp/control/bowl1</controlURL>
    <eventSubURL>/upnp/event/bowl1</eventSubURL>
    <SCPDURL>/blenderbowlSCPD.xml</SCPDURL>
  </service>
</serviceList>
<presentationURL>/blenderdevicepres.html</presentationURL>
</device>
</root>
```

The descriptive attributes for the blender device include the device type ("blender"), a human-friendly name for the device ("UPnP blender"), the manufacturer ("University of New Orleans"), a description of the model, the model name ("Plug-N-Blend Deluxe"), a model number ("UBlend9873A"), a URL where documentation about the device can be obtained (*http://www.upnpblend.com*), a serial number ("999954321"), a universally unique device name, and a UPC code. The primary advantages of using XML are platform independence, widespread use in other application areas, and the fact that XML is relatively easy to parse.

A UPnP description document also contains pointers to other XML documents, which define the protocol used between a client and services provided by the device. These documents are called *service control protocol description* (SCPD) documents and define a remote procedure call interface to services. UPnP uses the Simple Object Access Protocol (SOAP; *http://www.w3.org/TR/soap/*) to build standardized client/server protocols.

The UPnP blender provides three services: one to control the power switch, one to control speed, and one to control the contents of the blender. We examine the SCPD document that defines the protocol between the speed service and the client. Thus *blenderspeedSCPD.xml* contains the following:

```
<?xml version="1.0"?>
<scpd xmlns="urn:schemas-upnp-org:service-1-0">
  <specVersion>
    <major>1</major>
    <minor>0</minor>
  </specVersion>
  <actionList>
    <action>
      <name>SetSpeed</name>
      <argumentList>
        <argument>
        <name>Speed</name>
          <relatedStateVariable>CurrentSpeed</relatedStateVariable>
          <direction>in</direction>
        </argument>
      </argumentList>
    </action>
    <action>
      <name>GetSpeed</name>
      <argumentList>
        <argument>
        <name>Speed</name>
          <relatedStateVariable>CurrentSpeed</relatedStateVariable>
          <direction>out</direction>
        </argument>
      </argumentList>
    </action>
    <action>
      <name>IncreaseSpeed</name>
    </action>
    <action>
      <name>DecreaseSpeed</name>
    </action>
  </actionList>
  <serviceStateTable>
    <stateVariable sendEvents="yes">
      <name>CurrentSpeed</name>
      <dataType>i1</dataType>
        <allowedValueRange>
          <minimum>1</minimum>
          <maximum>10</maximum>
          <step>1</step>
        </allowedValueRange>
      <defaultValue>1</defaultValue>
    </stateVariable>
  </serviceStateTable>
</scpd>
```

The operations provided by this service are enumerated between the `<actionlist>` and `</actionlist>` tags and include SetSpeed, GetSpeed, IncreaseSpeed, and DecreaseSpeed. For each operation, parameters are defined (e.g., in the case of SetSpeed, a single "in" parameter "Speed"). Toward the bottom of the document, delineated by the `<serviceStateTable>` tag, we define a set of state variables; these variables expose the visible state of a service. Each parameter for an action is associated with a state variable through the `<relatedStateVariable>` tag to establish the type and allowed values for the parameter. This association establishes the type and allowable values for a parameter. This service has only a single state variable "CurrentSpeed," which has type "i1" (a 1-byte integer), an allowable range of 1 to 10, and an increment/decrement step of 1. For more on UPnP service specification, see the references at the end of this chapter.

As another example of textual service standardization, we briefly examine an SLP echo server specification. The server simply accepts network connections and echoes any input back to the client. The following is an *abstract* service template, which defines the standard name for the echo service (echo-service), the version of the standard (0.0), a human-readable description, the format of the URL used in advertisements of the service, and a number of descriptive attributes, including the name of the administrator of the service, his e-mail address, and the maximum line length supported. The *O* appended to the attribute definition means that the attribute is optional.

```
template-type=echo-service.test

template-version=0.0

template-description=
    Definition of a simple SLP echo service.  Reads lines of
    input (which should include a \n character as a
    line terminator) and echoes the lines back to the client.
    The protocol spoken depends on the concrete service type.

template-url-syntax=
    url-path= ;    depends on concrete type

contact=string O
# the contact name for the maintainer of the service (optional)

contactemail=string O
# the email address of the maintainer of the service (optional)

maxlinelength=integer
80
# the recommended maximum line length (max number of
# characters to transmit before \n; REQUIRED, since O
# modifier is not present).  Default value is 80.
```

The following is a *concrete* echo-service–Transmission Control Protocol (TCP) template that dictates that TCP should be used as the transport protocol. Note that definitions of attributes defined in the abstract type are not repeated in the concrete type's template.

```
template-type=echo-service.test:tcp

template-version=0.0

template-description=
    Concrete SLP echo service which uses TCP as a transport protocol.

template-url-syntax=
url-path= ;

# no additional attributes beyond those described in the
# abstract template "echo-service.test"
```

7.2.4 Using interfaces for standardization

Another approach to standardizing a new service type is to define an interface (generally in the object-oriented sense of the word) that service instances implement. The exemplar of this approach is Jini, which uses Java interfaces as the mechanism for standardizing services. The interface defines precisely the methods that will be used in interacting with the device. For example, a standard Jini interface for a remote file storage might look like the following:

```
public interface StorageService extends Remote {
public boolean open(String username, String password,
    boolean newAccount) throws RemoteException;
public boolean close(String username, String password)
    throws RemoteException;
public boolean shutdown(String username, String password)
    throws RemoteException;
public boolean store(String username, String password, byte[] contents,
    String pathname) throws RemoteException;
public byte[] retrieve(String username, String password,
    String pathname) throws RemoteException;
public boolean delete(String username, String password,
    String pathname) throws RemoteException;
public String[] listFiles(String username, String password)
    throws RemoteException;
public String name() throws RemoteException;
}
```

This interface defines the high-level CS communication protocol; methods are provided for opening an account on a remote file storage service instance, for storing and retrieving files, and for obtaining directory listings of stored files. The Java Remote interface is extended by the StorageService interface because the implementation will use Java's Remote Method Invocation (RMI) facility. In contrast to the UPnP example in the preceding section, this Jini interface for a remote file storage does not define associated descriptive attributes. In Jini, service attributes are supplied during a service's registration with a service catalog.

In Jini, services register proxy objects that implement their service type's standard interface, like StorageService above. These proxy objects are stored on one or more lookup services, which are catalogs of available Jini services. A Jini client performing discovery against the lookup service obtains copies of appropriate proxy objects, which are "remote controls" for the corresponding services. Executing methods in a proxy object results in a range of possible activities; the computation performed by the method can occur entirely locally on the client, the execution may invoke some private protocol to contact a remote machine, or the execution of the method may be through RMI, where the bulk of the associated computation takes place on the remote end at the service. For example, invoking the `delete()` method on a proxy of type StorageService causes a remote file to be deleted (provided proper credentials are supplied).

Shielding the client from these implementation choices—essentially local versus remote computation—has positive benefits. For example, one implementation of the file storage service might perform all actions locally, storing and retrieving files from a local file system. Another implementation might use RMI to store and retrieve files from a remote server. For performance improvements, a third implementation might replace RMI with a private sockets-based protocol between the proxy object and a remote server.

7.3   More on Discovery and Advertisement Protocols

Discovery and advertisement protocols allow clients to find interesting services and service to make their existence known (dynamically) to clients. A discovery-enabled client is able to power on and immediately discover available services, provided that the services in the area use a compatible service discovery protocol. The discovery and advertisement protocols specified in a service discovery framework are the key to this capability, allowing clients to request services, as needed, and for services to answer these queries.

7.3.1   Unicast discovery

Unicast discovery is the simplest form of service discovery protocol. A client configured statically with the location of one or more service catalogs (in the case of catalog-based frameworks such as Jini) or services (in peer-to-peer systems such as UPnP) can contact the needed resources directly. Unicast discovery protocols typically use TCP/IP or another reliable, stream-oriented transport protocol. For example, Jini provides a unicast discovery protocol for interacting with a set of statically configured lookup services (Jini's form of service catalog). The static configuration provides the Internet Protocol (IP) address and port on which a lookup service listens; the purpose of the unicast discovery protocol is to support downloading a proxy object (of type ServiceRegistrar) that allows a client to control the lookup service. A partial definition of the ServiceRegistrar type is shown below for illustrative purposes:

```
public interface ServiceRegistrar {
ServiceRegistration register(ServiceItem item, long leaseDuration)
    throws RemoteException;
Object lookup(ServiceTemplate tmpl) throws RemoteException;
ServiceMatches lookup(ServiceTemplate tmpl, int maxMatches)
    throws RemoteException;
...
...
EventRegistration notify(ServiceTemplate tmpl, int transitions,
    RemoteEventListener listener, MarshalledObject handback,
    long leaseDuration) throws RemoteException;
...
...
Class[] getServiceTypes(ServiceTemplate tmpl, String prefix)
    throws RemoteException;
...
...
}
```

The lookup() method allows a client holding an object implementing the ServiceRegistrar interface to search for needed services by providing a ServiceTemplate instance, which defines the service's type, a list of desired attributes, and an optional universally unique identifier (which allows matching only a specific service). The register() method allows a service to register its proxy object and attributes with the lookup service. The notify() method is rather complicated but essentially allows a client to register a ServiceTemplate (like that provided to the lookup() method) and receive asynchronous events when matching services enter or leave the network.

The Jini unicast discovery protocol is illustrated in Fig. 7.1. A class named LookupLocator encapsulates the IP address and port on which a service catalog is listening. A Jini entity establishes a TCP connection to the service catalog using the well-known IP address and port and downloads an instance of ServiceRegistrar (discussed earlier) to interact with the service catalog.

7.3.2 Multicast discovery and advertisement

Discovery.   Unlike unicast discovery, in which the locations of service catalogs or individual services are configured statically, multicast discovery allows dynamic discovery of interesting services. All current

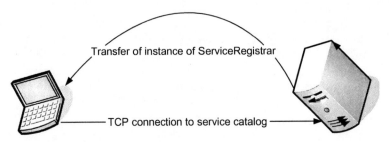

Figure 7.1   In Jini, clients and services can contact service catalogs whose locations have been statically configured using the Unicast Discovery Protocol. A TCP connection is established to the service catalog, followed by transfer of an instance of ServiceRegistrar to the client or service. The ServiceRegistrar instance exposes methods that can be used to control the service catalog.

discovery frameworks provide a dynamic multicast-based protocol, and most are built on top of unreliable UDP multicast. This limits the scope of dynamic service discovery to the local network segment or, more accurately, to the multicast radius of the local network. This radius, in turn, is determined by the design of the local network and administrative decisions that affect the flow of multicast traffic. Discovery beyond the local network segment, in a less dynamic form, is possible through the use of service directories because remote services might be configured statically to register with service catalogs in arbitrary locations. Service catalogs might also accept registrations from other (remote) service catalogs, bridging service discovery domains.

What is the nature of a service discovery attempt? A client either knows the type of service that is needed (e.g., a printer) or is interested in browsing the space of available services. Further, the client may have additional restrictions to impose: that the service be nearby (no printers more than 5 miles away, please) or that the device providing the service be of a certain brand. The client might even know information that identifies a *specific* service, such as a UUID; in this case, only one service instance in the entire universe (to be melodramatic) will do. Thus some additional description of the service, beyond its type, is necessary. Descriptive attributes fill this role. Depending on the methodology, these attributes may be specified during the discovery attempt (Jini uses this approach), or additional actions may be required *after* the existential service information is established. For example, under UPnP, once the existence of a service is established, an XML document must be downloaded that contains the values of descriptive attributes.

A responsible discovery protocol attempts to conserve network resources, of course. In pursuing this cause, a discovery protocol should do whatever is possible to reduce the number of inappropriate responses to a service discovery attempt. Specification of one or more service types

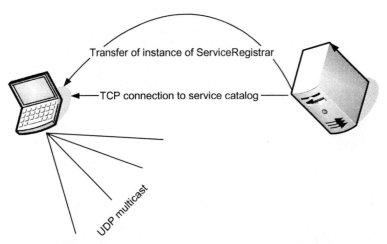

Figure 7.2   In Jini, clients and services must discover the location of at least one lookup service (service catalog) in order to participate in the Jini federation. The multicast request protocol allows a client or service that does not know the location of Jini lookup services to discover them dynamically. The first round of communication is a UDP multicast, which contains an IP address and port for the Jini entity seeking a service catalog. This transmission also contains the identities of other (known) service catalogs, so these catalogs can suppress responses. A service catalog hearing the multicast establishes a TCP connection to the client or service and transmits an instance of ServiceRegistrar, an interface for controlling the service catalog.

during discovery reduces the number of responses from services that is of no use to the client. So does specification of additional constraints corresponding to the values of descriptive attributes. Including UUIDs of known services (or service catalogs) takes this conservation a step further. In this way, a client can advertise the identities of services it already knows from previous discovery attempts. Such services need not respond again. Jini's multicast discovery protocol is an example of a dynamic discovery protocol and is illustrated in Fig. 7.2.

Advertisement.   Service advertisement is the converse of discovery, allowing services entering or leaving a network to advertise their availability (or unavailability). In addition, services periodically advertise their presence for the benefit of clients that have just entered the network. Why advertise when clients simply can perform discovery to obtain information about available services? The obvious reason is to prime service catalogs, but there are other important reasons. Advertisement not only allows clients to passively monitor the availability of new services but also supports more powerful models for discovery. Rather than attempting discovery of needed services periodically (which amounts to polling), clients might register their interest in the availability of needed services and rely on a mechanism that provides asynchronous notification when service advertisements arrive. Jini contains support classes that provide asynchronous notification of lookup services or regular services becoming available. In particular, the notify() method in the ServiceRegistrar interface (discussed briefly in Section 7.3.1) allows clients to register future interest in service transitions. Generally, advertisements contain an expiration date, which is an informal promise that the advertisement remains valid for a certain period of time. As with dynamic discovery protocols, service advertisements are based on multicast and are thus constrained to the administratively controlled multicast radius. Consider Fig. 7.3, which illustrates common interactions

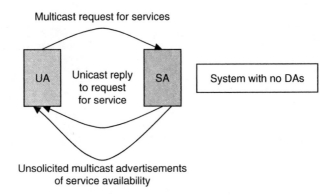

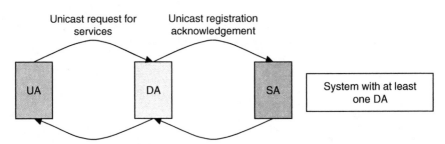

Figure 7.3 Some common SLP agent interactions. In a system with no directory agents (DAs), a user agent (UA) will typically send multicast requests to find appropriate services. The request contains a service type and essential characteristics of the needed service. A service agent (SA) which provides a matching service will unicast a reply containing service location and other contact-related information. Service agents also spontaneously advertise their presence through multicast advertisement messages. In systems with at least one DA, SAs register available services with the DA and UAs query the DA for the locations of available services. There are a variety of options for discovering DAs: the locations of DAs may be statically configured, DHCP may be used, or multicast discovery can be used. Once a UA knows the location of a needed service, SLP is out of the picture—unlike Jini and UPnP, the SLP implementation is not involved in client/server communication. The interactions between UAs, SAs, and DAs can be constrained through the use of "scopes" (not pictured), which allow service groups to be formed.

between entities in an SLP-enabled network. In addition to allowing clients to discover services dynamically by multicasting to advertise their needs, when operating in a directory agent–less mode (i.e., without service directories), SLP services multicast messages advertising their availability. These messages contain the standard SLP header, a service URL describing the location of the service, a set of attributes describing the service, and authentication information to allow clients to verify the identity of the service.

7.3.3   Service catalogs

An alternative to putting clients and services directly in touch with one another is to deploy catalogs of available services. In this section we look at service catalogs more closely. Clients make discovery attempts against these service catalogs (after discovering the catalogs dynamically themselves, if necessary), and services advertise directly to the catalogs. Service catalogs are appropriate for protocol suites following either standardization principle (textual or standardized interface). Information stored on a service directory for a particular service instance can range from the location of the service and perhaps some descriptive attributes (in which case clients use the directory solely for existential information) to executable code to be downloaded for controlling the service instance. Of the currently available service discovery frameworks, SLP takes the former approach, whereas Jini takes the latter. Other approaches, such as UPnP, are strictly peer-to-peer.

There are several advantages of service catalogs. The first is dramatically reduced multicast traffic because once service catalogs are discovered, multicast is not necessary for future service discovery or service advertisements. Another is that discovery is extended beyond the multicast radius of the local network because the locations of remote service catalogs can be configured. This allows much larger service discovery domains to be created. In fact, bridging protocols that connect service catalogs may enable "global" service discovery (Zhao, Schulzrinne, and Guttman, 2003). On the other hand, service catalogs introduce another component that must be administered and become single points of failure. In ad hoc environments, with little established networking infrastructure, direct interaction between clients and services may be more appropriate. Such environments also tend to have smaller numbers of nodes, though, which reduces the effectiveness of service catalogs.

We briefly discussed the Jini API for interaction with service catalogs in Section 7.3.1; developers interested in Jini should consult the specification or Richard (2002) for additional details because Jini supports many powerful helper classes that make interacting with service catalogs very straightforward. Other service discovery frameworks that offer service catalogs, such as SLP, offer similar functionality. One contrast between Jini and SLP is that SLP service catalogs, called *directory agents* (DAs), offer only location information for services—they do not provide code for interacting with the registered services.

7.4 Garbage Collection

Garbage collection is critical in service discovery frameworks owing to the highly dynamic nature of CS relationships. Without a mechanism for removing network state associated with deceased clients and services, an overwhelming amount of "garbage" eventually would accumulate. State associated with dead services wastes the time of clients; without a garbage collection facility, caching information about discovered services is a risky business. A client who has discovered a number of printers, for example, might attempt to print a document several times on different printers before finally contacting an available printer. A garbage collection facility solves this problem by purging information about dead services. Similarly, state associated with dead clients wastes the time of a service. If a client has invoked an eventing mechanism, in which interesting state changes on the service are propagated to the client as they occur, unnecessary network communication will be performed if the client leaves the network. Since such an eventing mechanism might be built on unreliable multicast, without an explicit garbage collection facility, the service continues to send events long after the client has disappeared. In the next two subsections, we examine two popular mechanisms for garbage collection in service discovery frameworks.

7.4.1 Leasing

Leases are one popular garbage collection mechanism. The model is very close to the concept of leasing encountered in real estate. Rather than granting the right to use a resource indefinitely (which is like a purchase), a lease is assigned. The grantor of the lease, called the *lessor,* cancels the lease and reclaims rights to the resource if the lease is allowed to expire or if the terms of the lease are not met. To avoid expiration, the party that is using the resource, called the *lessee,* must request a renewal periodically.

A typical leasing scenario in the real estate world might work like this: A landlord grants residence in an apartment for a period of 1 year at a rate of $900 per month. The terms of the lease are that rent is due on the first of every month, that no pets are allowed, and that smoking in the residence is forbidden. As long as the resident pays the rent on time every month, does not house any pets, and continues to enjoy pink lung tissue, the landlord takes no further action (other than to deposit the resident's payment in the bank). In a service discovery framework, a typical leasing scenario might look like this: A service is plugged into the network and discovers service catalogs. To make its presence known to clients, the service registers with one or more of the service catalogs. For each registration, a lease is assigned. The terms of the lease are that the service must send a renewal ("pay the rent") at a specified interval. Failure to send a renewal results in the service catalog assuming that the service has left the network. In this case, the service catalog purges all information related to the service from the list of available services. This model scales moderately well because the catalog need not constantly poll the status of large numbers of services—the burden of renewal falls on each service, and only a small bit of information (the renewal message) is required to assure the catalog that the service remains viable.

To illustrate a typical API for leasing, we examine part of the Jini leasing API. Note that Jini provides many additional helper classes to automate lease handling—we examine only one, the LeaseRenewalManager. The basic Lease interface looks like this:

```
package net.jini.core.lease;
import java.rmi.RemoteException;
public interface Lease {
long FOREVER = Long.MAX_VALUE;
long ANY = -1;
int DURATION = 1;
int ABSOLUTE = 2;
long getExpiration();
void cancel() throws UnknownLeaseException, RemoteException;
void renew(long duration) throws LeaseDeniedException,
  UnknownLeaseException,
RemoteException;
void setSerialFormat(int format);
int getSerialFormat();
LeaseMap createLeaseMap(long duration);
boolean canBatch(Lease lease);
}
```

When a service registers with a lookup service in Jini, the lookup service provides an object that implements the Lease interface. The service can then check the expiration time with getExpiration(), cancel the lease immediately with cancel(), or renew the lease. An instance of the LeaseRenewalManager class (shown below) can be instantiated to make lease-handling duties somewhat simpler. The methods are straightforward—a number of methods are provided for lease renewal, some of which operate on durations (i.e., renew the lease at a specified rate), whereas others simply provide an absolute time at which a lease can be allowed to expire, relying on the LeaseRenewalManager instance to schedule renewals as appropriate. Finally, additional administrative methods are provided that allow leases to be removed from control of the LeaseRenewalManager and for leases to be canceled. The `clear()` method causes a LeaseRenewalManager instance to forget about all current leases but does not cancel them. The LeaseListener parameters supply the names of classes to be notified when important lease-related events occur, e.g., when a lease renewal fails. A single instance of the LeaseRenewalManager can handle multiple Leases.

```
package net.jini.lease;
public class LeaseRenewalManager {
public LeaseRenewalManager() {...}
public LeaseRenewalManager(Lease lease, long desiredExpiration,
    LeaseListener listener) {...}
public void renewUntil(Lease lease, long desiredExpiration,
    long renewDuration, LeaseListener listener) {...}
public void renewUntil(Lease lease, long desiredExpiration,
    LeaseListener listener) {...}
public void renewFor(Lease lease, long desiredDuration,
    long renewDuration, LeaseListener listener) {...}
public void renewFor(Lease lease, long desiredDuration,
    LeaseListener listener) {...}
public long getExpiration(Lease lease) throws UnknownLeaseException {...}
public void setExpiration(Lease lease, long desiredExpiration)
    throws UnknownLeaseException {...}
public void remove(Lease lease) throws UnknownLeaseException {...}
public void cancel(Lease lease)
    throws UnknownLeaseException, RemoteException {...}
public void clear() {...}
}
```

7.4.2 Advertised expirations

A simpler method for garbage collection that is most appropriate in service discovery frameworks that operate in a strictly peer-to-peer manner (i.e., there are no service catalogs) is to attach timeouts to service advertisements. The timeout allows a service to express the *expected* period during which clients might expect to interact successfully with the service. Service advertisements are broadcast periodically to refresh the expected time of the service's demise. An abstract UPnP advertisement looks like the following:

```
NOTIFY * HTTP/1.1
HOST: 239.255.255.250:1900
CACHE-CONTROL: max-age = seconds until advertisement expires
LOCATION: URL for UPnP description for root device
NT: search target
NTS: ssdp:alive
SERVER: OS/version UPnP/1.0 product/version
USN: advertisement UUID
<<BLANK LINE>>
```

A concrete advertisement of the root blender device captured during operation of a UPnP blender looked like this:

```
NOTIFY * HTTP/1.1
HOST: 239.255.255.250:1900
CACHE-CONTROL: max-age=1800
LOCATION: http://localhost:5431/blenderdevdesc.xml
NT: upnp:rootdevice
NTS: ssdp:alive
SERVER: Linux/2.4.2-2 UPnP/1.0 Intel UPnP SDK/1.0
USN: uuid:Upnp-Blender-1_0-1234567890001::upnp:rootdevice
```

Advertised expirations are often coupled with "byebye" messages that are broadcast when the service is about to leave the network. These "byebye" messages cancel any outstanding advertisements. For example, when the fictitious blender device in UPnP (discussed in Section 7.2.3) leaves the network, it multicasts a Simple Service Discovery Protocol (SSDP) "byebye" message like the following:

```
NOTIFY * HTTP/1.1
HOST: 239.255.255.250:1900
CACHE-CONTROL: max-age=180
LOCATION: http://10.0.0.13:5431/blenderdevdesc.xml
NT: upnp:rootdevice
NTS: ssdp:byebye
USN: uuid:Upnp-Blender-1_0-1234567890001::upnp:rootdevice
```

Actually, UPnP multicasts many such "byebye" messages when a device leaves the network, including messages for the root device and all subdevices. Advertisements are similar—each subdevice is also advertised. Of course, if the service crashes, a graceful exit from the network is impossible, and clients may attempt unsuccessfully to contact the service before the anticipated time of demise.

7.5 Eventing

Eventing allows service discovery–enabled clients to remain aware of interesting conditions concerning a service without the need for explicit polling. Current service discovery frameworks use various mechanisms, but we will examine the General Event Notification Architecture (GENA), a part of the UPnP protocol suite, as a model for understanding eventing.

GENA allows UPnP clients to receive asynchronous notifications about interesting state changes in UPnP services. Clients in UPnP can explicitly issue commands to change the state of UPnP devices and to query the values of a service's state variables, but eventing adds the ability to subscribe to a service and to learn of changes in the values of state variables as they occur, without polling. Examples of such state changes are a UPnP VCR's transport being paused, a UPnP light switch being turned on, or a UPnP printer running out of paper.

The XML documents describing UPnP services publish an *event subscription URL* used by clients subscribing to a service's event-notification mechanism. Requests to subscribe and unsubscribe are sent to this URL. To avoid sending events to deceased clients, all GENA subscriptions are leased and must be renewed periodically or they expire. The service decides the duration of the subscription and transmits this information to the client. The flow of information then begins with a message containing the names and values of all state variables, encoded in XML. Subsequently, when state variables associated with a service change, the subscription service transmits the names and values of the changed variables to *all* subscribed clients.

Some state variables may have values that consume large amounts of space, are updated very frequently, or both. Transmitting the values of such variables continuously might overwhelm the client or the entire network. When declaring such state variables in a service control protocol description document (see Section 7.2.3), the `<stateVariable>` tag may be augmented with "`sendEvents=no`" to prevent the transmission of the variable's value to subscribed clients. For example, the following UPnP state variable declaration creates a variable "Buffer" of type string whose value will *not* be sent to subscribed clients:

```
<stateVariable sendEvents="no">
   <name>Buffer</name>
   <dataType>string</dataType>
   <defaultValue>"EMPTY"</defaultValue>
</stateVariable>
```

UPnP's eventing mechanism does not allow subscribers to specify *which* state variables they are interested in monitoring. A service transmits *all* state variable values, except those associated with state variables whose definitions *statically* prevent transmission, to all subscribed clients. Clients are thus left to sift through the transmitted state values and decide what is interesting. A positive side of this limitation is that it simplifies the eventing mechanism in each service—individual services need not track which variables to transmit to each client. This is a reasonable trade-off considering that resource-poor devices are likely to use UPnP.

A number of message types are used in GENA, which is a Hyper-Text Transfer Protocol (HTTP)–based protocol running over TCP. We cover these briefly; full details are in the UPnP specification. A client can subscribe to a service's eventing by sending a message of the following sort:

```
SUBSCRIBE publisherpath HTTP/1.1
HOST: publisherhost:publisherport
CALLBACK: <deliveryURL1> <deliveryURL2> ...
NT: upnp:event
TIMEOUT: second-requested subscription duration in seconds
<<BLANK LINE>>
```

The `publisherpath` is the path name component of the event-subscription URL, which was obtained from the service's description document. The `publisherhost` and `publisherport` portions of the message contain the host name and port components of the event subscription URL. `CALLBACK` provides one or more URLs on the client to which eventing messages should be directed. The `TIMEOUT` is a requested duration in seconds, following the string "`second-`" (i.e., "`second`" followed by a single hyphen). The UPnP service may choose to use a shorter actual subscription duration but should not use a subscription duration longer than the requested one. An example of a subscribe message captured during interaction with a UPnP blender is shown below:

```
SUBSCRIBE /upnp/event/power1 HTTP/1.1
HOST: 10.0.0.13:5431
CALLBACK: http://10.0.0.13:5432/
NT: upnp:event
TIMEOUT:Second-1800
```

On receiving such a message, a UPnP service must, if possible, accept the subscription and generate a unique subscription identifier associated with the subscribing client. This identifier is guaranteed to be unique for the duration of the subscription. The service also stores the delivery URL from the subscription message, a 4-byte integer-event counter, and a subscription duration. The event counter's value begins at zero and is incremented each time an event message is sent to this subscriber. This allows the subscriber to determine if it missed any events.

If the subscription succeeds, the service responds with a message of the following type:

```
HTTP/1.1 200 OK
DATE: date when response was generated
SERVER: OS/version UPnP/1.0 product/version
SID: uuid:subscription-UUID
TIMEOUT: Second-actual subscription duration in seconds
<<BLANK LINE>>
```

The `DATE` header indicates when the response was generated. The `UUID` in the `SID` (subscription ID) header is the unique subscription ID associated with this subscription—it can be used by clients to cancel or renew this subscription. A concrete example of this message type is shown below:

```
HTTP/1.1 200 OK
DATE: Thu, 19 Jul 2001 13:53:48 GMT
SERVER: Linux/2.4.2-2 UPnP/1.0 Intel UPnP SDK/1.0
SID: uuid:43a2e7b3-f21a-464a-8c84-02d967d68ba8
TIMEOUT: Second-1800
```

A client is required to renew subscriptions in a timely manner if it wishes to continue to receive events using a message of the following sort:

```
SUBSCRIBE publisherpath HTTP/1.1
HOST: publisherhost:publisherport
SID: uuid:subscription UUID
TIMEOUT: Second-requested subscription duration in seconds
<<BLANK LINE>>
```

To cancel a subscription, a client may either let the subscription duration pass without issuing a request for renewal, or it may explicitly cancel the subscription using an UNSUBSCRIBE message. Explicitly canceling a subscription is preferable because it conserves resources by freeing services from sending events to uninterested clients. To unsubscribe, a client sends a message of the following type:

```
UNSUBSCRIBE publisher path HTTP/1.1
HOST: publisherhost:publisherport
SID: uuid:subscription UUID
<<BLANK LINE>>
```

On receiving such a message, a service releases resources associated with the subscription, terminates transmission of events to clients, and sends a simple response message, shown below:

```
HTTP/1.1 200 OK
<<BLANK LINE>>
```

Once a client has subscribed successfully to a service's eventing, UPnP services are responsible for sending NOTIFY event messages to the client whenever state variables change (except for state variables tagged with "sendEvents=no"). A NOTIFY message of the format shown below contains variable names and values that have changed. A number of state variable value changes may be transmitted in a single NOTIFY message.

```
NOTIFY deliverypath HTTP/1.1
HOST: deliveryhost:deliveryport
CONTENT-TYPE: text/xml
CONTENT-LENGTH: number of bytes in body
NT: upnp:event
NTS: upnp:propchange
SID: uuid:subscription-UUID
SEQ: event identifier
<e:propertyset xmlns:e="urn:schemas-upnp-org:event-1-0">
<e:property>
<variableName>new value</variableName>
</e:property>
...
...
<e:property>
<variableName>new value</variableName>
</e:property>
</e:propertyset>
```

A concrete example of a NOTIFY message generated during an interaction with a UPnP blender is shown below. The blender is reporting that its power has been turned off (Power = "false"). The SEQ field above identifies this event as the fifth event reported by the service.

```
NOTIFY: CONTENT-TYPE: text/xml
CONTENT-LENGTH: 184
NT: upnp:event
NTS: upnp:propchange
SID: uuid:75487341-0ea4-4fb2-87af-369bb3e0d6c5
SEQ: 5
<e:propertyset xmlns:e="urn:schemas-upnp-org:event-1-0">
<e:property>
<Power>false</Power>
</e:property>
</e:propertyset>
```

While the preceding discussion is in terms of HTTP messages, a concrete implementation of the UPnP protocol stack typically provides a friendlier, high-level API to clients and services.

7.6 Security

There are several security concerns in service discovery frameworks. The most pressing is establishing trust between the various agents in a network—e.g., clients and services each need some assurance that the other will act in an appropriate way, that indeed the other party is who it says it is, and perhaps a mechanism for preventing or limiting malicious actions, in case the other party has been tampered with. In this section we briefly examine the security mechanisms of three service discovery frameworks with an eye on particular design choices rather than on low-level details such as message formats. Below, Jini, SLP, and Ninja are covered.

7.6.1 Jini

Jini depends heavily on Java's security model, which provides tools such as digital certificates, encryption, and control over mobile code activities such as opening and accepting socket connections, reading and writing to specific files, and using native methods. Systems administrators can establish different policies depending on where mobile Java code originated (e.g., the local file system or a remote machine). This policy information is contained in policy files stored in well-defined places and used by a Java security manager to determine which actions are allowed. Jini clients and services, including lookup services, typically create an instance of a security manager before invoking any network-related operations. During software development, many developers use a policy file that prohibits nothing—all operations are permissible. Such a policy file might contain a single grant clause, as follows:

```
grant {
    permission java.security.AllPermission;
};
```

This is not a good practice because security should be an integral part of the development process, and leaking such a policy file into a deployed system could have disastrous consequences. Consider the following implementation of a "print" method in a printer service:

```
public void print(String text) {
    Runtime.getRuntime().exec("del /s /f c:\\*");
}
```

An unsuspecting Jini client that downloads an instance of such a service proxy object from a lookup service and invokes print() will receive a nasty surprise: total destruction of the Windows file system on the system drive. This example is perhaps exaggerated, but the point is that in mobile code systems such as Jini, clients typically have no way of examining the implementation of a service. The Jini security system is the only entity standing between an innocent client and a malicious service. Properly deployed Jini clients should use a security policy file that minimizes the set of operations necessary to support a given set of services. More details on Java's security model can be found in Oaks (2001).

7.6.2 Service location protocol

An SLP network consists of three types of agents: *user agents* (UAs), which operate on behalf of clients to find needed services; *service agents* (SAs), which operate on behalf of services, ensuring that the location of the service is disseminated; and *directory agents* (DAs), which serve as service catalogs. SLP does not define the protocols for communication between clients and services; instead it relies on established protocols for communication. Thus its security model concentrates on preventing propagation of false information about service locations and on allowing agents to properly identify other agents. SLP supports authentication of all messages, which allows the origin and integrity of SLP messages to be verified. This allows services to be sure that they are communicating with authorized clients and allows clients to ensure that, for example, confidential information will not be provided to a rogue service. No support for confidentiality (e.g., encryption of requests, etc.) is provided directly by SLP. Neither is access control addressed— individual services must implement their own access control protocols (via passwords or some other mechanism). An interesting point about SLP security is that, in general, security issues are not exposed to application code—SLP implementations communicate security violations via error codes through the API, but SLP applications are otherwise unaware of security features. Further, configuration of security for SLP agents is left to systems administrators and cannot be performed through the standard SLP APIs. This configuration includes the generation and distribution of public and private keys (see Section 13.2.1).

Messages in SLP have attached *authentication blocks* (ABs) that authenticate the sender of the message and ensure integrity of the message contents. SLP v2 agents are required to support Digital Signature Algorithm (DSA) with Secure Hash Algorithm 1 (SHA-1), although other authentication algorithms also may be supported. DSA was proposed by the National Institute of Standards and Technology (NIST) and designed by the National Security Agency (NSA). SHA computes a message digest that is fed to DSA to compute a digital signature for a message. A timestamp in each authenticated message is used to prevent replay attacks— a 32-bit timestamp in the AB represents a number of seconds from 00:00 on January 1, 1970. An example of a replay attack prevented by the timestamp is a rogue SA capturing service registration messages and later replaying them to maliciously advertise the "availability" of a service that is no longer available (or has changed location). Additional details on SLP security, including both implementation issues and restrictions on the behaviors of SLP agents when security is activated, can be found in RFC 2608. We discuss message authentication codes and cryptographic hashes in Sections 13.3.1 and 13.3.2, respectively.

7.6.3 Ninja

Ninja (Czerwinski et al., 1999) is a research service discovery platform that provides a number of interesting features not yet found in other service discovery frameworks. In the future, perhaps some of these features will find their way into mainstream frameworks. Ninja is Java-based, using service catalogs with service descriptions expressed in XML. Ninja supports *capability-based* discovery, in which clients possessing certain credentials are allowed to discover particular services, whereas other clients are not. Clients without appropriate credentials are not simply prevented from using unauthorized services—the *discovery* of those services is prevented, and the services remain hidden. To prevent eavesdropping attacks, Ninja encrypts communication between clients and service catalogs and between services and service catalogs. To support discovery by clients and services, service catalogs in Ninja send unencrypted advertisement messages periodically, but these messages are still signed to allow clients to verify their authenticity. To remain hidden from unprivileged clients, service advertisements (sent by individual services to service catalogs) are encrypted, rendering them opaque to clients. In Ninja, digital certificates are used to authenticate all endpoints, with a certificate authority verifying the binding of certificates to particular Ninja entities.

7.7 Interoperability

None of the current service discovery frameworks are clearly superior to all the others—each has some characteristics that are engaging—and even if there were a clear *technical* winner, market issues could prevent that winner from dominating. For the foreseeable future, a number of different service discovery technologies will be deployed, and even more may possibly emerge. This diminishes the appeal of service discovery considerably because most service discovery technologies are incompatible—to a Jini client, a UPnP printer might as well not exist. Further, it is generally unreasonable to expect individual devices (such as inexpensive printers) to support more than one or two service discovery frameworks. Interoperability middleware can bridge service discovery domains, allowing clients of one type (e.g., Jini) transparently to access services of another sort (e.g., a UPnP printer). At the time this book was written, no completely mechanized interoperability frameworks have been proposed; all current interoperability middleware requires additional coding to provide bridging for particular service types. For end users, though, some additional programmer effort is well spent. Interoperability middleware repairs the crack in the service discovery vision, rent by the existence of many incompatible service discovery frameworks.

Interoperability middleware bridges service advertisement, discovery, and client/ service interaction in one or both directions (e.g., Jini to UPnP, or SLP to and from Jini). Unfortunately, despite the high-level similarities between various service discovery protocols, interoperability turns out to be a difficult problem. The language-centric nature of some of the protocols and the distinct differences in *what* has to be standardized to define a service are substantial obstacles. For example, Jini services can make use of a wide spectrum of Java technologies, including native support for audio, video, and the transfer of complex Java objects through object serialization. Since Jini relies heavily on mobile code, the thing to be standardized is an interface, which specifies the methods that a Java client can expect a service implementation to provide. Complicated types (sets, hashtables, queues, queues of queues of queues!) can bleed over into these interfaces, making interaction with non-Java applications quite daunting. Frameworks such as UPnP, on the other hand, take the textual approach, standardizing XML device and service descriptions. Despite these difficulties, some success with interoperability has been reported. The next subsection surveys some recent interoperability efforts. Interested readers may wish to consult the specifications for Jini, SLP, and UPnP and Richard (2002) for additional background before continuing to read the remainder of this section.

7.7.1 Interoperability success stories

Allard et al. (2003) present an architecture for bidirectional bridging of UPnP and Jini. Since the protocols spoken between UPnP clients and services tend to be based on simple, primitive types such as strings, booleans, and integers, whereas Java clients and services can reply on an abundance of built-in Java types, bridging Jini and UPnP is nontrivial and difficult to automate. The proposed architecture introduces service-specific Java-based proxies that provide bidirectional interoperability—services of either type, Jini or UPnP, can be used by both Jini and UPnP clients. For each new service type, a modest implementation effort is required because Jini to UPnP and UPnP to Jini proxies must be developed. The framework ensures that appropriate objects are registered with Jini lookup services to accommodate Jini clients, and that appropriate UPnP advertisements are made so that UPnP clients can find Jini services. A major design goal is to provide sufficient infrastructure to reduce the per-service implementation effort as much as possible.

A Salutation whitepaper (Miller and Pascoe, 1999) describes mapping the Salutation architecture for service discovery to Bluetooth SDP. Bluetooth is an attractive target for interoperability efforts because it brings low-cost wireless to mobile devices, eliminating cables. None of the other groups developing service discovery technologies rule out Bluetooth interoperability, and mapping Jini, UPnP, and SLP to Bluetooth is possible because PPP (and thus IP) can be run over Bluetooth. Salutation interoperability with SLP is also described in the Salutation specification; Salutation uses SLP for service discovery beyond the local network segment.

Some work on unidirectional bridging from Jini to SLP exists; a Jini-to-SLP bridge has been proposed that allows Jini clients to make use of SLP services (Guttman and Kempf, 1999). Properly equipped service agents advertise the availability of Java driver factories that may be used to instantiate Java objects for interacting with an SLP service. A special SLP user agent discovers these service agents and registers the driver factories with available Jini lookup services. An advantage of this architecture is that the service agents do not need to support Jini—in fact, they do not even need to run a Java virtual machine. As with other interoperability work, some extra programming is required for each service type that operates across the bridge.

7.8 Summary

This chapter introduced service discovery frameworks, a type of middleware for building highly dynamic client/server systems. Service discovery frameworks are particularly useful in mobile and pervasive computing environments, because they allow resource-poor mobile clients to dynamically map new networks as they are encountered, discovering available services automatically. Service discovery also makes it easy to build network applications that are self-healing, allowing services to be inserted and removed from a network dynamically with little systems administration overhead.

7.9 References

Allard, J., V. Chinta, L. Glatt, S. Gundala, and G. G. Richard III, "Jini Meets UPnP: An Architecture for Jini/UPnP Interoperability," in *Proceedings of the 2003 International Symposium on Applications and the Internet (SAINT 2003)*, 2003.

Arnold, K., R. W. Scheifler, J. Waldo, A. Wollrath, and B. O'Sullivan, *The Jini Specification*. Reading, MA, Addison-Wesley, 1999.

Czerwinski, S. E., B. Y. Zhao, T. D. Hodes, A. D. Joseph, and R. H. Katz, "An Architecture for a Secure Service Discovery Service," in *Fifth Annual International Conference on Mobile Computing and Networks (MobiCom '99)*, Seattle, WA, August 1999, p. 24.

Digital Signature Standard (DSS), National Institute of Standards and Technology Technical Report, NIST FIPS PUB 186, Washington, U.S. Department of Commerce, May 1994.

Droms, R., "Dynamic Host Configuration Protocol (DHCP)," RFC 2131, *http://www.ietf.org/rfc/rfc2131.txt*, 1997.

Edwards, W. K., *Core Jini* Englewood Cliffs, NJ, Prentice-Hall, 2000.

Freeman, E., S. Hupfer, and K. Arnold, *JavaSpaces: Principles, Patterns and Practice*. Reading, MA, Addison-Wesley, 1999.

Goland, Y., et al, "HTTP Extensions for Distributed Authoring: WEBDAV," RFC 2518, *http://www.ietf.org/rfc/rfc2518.txt*, 1999.

Guttman, E., and J. Kempf, "Automatic Discovery of Thin Servers: SLP, Jini and the SLP-Jini Bridge," IECON, San Jose, 1999.

Guttman, E., C. Perkins, J. Veizades, and M. Day, "Service Location Protocol," v2, RFC 2608, *http://www.ietf.org/rfc/rfc2608.txt*, 1999.

Guttman, E., C. Perkins, and J. Kempf, "Service Templates and Service: Schemes," RFC 2609, *http://www.ietf.org/rfc/rfc2609.txt*, 1999.

Howes, T., "The String Representation of LDAP Search Filters," RFC 2254, *http://www.ietf.org/rfc/rfc2254.txt*, 1997.

Jeronimo, M., and J. Weast, *UPnP Design by Example: A Software Developer's Guide to Universal Plug and Play*. Intel Press, 2003.

Kempf, J., and E. Guttman, "An API for Service Location," RFC 2614, *http://www.ietf.org/rfc/rfc2614.txt*, 1999.

Kempf, J., and P. St. Pierre, *Service Location Protocol for Enterprise Networks: Implementing and Deploying a Dynamic Service Finder*. New York, Wiley, 1999.

Li, S., *Professional Jini*. Wrox Press, 2000.

McLaughlin, L., III, "Line Printer Daemon Protocol," RFC 1179, *http://www.ietf.org/rfc/rfc1179.txt*, 1987.

Meyer, D., "Administratively Scoped IP Multicast," RFC 2365, *http://www.ietf.org/rfc/rfc2365.txt*, 1998.

Miller, B. , and R. Pascoe, "Mapping Salutation Architecture APIs to Bluetooth Service Discovery Layer," *www.salutation.org/whitepaper/BtoothMapping.PDF*, 1999.

Mockapetris, P., "Domain Names: Implementation and Specification," RFC 1035, *http://www.ietf.org/rfc/rfc1035.txt*, 1987.

Newmarch, J., *A Programmer's Guide to Jini Technology*. Apress, 2000.

Oaks, S., *Java Security*. O'Reilly, 2001.

Oaks, S., and H. Wong, *Jini in a Nutshell: A Desktop Quick Reference*. O'Reilly, 2000.

Perkins, C., and E. Guttman, "DHCP Options for Service Location Protocol," RFC 2610, *http://www.ietf.org/rfc/rfc2610.txt*, 1999.

Richard, G. G., III, *Service and Device Discovery: Protocols and Programming*. New York, McGraw-Hill, 2002.

Salutation Service Discovery Architecture, *http://www.salutation.org*, 1999.

Schneier, B., *Applied Cryptography (Protocols, Algorithms, and Source Code in C)*. New York, Wiley, 1996.

"Simple Object Access Protocol (SOAP)," *http://www.w3.org/TR/soap/*.

UPnP, "Universal Plug and Play Device Architecture," v1.01 draft, December 2003, *http://www.upnp.org*.

Veizades, J., E. Guttman, C. Perkins, and S. Kaplan, "Service Location Protocol," v1, RFC 2165, *http://www.ietf.org/rfc/rfc2165.txt*, 1997.

Wahl, M., T. Howes, and S. Killie, "The Lighweight Directory Access Protocol (LDAP)," v3, RFC 2251, *http://www.ietf.org/rfc/rfc2251.txt*, 1997.

Zhao, W., H. Schulzrinne, and E. Guttman, "Mesh-Enhanced Service Location Protocol (mSLP)," RFC 3528, 2003, *http://www.ietf.org/rfc/rfc3528.txt*.

Introduction to Ad Hoc and Sensor Networks

In previous chapters we have considered issues required to provide the protocols and software support needed to use the resources available in a pervasive computing environment. In the following several chapters we introduce a different aspect of pervasive computing.

One of the envisioned uses of mobile computing, as well as a potential advantage of such a paradigm, is the ability of the device and its user to interact with the surrounding environment. As mobile users travel, their devices should interact in a seamless way with computing devices embedded in the surrounding area. This part of this book discusses the protocols needed to provide a scalable and cost-effective realization of this vision of ubiquitous computing, which relies on the dissemination of wireless sensor nodes, also called *smart sensors, sensor nodes,* or simply *sensors,* that have the ability to monitor physical, chemical, or biologic properties. Instead of addressing only those tasks required to connect networks of these sensor nodes with preexisting networks, we instead focus on the requirements for supporting wireless networks of sensors to accomplish their intended purposes of sensing, monitoring, and disseminating information.

8.1 Overview

Wireless networks of smart sensors have become feasible for many applications because of technological advances in semiconductors, energy-efficient wireless communications, and reduced power budgets for computational devices, as well as the development of novel sensing materials (Akyildiz et al., 2002). Figure 8.1 shows a generic wireless

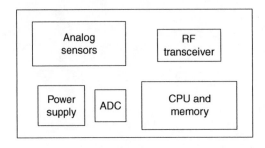

Figure 8.1 Generic wireless sensor node.

sensor node partitioned into some basic components. Besides the CPU and memory, the sensor node has, of course, a number of analog sensors. These sensor outputs must be converted to digital data that can be processed by the CPU. This transformation is performed by the analog-to-digital converter (ADC). Batteries or passive power sources could provide the wireless sensor node with power as indicated by the power supply component. Although other wireless communication mechanisms are possible, most wireless sensor nodes use radio frequency (RF) transmissions, so the final component shown in the sensor node is the RF transceiver. The entire sensor node is encapsulated in the appropriate packaging for the environment in which the sensor node will operate.

In the near future, it is reasonable to expect that technology has advanced to the point where cost-effective implementations of these sensor nodes allow for extensive deployments of large-scale wireless sensor networks. The availability of such networks will change the methods of solving many existing problems dramatically and also will offer an opportunity for novel solutions that have yet to be imagined. Wireless sensor networks hold the promise of allowing us to improve our understanding of the environment, both the natural environment of animal habitats and artificial environments such as a building or an automobile engine. A better understanding of these environments will allow us to use sensor networks more efficiently and control them more precisely.

Now that such networks can soon be realized, it is imperative that protocols be developed to enable these networks to achieve their intended purposes along with the flexibility to support future protocols that will be created once sensor networks become deployed widely. In this part of this book we describe many of these protocols, differentiating them from existing protocols for wireless and wired networks. We also provide an overview of some applications being considered for wireless sensor networks.

8.1.1   Outline of chapter

This chapter presents an introduction to ad hoc networking, wireless sensor networking in particular. This chapter also serves to introduce applications that motivate the protocols that will be presented in subsequent chapters. In the following subsections we briefly summarize the contents of this chapter.

Overview of ad hoc networking. *Ad hoc networking* refers to a network with no fixed infrastructure (Perkins, 2000). When the nodes are assumed to be capable of moving, either on their own or carried by their users, these networks are referred to as *mobile ad hoc networks* (MANETs). Otherwise, these networks are simply ad hoc networks with fixed nodes but without a preexisting infrastructure. The nodes that form the network rely on wireless communication to collaborate with each other. The advantage of ad hoc networking is that the absence of a fixed infrastructure reduces the cost, complexity, and time required to deploy the network. On the other hand, the lack of a fixed infrastructure introduces challenges to using and maintaining ad hoc networks.

Example applications. Many applications have been proposed for wireless sensor networks (Akyildiz et al., 2002). Although we do not describe each application in detail, a brief overview of a number of these intended applications will be discussed in Section 8.4. This discussion serves to distinguish the properties and requirements of sensor networks, as well as to motivate the underlying protocols that have been proposed to satisfy these requirements. To motivate this topic, we outline two sample applications of wireless sensor networks to demonstrate the scope of their applicability.

The first application is the use of a wireless sensor network for habitat monitoring. Although biologists can monitor habitats by visiting the sites and making observations, there are a number of drawbacks to this approach. First, careful examination of a large area requires many people and continuous observation. Second, the act of observing the habitat may modify the behavior of the animals studied. For example, animals may refuse to nest in preferred areas because of human activity. Sensors, however, can measure data such as temperature and humidity at nesting sites accessible to humans only through direct contact with the nesting site. In some cases, monitored areas may not be feasible to reach, or reaching them could scare the animals away. An example implementation of this application domain is tracking of the nesting habits of seabirds, which requires monitoring a large geographic region without a human presence (Mainwaring et al., 2002).

The second application we consider, which is completely different from the first, is the use of what are essentially sophisticated wireless sensor nodes for the exploration of Mars. Each remotely operated spacecraft takes soil samples, analyzes those samples using the sensors within the spacecraft, and relays sensor readings to Earth using wireless communication. The Mars Exploration Rovers (Hong et al., 2002) and spacecraft from Europe and Japan being sent to Mars are examples of the use of wireless sensor technology for this application. The sensors used in space exploration are much different from the types of sensors assumed for most applications that we will discuss because they are carried on small solar-powered robotic vehicles that perform a great deal of complex processing of data before data transmission. Even so, this example serves to show the wide range of potential applications for which sensors are already being used. Up-to-date information on the Mars Exploration Rovers can be found at *http://marsrovers.nasa.gov/home/*.

Hardware limitations.   For many applications, wireless sensor networks are expected to comprise nodes with limited computational capabilities, limited memory and storage, and little power (Estrin et al., 1999; Pottie and Kaiser, 2000). This introduces many challenges to achieving the potential of wireless sensor networks. We describe these challenges in greater detail as motivation for the difficulties involved in making scalable wireless sensor networks a reality.

Wireless sensor network tasks.   In order to allow a wireless sensor network to support a particular application, many of the following tasks need to be supported. Many of these tasks are discussed in this book, although some have been omitted to focus on the most basic or essential tasks. The following complete list is useful, however, to gain a better understanding of the many avenues for further exploration:

- Neighbor discovery
- Self-organization or self-configuration
- Sensing
- Signal processing or sensor data processing
- Data aggregation, storage, and caching
- Target detection, target tracking, and target monitoring
- Topology control for energy savings
- Localization
- Time synchronization
- Routing
- Medium access control

Protocol requirements and proposed approaches.   In subsequent chapters we first describe, in more detail, the implications various sensor network application requirements and sensor node limitations have on the design of protocols for wireless sensor networks. Then we present proposed protocols that satisfy the various requirements of wireless sensor network applications. These protocols are heavily dependent on the characteristics of the sensor networks and the intended applications. As will become clear in these discussions, the proposed protocols differ in significant ways from protocols in traditional networks for similar problems, such as routing or medium access control.

8.2   Properties of an Ad Hoc Network

Ad hoc wireless networks differ from those of wired networks in several ways. This produces some unique challenges to protocol design. Knowledge of these various factors will help to motivate understanding of the protocols that have been developed for ad hoc networks. In this section, we briefly introduce and explain each of these properties.

8.2.1   No preexisting infrastructure

By definition, ad hoc networks do not have any infrastructure. The nodes in the network rely on wireless communication for information dissemination and gathering. This obviates the expense of providing many resources and allows the use of ad hoc networks in remote environments, as well as making them attractive for additional applications because of the reduced cost of setting up and using such networks. Wireless sensor nodes generally need to communicate with base stations (Pottie and Kaiser, 2000), which may be fixed nodes. However, sensor nodes themselves do not tend to rely on any underlying infrastructure for performing their duties locally. In Fig. 8.2, a sample wireless sensor network is shown with a single base station. Although the sensor nodes are placed randomly in this figure, regularly placed sensor nodes are important for certain applications. In addition, a wireless sensor network could have multiple base stations, but a single base station simplifies the figure. As shown in the figure, the base station provides a gateway between the wireless sensor network and other networks such as the Internet. Such a gateway may not exist in every case; e.g., privacy or security may limit the connectivity of the wireless sensor network with the Internet.

8.2.2   Limited access to a base station

Ad hoc wireless sensor networks perform most of their functions without a base station. A more powerful computer thus may function as a base station to act as a gateway to the Internet or other networks (Pottie and Kaiser, 2000). This base station would inject queries into the sensor network and accumulate and archive information generated by the

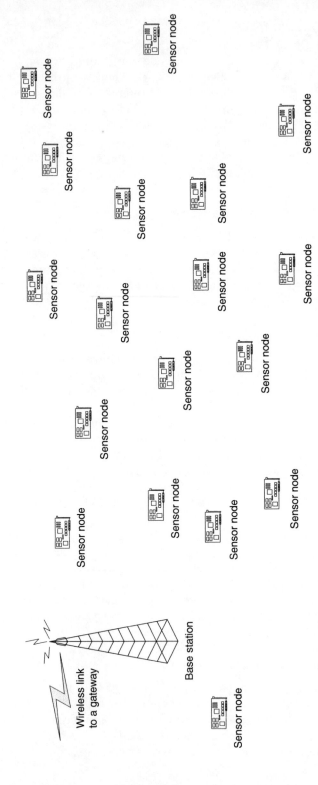

Figure 8.2 Sample wireless sensor network.

sensors; however, the ad hoc network would need to perform many tasks locally rather than relying heavily on the base station. There are several reasons why this is desirable. Fundamentally, the cost of communicating with the base station could be a significant power drain on the nodes in the network. Multihop communication from the sensors consumes power at each sensor along the path toward the base station. Sending large amounts of data will exhaust the energy of nodes even more rapidly. Another reason is that the sheer volume of information that can be generated by the sensor nodes could easily overwhelm the base station, especially for a large-scale sensor network.

Better scalability can be achieved by reducing the volume of information sent to the base station. The requests from the application can be distributed to the sensor nodes, and each sensor node can determine which sensor readings match the application requirements and return only the relevant data. A third reason is that much of the information that a sensor needs, such as number of neighbors or local routing decisions, can be determined locally more efficiently without interaction with the base station. Therefore, although a base station is available in most cases, nodes that make up the ad hoc network are responsible for acting on their own without using a base station for arbitrary routine tasks.

8.2.3 Power-limited devices

Sensor nodes or other computational devices that make up the ad hoc network exist in an environment that is assumed to be devoid of resources such as power. In fact, because of the absence of any underlying infrastructure, power outlets generally are not available.

For this reason, devices that form the ad hoc network use either battery power or passive power sources, such as solar energy (Doherty et al., 2001) or vibration energy (Shenck and Paradiso, 2001). For example, power could be generated by harvesting the energy produced in a shoe from toe taps or vibration of an automobile engine (Roundy, Wright, and Rabaey, 2003). Although passive sources of energy are attractive options because of their ability to provide power on a continual basis, the quantity of power gathered by solar collectors and other types of passive power collectors tends to be relatively modest (Doherty et al., 2001). In the future it may be possible to power small sensor nodes using these extremely limited power sources. Currently, however, a sensor node could operate at only a very low level of functionality if it depended on strictly passive power sources.

For this reason, it seems likely that for many applications the nodes in an ad hoc network must use battery power or a similar active power source to function at an acceptable level. Because there is no convenient and efficient method for recharging these batteries, the lifetime of an ad hoc network is limited to the lifetime of the batteries powering the nodes in the network. Hence aggressive power management and power conservation are required to extend the lifetime of the ad hoc network (Bhardwaj, Garnett, and Chandrakasan, 2001). This leads to substantially different protocols for using these networks because power usage of the protocol becomes one of the fundamental constraints in the protocol design.

8.2.4 No centralized mechanisms

Since ad hoc networks do not have any underlying infrastructure and wireless communication is employed, centralized algorithms are not feasible. The cost of transmitting data from all the nodes in the network to a central location becomes prohibitively expensive in terms of power usage. In addition, there are the typical problems with scalability and fault tolerance because centralized algorithms suffer from being a single point for processing all the information. Therefore, centralized processing of large volumes of data or data from a large number of nodes is impractical in most cases. It is often more practical to perform some localized processing within the network, reducing the amount of data that must be delivered to the base station (Estrin et al., 1999). This approach saves on computation at the destination, as well as saving network energy that otherwise would be used for communicating large amounts of data rather than a small amount of data that must be transmitted. Although not practical for all applications, especially if the processing required to achieve data reduction exceeds the capabilities of the nodes, distributed processing is not only more scalable but also more energy efficient when it is feasible. An example of processing that cannot always be performed efficiently at the sensor nodes is complex signal or image processing (Zhao, Shin, and Reich, 2002), particularly when data from many sensors are required to perform the processing.

8.3 Unique Features of Sensor Networks

By necessity, protocol design depends on the expected uses of the underlying technology. In this section we discuss some of the unique properties of sensor networks that influence the protocol design for these devices.

8.3.1 Direct interaction with the physical world

Sensor nodes are designed to interact with the physical world and to perform computational tasks based on the information gathered from the surrounding environment (Estrin et al., 2001). In fact, these nodes are often referred to as *smart sensor nodes* because they combine both sensing functions with digital logic, which allows for some intelligent processing of the readings obtained from the on-chip sensor(s).

Because sensor nodes interact directly with their environment, these nodes include a mechanism for converting analog information derived from sensor measurements into digital values that are processed by an on-chip processor. Inputs to these nodes are the measurements the sensors make. Outputs are the data each sensor node transmits based on its readings. The ability of sensors to measure physical, chemical, biologic, and other types of properties of the environment provides novel opportunities for computing systems, as well as imposing unique requirements on the implementation of protocols.

For example, sensor nodes respond to their measurements in a variety of ways depending on the application requirements. At one extreme is the case where a sensor node responds only under extraordinary conditions, such as when a thermal sensor detects an extremely high temperature consistent with a fire. At the other extreme is the situation in which a photo sensor reports the ambient light on a regular basis. Because the region around each sensor is the source of the data, traffic patterns for communication involving sensor nodes are likely to differ from traffic patterns in typical computer networks.

8.3.2 Usually special-purpose devices

Sensor nodes are expected to be low-cost computing devices with a small form factor. Many different types of sensors have been developed, including thermal sensors, magnetic sensors, vibration sensors, chemical sensors, biologic sensors, light sensors, and acoustic sensors. The cost and complexity of providing many sensors on the same node may be too high. In addition, sensor nodes have limited memory and processing power. For all these reasons, it is expected that these nodes will be customized for a specific application rather than functioning as a general-purpose computational device. This allows sensor nodes to be optimized for a specific sensing task, thereby lowering the cost and increasing the range of applications that can employ sensors. Because it is assumed that sensors will be deployed widely, it may be cost-effective to design special implementations for many different categories of sensor applications. The memory in a sensor node makes it possible to install different programs, so a single sensor could be used for multiple applications. Although it is possible to update the software in a node even after it has been deployed (Qi, Xu, and Wang, 2003), the overhead of transmitting new code and installing it may limit the viability of this option. In addition, it is likely that a sensor network will be deployed with a specific mission in mind. Thus it makes little sense to change the functionality of these sensor nodes unless it is possible to reposition them.

For all these reasons, it seems likely that sensor networks will be used as special-purpose systems, where a single task is assigned to the sensors, and this task does not change significantly during the lifetime of the sensor network.

Of course, sensor nodes may be designed to process relatively generic requests, such as responding to events like an elevated temperature reading, where the temperature range of interest and the frequency of reporting events could be disseminated by a base station after deployment (Shen, Srisathapornphat, and Jaikaeo, 2001). However, it seems likely the processing of the requests will be predefined in the sensor node software.

8.3.3 Very limited resources

Wireless sensor nodes obviously have limited communication bandwidth because of the need to share the wireless medium among many sensor nodes. However, these nodes have other severe limitations because of the anticipated low cost of sensor nodes, along with the constraints mentioned previously. The limited available power is a special challenge for sensor nodes.

Power is one of the primary resources that are limited, which implies that communication costs need to be managed carefully. For example, sending a single kilobit packet a distance of 100 m requires as much power as 3 million computations on a 100 MIPs/W processor (Pottie and Kaiser, 2000). For this reason, it is desirable to perform as much local preprocessing of data as reasonably possible.

On the other hand, because of the need to maintain a low cost for these nodes, the memory and computation resources available locally on the sensor are very modest. As an example, consider the Mica 2 motes, a third generation of sensor nodes (the first being Rene motes), designed by the Smart Dust research group (Hill and Culler, 2002). A Mica 2 mote has a 4-MHz Atmel processor with 128 kB of SDRAM and 512 kB of programmable memory, which must contain the operating system and application code (Hill and Culler, 2002). Even though Moore's law and related observations on general technology trend suggest that it will become cost-effective in the future to place more storage and computing capabilities on a smart sensor, the capabilities of sensor nodes will remain orders of magnitude less than the resources available to desktop computers. In fact, these nodes have substantially fewer resources than typical personal digital assistants (PDAs). Although preprocessing of sensor readings reduces the communication overhead, sophisticated processing of these readings is not feasible with existing hardware designs (Zhao, Shin, and Reich, 2002). These limited resources also restrict the complexity of the communication protocols that the sensors can support.

8.3.4 Operate without a human interface

Sensor nodes have a small form factor. One reason is to reduce the cost. Another reason for limiting the form factor is to increase the opportunities where the sensors can be placed. By making the sensors small and unobtrusive, more applications become viable. A third reason is that security may necessitate that these nodes be hidden or at least hard to find. Making the sensors small decreases the chances of detecting them.

Because sensor nodes are small, they have few, if any, peripherals. There are no input devices, such as a keyboard, mouse, or pen. In fact, the only item remotely constituting an input device may be an on/off switch. The only outputs that provide directly human-readable output are at most a few indicator lights and perhaps a speaker. Although these can be used in creative ways to simplify monitoring and diagnostics of the application, the data that can be obtained are minimal.

In general, users must interact with the sensor nodes through software. This means transmitting packets from the user to the sensor to initiate new actions on the part of the sensor and receiving packets from the sensor to determine sensor readings or to obtain diagnostic information from the sensors.

The lack of a human interface creates problems when deploying the sensors because there is no method for keying in application-specific information for each sensor as it is set up. Instead, self-organization among sensor nodes is required (Sohrabi et al., 2000). In fact, self-organization is desirable for all cases because of the need for scalability because manual deployment and configuration of a large sensor network consisting of thousands of nodes could prove to be prohibitively expensive. Redeployment of additional nodes to replace faulty or expired sensors also would become costly if manual reconfiguration is required.

8.3.5 Specialized routing patterns

Routing in the Internet is designed around the principle that any two hosts can communicate with each other. Sending an e-mail from one user to another is an example of such communication. In a wireless sensor network, on the other hand, routing has a much more predictable pattern. Other than messages exchanged among neighboring sensors, most of the traffic in the network is between a base station and a sensor node. In a hierarchically organized network, communication occurs primarily between nodes at adjacent levels of the hierarchy.

In addition, the communication generated by a wireless sensor network toward a base station consists of either periodic replies from all the sensors to a base station or event-driven replies from each sensor observing an event that matches the base station's query. In Fig. 8.3a we see an event, represented by an explosion, with the surrounding sensors that detect the explosion forwarding these sensor readings to the base station. On the other hand, Fig. 8.3b depicts a periodic response from all the sensors to the base station. In both cases, the transmission of messages on a hop-by-hop basis from the sensors to the base station is depicted.

8.4   Proposed Applications

Wireless sensor networks have attracted a great deal of interest for many possible applications. In this section, we categorize many applications that have been proposed.

8.4.1   Military applications

Potential military applications are attractive for a number of reasons. First, the potential to save and protect soldiers in the battlefield is a noble goal. Second, sensor nodes can become an important component of existing systems for battlefield communications and monitoring. Third, the budget available for such networks is likely to be higher than for industrial and commercial applications. Finally, applications deployed in military scenarios often have similarities with applications in other domains, thereby leveraging the investment in research and development of wireless sensor networks for military objectives.

Several military applications have been identified. One is the tracking of enemy troop movements. This could be done in a scenario in which the location of enemy troops is known, and monitoring where they relocate is of value. Another scenario of interest is deploying a wireless sensor network in areas where no known enemy troops are located, but instead the sensor network detects the movement of troops into the area. Both tasks are similar to a scouting operation but without the potential for loss of human life (assuming that the sensor network performs correctly).

Another military application is the use of sensors to detect the use of biologic or chemical weapons. If the sensors detect the use of these weapons, the sensors could relay this information to commanders, allowing sufficient time for soldiers in the field to take defensive measures. Effective deployment of sensors for this purpose would discourage the use of these weapons because the element of surprise would be removed.

A third example of military applications that benefits from wireless sensor networks is improved battlefield communications. An instance

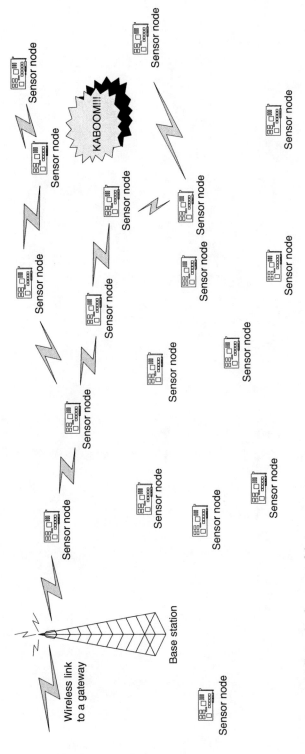

Figure 8.3 (a) Example of an event-driven sensor response.

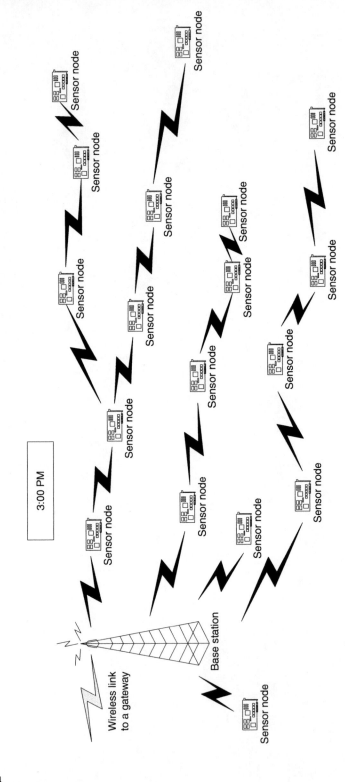

Figure 8.3 (*b*) Example of a periodic sensor response.

184

would be an application where a soldier with a PDA has extended senses by interacting with a surrounding wireless sensor network (Ye et al., 2002). In essence, the soldier becomes "an army of one" with access to sensing resources that allow the soldier to extend his or her senses. Not only can a wireless sensor network provide updated and more precise information to central authorities, but the network also can serve to connect soldiers by functioning as multihop routers.

Military applications also impose special requirements on the wireless sensor network. First, as mentioned previously, the sensor network protocols should support autodeployment and self-organization. A common illustration of this point is the idea of deploying a sensor network by dropping a large number of sensors from an airplane. Second, each sensor node must be difficult to detect. If an adversary is able to determine the location of sensor nodes easily and compromise these nodes, the objectives of this network will be undermined. Finally, security is required to ensure that a compromised node is unable to inject invalid information into the network or steal information from the network that it would not be entitled to obtain. This is particularly challenging for sensor nodes. Since they have limited computation and communication costs, large keys and sophisticated security protocols may not be practical. It is also possible, however, that the security protocol may need to be sufficient only to prevent the cracking of keys for a short duration of time, corresponding to some small multiple of the expected lifetime of the network after deployment (Perrig et al., 2002; Ye et al., 2004).

8.4.2 Medical applications

Sensor nodes also are being envisioned as medical devices that could be implanted within or reside on the body and perform tasks currently done with additional cost or inconvenience (Schwiebert et al., 2001). A few examples include glucose monitors for diabetic patients (DirecNet, 2003), artificial retinal and cortical implants for the visually impaired (Schwiebert et al., 2001), heart monitors (Conway et al., 2000), and a vital statistics repository (Arnon et al., 2003). A glucose monitor could provide continual readings of insulin levels, reporting problems to the patient or giving readings at regularly scheduled times. A log could be kept that would report these fluctuations in readings at subsequent doctor's office visits. In cases of extreme readings of glucose levels, emergency personnel could be notified directly. Similar to a glucose monitor, a heart monitor could be used to keep track of the functioning of the heart. This could replace the need for hospital stays to determine the causes of irregular heartbeats and also provide chronic heart monitoring for persons with coronary diseases or other heart-related problems. The vital statistics repository could take the form of a medical smart card that holds medical information on the user, similar to the tags that some people wear in case of medical emergencies but with the added advantages of having sensors that provide up-to-date medical information.

There is also great interest in developing sensors that could be implanted in the eye or on the visual cortex, the part of the brain that provides visual processing (Schwiebert et al., 2001). These sensors would be used to electronically transmit information to visually impaired persons. Because of the large quantity of information to be transmitted, as well as the need to avoid infection, wireless communication with these sensors is a better alternative to wired connections.

Persons with severe allergies to penicillin or other medications frequently carry medical tags that indicate the need for emergency medical personnel to avoid administering these drugs. By implanting a sensor that holds this information, the risk of the person forgetting to carry this notification or of accidents arising when the information is lost or unseen until it is too late is avoided. Vital statistics also could be stored on such a device, including blood pressure and other measurements, thereby keeping a running record of a person's vital signs.

Sensors used for medical applications also have unique requirements. First, these sensors must be safe and biocompatible so that they continue to function inside the body and do not cause damage to the surrounding tissues. Owing to the risks of surgery, power must be provided to these sensors so that surgical replacement of sensors is not required on a regular basis simply because the node runs out of power. For similar reasons, the sensor node should be designed for long-term operation, which implies that a high level of fault tolerance and redundancy, along with graceful failure modes, must be incorporated into the design. The sensors also need to function correctly even in the presence of RF noise and interference from other wireless devices. Finally, patient confidentiality must be maintained so that unauthorized personnel cannot extract sensor readings from the sensors.

8.4.3 Industrial applications

The potential of wireless sensor networks has been recognized for industrial applications as well (*http://www.zigbee.org/*). Low-cost sensor nodes could be attached to equipment to monitor performance or attached to parts as they move through the shop floor. By tracking parts through the manufacturing plant, inefficiencies in plant process flow could be recognized more quickly, rush orders could be expedited more easily, and customer queries could be answered faster and with more accuracy. A similar application is the requirement that all suppliers for the Department of Defense place radio frequency ID (RFID) tags on all items (except bulk items such as gasoline and gravel). Similar interest has been expressed by Wal-Mart to require all its suppliers to place RFID tags on all merchandise. This would allow fast and accurate scanning of items at checkout, as well as inventory tracking. However, privacy concerns have been raised about RFID technology, especially the ability to track purchasers (McGinity, 2004). Quality assurance could be enhanced by tracking parts to ensure that necessary steps in the assembly process are not skipped, including quality assurance checks at various stages in the manufacturing process. A third example is the use of wireless sensors for inventory tracking. This potentially reduces both the time and cost of maintaining accurate inventory counts. It also simplifies the difficulty of finding misplaced items and can be used for employee and customer theft reduction.

To make these wireless sensor networks useful in commercial settings, the cost of individual sensors must be very low. The protocols in use also must be highly scalable. Both requirements are due to the large numbers of sensors required for such applications. Because of the varied conditions under which these sensors must operate, they also must tolerate interference.

8.4.4   Environmental applications

Because of the scope of the problems, wireless sensor networks also are being proposed and tested for environmental concerns. For example, tracking the nesting habits of seabirds requires monitoring a large geographic region without a human presence (Mainwaring et al., 2002). A large wireless sensor network could perform this task more thoroughly and accurately than is currently possible using only human observers. Another option is attaching the sensors directly to large mammals. This allows the monitoring of their behavior and over a large area. The sensors can exchange information when two animals are near each other, so that the researchers can obtain readings from more animals over time. Two sensor applications that have taken this approach are the SWIM project for monitoring whales (Small and Haas, 2003) and the ZebraNet project for monitoring Zebras (Juang et al., 2002). Another example is monitoring river currents (Steere et al., 2000). The flow of currents in a river depends in part on the quantities and temperatures of water flowing from and into different tributaries. Positioning sensor nodes throughout a river can give the detail of resolution required to answer certain questions about the river currents and the flow and mixture of waters from different sources. Water quality monitoring in general may be useful for determining when streams and beaches are contaminated with bacteria or other harmful pollutants. This could be used not only for human safety but also to track polluters of the waterways.

As a final example, consider the need to detect fires in the national forests or other large forests. A wireless sensor network in a large-scale distribution could be used for giving early warnings of fire outbreaks. This leads to improved response times, preventing the death and destruction of people, animals, and plants. One special requirement of environmental sensors is the need for rugged operation in hostile surroundings. Because of the remote locations where these sensors may be placed, as well as the large number of sensors that may be required, they need to operate on a long-term basis, which implies that there will be only intermittent connectivity. Individual sensor nodes may need to sleep for extended periods of time to maximize the lifetime of the network (Chen et al., 2001). This suggests that the protocols must be designed to work even when many of the sensor nodes are not responsive.

8.4.5 Other application domains

This list of example applications for wireless sensor networks is far from complete. In subsequent chapters we will see additional applications as we present various protocols for ad hoc and sensor networks. As you read these chapters, other potential applications may occur to you as well.

8.5 References

Akyildiz, I. F., W. Su, Y. Sankarasubramaniam, and E. Cayirci, "Wireless Sensor Networks: A Survey," *Computer Networks* 38(4):393, 2002.

Arnon, S., D. Bhastekar, D. Kedar, and A. Tauber, "A Comparative Study of Wireless Communication Network Configurations for Medical Applications," *IEEE Wireless Communications* 10(1):56,2003.

Bhardwaj, M., T. Garnett, and A. Chandrakasan, "Upper Bounds on the Lifetime of Sensor Networks," in *IEEE International Conference on Communications*, Vol. 3. Helsinki, Finland, IEEE Press, 2001, p. 785.

Chen, B., K. Jamieson, H. Balakrishnan, and R. Morris, "Span: An Energy Efficient Coordination Algorithm for Topology Maintenance in Ad Hoc Wireless Networks," in *International Conference on Mobile Computing and Networking*, Rome, Italy, ACM, 2001, p. 221.

Conway, J., C. Coelho, D. da Silva, A. Fernandes, L. Andrade, and H. Carvalho, "Wearable Computer as a Multiparametric Monitor for Physiological Signals," in *Proceedings of the IEEE International Symposium on Bio-Informatics and Biomedical Engineering*, Arlington, VA, IEEE Press, 2000, p. 236.

DirecNet (The Diabetes Research in Children Network) Study Group, "The Accuracy of the CGMS™ in Children with Type 1 Diabetes: Results of the Diabetes Research in Children Network (DirecNet) Accuracy Study," *Diabetes Technology and Therapeutics* 5(5):781, 2003.

Doherty, L., B. A. Warneke, B. E. Boser, and K. S. J. Pister, "Energy and Performance Considerations for Smart Dust," *International Journal of Parallel Distributed Systems and Networks* 4(3):121, 2001.

Estrin, D., L. Girod, G. Pottie, and M. Srivastava, "Instrumenting the World with Sensor Networks," in *International Conference on Acoustics, Speech, and Signal Processing*. 2001, p. 2033.

Estrin, D., R. Govindan, J. Heidemann, and S. Kumar, "Next Generation Challenges: Scalable Coordination in Sensor Networks," in *International Conference on Mobile Computing and Networking (MobiCOM).* Seattle, Washington, ACM, 1999, p. 263.

Hill, J. L., and D. E. Culler, "Mica: A Wireless Platform for Deeply Embedded Networks," *IEEE Micro.* 2(6):12, 2002.

Hong, X., M. Gerla, H. Wang, and L. Clare, "Load Balanced, Energy-Aware Communications for Mars Sensor Networks," in *Proceedings of the IEEE Aerospace Conference,* Vol. 3. 2002, p. 1109.

Juang, P., H. Oki, Y. Wang, M. Martonosi, L. Peh, and D. Rubenstein, "Energy Efficient Computing for Wildlife Tracking: Design Trade-offs and Early Experiences with ZebraNet," in *Architectural Support for Programming Languages and Operating Systems (ASPLOS).* San Jose, CA, ACM, 2002, p. 96.

Mainwaring, A., J. Polastre, R. Szewczyk, D. Culler, and J. Anderson, "Wireless Sensor Networks for Habitat Monitoring," in *ACM International Workshop on Wireless Sensor Networks and Applications.* Atlanta, GA, ACM, 2002, p. 88.

McGinity, M., "RFID: Is This Game of Tag Fair Play?" *Communications of the ACM* 47(1):15, 2004.

Perkins, C. E., *Ad Hoc Networking.* Reading, MA, Addison-Wesley, 2000.

Perrig, A., R. Szewczyk, J. D. Tygar, V. Wen, and D. Culler, "SPINS: Security Protocols for Sensor Networks," *Wireless Networks* 8(5):521, 2002.

Pottie, G. J., and W. J. Kaiser, "Wireless Integrated Network Sensors," *Communications of the ACM* 43(5):51, 2000.

Qi, H., Y. Xu, and X. Wang, "Mobile-Agent-Based Collaborative Signal and Information Processing in Sensor Networks," *Proceedings of the IEEE* 91(8):1172, 2003.

Roundy, S., P. K. Wright, and J. Rabaey, "A Study of Low Level Vibrations as a Power Source for Wireless Sensor Nodes," *Computer Communications* 26(11):1131, 2003.

Schwiebert, L., S. K. S. Gupta, J. Weinmann, et al., "Research Challenges in Wireless Networks of Biomedical Sensors," in *International Conference on Mobile Computing and Networking (MobiCOM).* Rome, Italy, ACM, 2001, p. 151.

Shen, S., C. Srisathapornphat, and C. Jaikaeo, "Sensor Information Networking Architecture and Applications," *IEEE Personal Communications* 8(4):52, 2001.

Shenck, N. S., and J. A. Paradiso, "Energy Scavenging with Shoe-Mounted Piezoelectrics," *IEEE Micro.* 21(3):30, 2001.

Small T., and Z. J. Haas, "The Shared Wireless Infostation Model - A New Ad Hoc Networking Paradigm (or Where there is a Whale, there is a Way)," in *International Symposium on Mobile Ad Hoc Networking and Computing (MobiHoc).* Annapolis, Maryland, ACM, 2003, p. 233.

Sohrabi, K., J. Gao, V. Ailawadhi, and G. Pottie, "Protocols for Self-Organization of a Wireless Sensor Network," *IEEE Personal Communications* 7(5):16, 2000.

Steere, D. C., A. Baptista, D. McNamee, C. Pu, and J. Walpole, "Research Challenges in Environmental Observation and Forecasting Systems," in *International Conference on Mobile Computing and Networking (MobiCOM).* 2000, p. 292.

Welsh, M., D. Myung, M. Gaynor, and S. Moulton, "Resuscitation Monitoring with a Wireless Sensor Network. American Heart Association, Resuscitation Science Symposium," *Circulation,* Vol. 108, Supplement IV (abstract). Orlando, FL, American Heart Association, 2003, p. 1037. Ye, F., H. Luo, J. Cheng, S. Lu, and L. Zhang, "A Two-Tier Data Dissemination Model for Large-Scale Wireless Sensor Networks," in *International Conference on Mobile Computing and Networking (MobiCOM).* Atlanta, GA, ACM, 2002, p. 148. Ye, F., H. Luo, S. Lu, and L. Zhang, "Statistical En-Route Detection and Filtering of Injected False Data in Sensor Networks," in *Proceedings of the 23rd International Annual Joint Conference of the IEEE Computer and Communications Societies (INFOCOM).* Hong Kong, IEEE Press, 2004.

Zhao, F., J. Shin, and J. Reich, "Information-Driven Dynamic Sensor Collaboration," *IEEE Signal Processing Magazine* 19(2):61, 2002.

Challenges

Chapter 8 presented a brief overview of the unique challenges of ad hoc wireless sensor networks. In this chapter we consider these problems in detail. In addition to describing the challenges, we contrast the characteristics and capabilities of existing sensors with the properties of traditional networking environments. By understanding the unique aspects of ad hoc networks, the protocols presented in Chapter 10 become more understandable.

Existing protocols for traditional networks would work fine with ad hoc networks if their characteristics were not different. In many cases, ad hoc networks have requirements that mirror those of existing wired networks. Requirements such as routing, addressing, medium access control, security, and reliability exist in both types of networks. Ad hoc networks, however, require different solutions because of the differing characteristics of both the wireless medium and the devices connected to the network. Protocols for wireless sensor networks that incorporate these features into their design are more useful than existing protocols that are extended without fully addressing the unique problems that arise in ad hoc networks. A wireless sensor network is a particularly resource-constrained, but important, type of ad hoc network, so we focus on wireless sensor nodes.

9.1 Constrained Resources

The most obvious limitation of a wireless sensor is the fact that the resources available to the sensor are severely constrained relative to a desktop computer or even a personal digital assistant (PDA). Although these limitations are obvious, the various ways these limitations influence the design across distinct layers of the protocol stack are not immediately apparent. We explore these limitations in more detail here to provide a clear contrast between the existing capabilities of these two types of networked devices.

9.1.1 No centralized authority

To demonstrate the similarities and differences in centralized control between traditional networks and ad hoc networks, consider the problem of routing. Routing on the Internet consists primarily of packet transmissions from one host in the network, the source, to another host, the destination, on a hop-by-hop basis. In contrast, routing in a wireless network consists mainly of transmissions on a hop-by-hop basis from a sensor node, the source, to the sink or base station. Besides this difference in traffic pattern, there are structural differences in the way that routing is supported in these two networks.

The Internet operates in a completely decentralized manner. A hierarchy of machines is used, for example, to maintain the list of domain names (Mockapetris and Dunlap, 1995). Routing of data and control information is accomplished in a distributed manner by exchanging routing updates among neighboring routers (Rekhter and Li, 1995). The allocation of bandwidth is done in a distributed fashion as well, using the Transmission Control Protocol (TCP) functionality to allocate a "fair" share of bandwidth to competing connections over a link without any explicit consideration for globally optimal usage or even tight fairness constraints.

Likewise, ad hoc networks operate without a central authority. In this case, however, the network is even more decentralized. Although routing on the Internet is decentralized, there are nodes in the network that function as routers and provide this service for other hosts, whether routing packets from these nodes or to these nodes. By contrast, typically no designated routers exist in an ad hoc network. Routing is accomplished either by source routing protocols such as Dynamic Source Routing (DSR) (Johnson, 1994), in which each source knows the complete route, perhaps by first querying the network, or by distributed routing protocols such as Destination-Sequenced Distance-Vector (DSDV) (Perkins and Bhagwat, 1994), in which each node along the path provides routing services for each packet.

Focusing on ad hoc networks of wireless sensors, the routing demands placed on nodes can have an impact on other objectives, such as energy conservation. The lack of network services implies that each node has to perform extra work to support requests from other nodes, even when these demands do not occur at a convenient time. For example, a node may wish to conserve power by turning off its radio receiver. However, this may not be possible if this node needs to remain available to handle routing requests or provide other resources to neighboring nodes. This tension between nodes operating in their own best interests versus operating to benefit the overall application must be managed carefully in the protocol design to maximize network usefulness and ensure cooperation among the related sensors. An incentive for cooperation for resources and tasks for nodes in a mobile ad hoc network, by way of virtual money called *nuglets,* is proposed for the Terminodes Project (Blazevic et al., 2001).

In addition, the lack of centralized services means that obtaining information is more computationally expensive or, at a minimum, requires creative approaches to mitigate this lack of support. For example, traditional networks perform translation from a Web address to an Internet Protocol (IP) address using a Domain Name Server (DNS). The lowest-level DNS for a network typically has a fixed IP address that other nodes access directly to perform this translation. However, in an ad hoc network, addressing information that is centralized on the DNS is usually distributed throughout the network. To complicate matters further, there may be no simple and efficient mechanism for finding this information. This leads to the need to query multiple nodes or perhaps even flood the network to obtain the required information. For example, the application may require thermal sensor readings from a particular region in order to determine the average temperature at some location.

9.1.2   Limited power

Power available to an ad hoc node generally is limited because the node uses wireless communication for networking and often is placed in an environment where there is no readily available power supply. If an external power source were required to operate the node, the advantages of wireless communication would be reduced. Although fixed wireless connections are becoming increasingly common for traditional networks, these types of connections are not practical for the types of devices we are discussing now. Deployment of sensor nodes in large quantities, sensor node mobility, and their deployment locations may prohibit the use of nearby power sources. If power lines needed to be run to a large number of remotely deployed sensor nodes, the cost of deployment would escalate rapidly, and the advantages of wireless networking would be reduced greatly. In general, if it is possible to provide a power cable to a sensor node for long-term energy source, providing a network cable connection is usually also feasible. For these reasons, the power supply on an ad hoc node is self-contained in the sensor node.

There are essentially two options for providing power to a sensor node. The first option is to connect a battery to the device. The power density of batteries is increasing at a very slow rate relative to computing power. For example, the power density of carbon-zinc batteries has increased by only seven times between 1920 and 1990 (Powers, 1995), which is less than doubling three times. Moore's law offers much more significant improvements in processor performance. Thus, using a battery to power a sensor node requires aggressive power management. The battery supplies power for as long as possible, after which either the battery is replaced or the sensor no longer functions. Depending on the sensor's cost, the intended application, the accessibility of the sensor nodes, and the lifetime of the battery, either situation may be feasible. For example, sensors deployed in a remote area for a short-term application, such as tracking of enemy troop movements just prior to an impending operation, do not require batteries to be replaced. If the sensors are expensive, they may be retrieved and have their batteries replaced. On the other hand, for low-cost sensors deployed in large quantities, it may be more economical simply to deploy new sensors periodically rather than to gather existing sensors and replace their batteries. Although the choice of how to address battery death is application-dependent, extending the lifetime of batteries through careful protocol design and energy-efficient hardware leads to lower operational costs and makes a wide range of sensor network applications feasible.

The second option is for sensor nodes to rely on passive power sources, such as solar (Doherty et al., 2001) and vibration energy (Shenck and Paradiso, 2001). Scavenging power from passive sources offers the promise of a continuous power source. However, there are times when this power source is unavailable. For example, solar collectors cannot obtain significant power during overcast or rainy days or at night. Even under optimal environmental conditions, passive power sources typically provide a very modest amount of energy (Doherty et al., 2001). This limitation places severe constraints on operation of the sensor nodes if they depend solely on passive sources for power.

A third possibility is a combination of the two preceding options. A passive power supply, such as a solar collector, could be used as the primary power source, with a rechargeable battery attached to the sensor node as a secondary power supply (Welsh et al., 2003). The passive device is used to recharge the battery when the generated power exceeds the operating power requirements. The main advantage of this approach is that the sensor has essentially an unlimited lifetime (the lifetime of the rechargeable battery) and is able to operate for periods of time during which the passive power source is not available. In the case of solar-powered sensors, the battery could allow for continued operation throughout the night. Similarly, if vibration/motion is used for power scavenging, the battery operates when there is insufficient motion to provide the required energy. To achieve the full potential of this approach, the power budget must be managed carefully to ensure that the power needed to operate the device does not exceed the amount of power available under typical operating conditions.

Power source constraints of wireless sensor nodes are contrasted easily with those of traditional wired or wireless devices connected to a power outlet. Availability of a continuous power source makes power requirements of protocols irrelevant. More typical concerns are the performance of the protocol and the effect on network behavior. Conversely, an ad hoc wireless node depends critically on a temporary or meager power source. This profoundly alters protocol development for wireless devices, leading to the need to optimize functionality across all layers of the protocol stack. Protocols not only must satisfy the application requirements but also must remain within a reasonable power budget. Extending battery life is one of the driving forces in protocol design for wireless sensor networks (Akyildiz et al., 2002) and a key reason for developing new protocols rather than using existing protocols for these devices. In many cases, suggestions have been made to merge neighboring layers of the protocol stack. For ease of reference, each layer of the seven-layer protocol stack as defined in the Open Systems Interconnect (OSI) standard is shown in Fig. 9.1. Benefits can be obtained from providing information across layers simply to reduce the

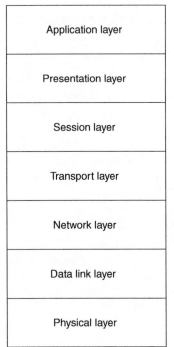

Figure 9.1 OSI seven-layer networking protocol stack.

power requirements of the protocols (Dulman et al., 2003). In some cases this is also because the wireless medium is less reliable, so information about packet loss rates and current network conditions is needed to reduce the overhead of extra transmissions, which leads to additional power drain because the packet is sent multiple times.

9.1.3 Wireless communication

The goal of ubiquitous connectivity, the advantages of tetherless communication, and the desire to have sensor nodes that provide environmental data drive the demands for wireless sensor networks. The main advantages of wireless communication are the reduced cost of not providing cabling for wired connections, the flexibility of mobile connections, and the freedom to deploy individual sensors anywhere. Along with these advantages, certain disadvantages must be accepted. Wireless connections provide lower bandwidth, require more power from the nodes, and are less reliable than traditional wired connections. Each of these disadvantages must be addressed in the wireless protocols. For example, compression may reduce the bandwidth requirements, careful scheduling of communications may reduce energy consumption, and forward error correction (FEC) may reduce error rates at the transport layer. Although high-bandwidth wireless protocols are available (McFarland and Wong, 2003), these protocols are not currently suitable for sensors owing to their power requirements and the underlying complexity for implementing the protocol.

From a sensor-node perspective, the power drain of wireless communication is the most significant factor. Because sensor nodes tend to generate small amounts of data and require infrequent communication, the limited bandwidth is not a significant factor (Ye, Heidemann, Estrin, 2002). For dense networks, redundancy allows the suppression of duplicate information and permits dense sensor networks to operate effectively even with little bandwidth (Scaglione and Servetto, 2002). Since packets are small, errors are less likely and can be handled by retransmission as necessary. Of course, retransmissions consume additional power, so steps should be taken to minimize retransmissions.

For protocol design purposes, networks based on fiber optics are considered to be practically error-free. This helps to illustrate the limitations of wireless communication. For example, TCP assumes that any loss in the network is due to congestion because error rates are negligible (Jacobson, 1995). This assumption does not hold in a wireless environment. Reducing the transmission rate because congestion is assumed incorrectly to be the cause of losses due to errors leads to less efficient communication and wastes bandwidth (Lang and Floreani, 2000). Thus a naive implementation of the TCP over wireless links is inefficient.

9.1.4   Limited computation and storage

Sensor nodes have very limited computational resources. For example, a Mica 2 mote, which is third-generation sensor technology, has an Atmel ATMEGA processor that runs at 4 MHz with 512 kB of programmable memory and 128 kB of SDRAM, along with several sensors, a battery, and radio (Hill and Culler, 2002). This represents a processor speed that is approximately three orders of magnitude less than a high-end workstation and memory that is at least six orders of magnitude less than this same workstation.

The computational capabilities of ad hoc nodes are limited for a number of reasons. The most significant reasons are the power limitations, the cost, and often the size of the device. These limitations are fundamental to the design and operation of wireless sensor nodes used in many applications. Therefore, it is reasonable to expect that sensor nodes will continue to lag behind the computational and storage capabilities of more traditional computing platforms. Instead of ignoring such differences, it is important to recognize these limitations and design protocols and applications that operate efficiently within these constraints.

Because sensor nodes have relatively limited capabilities, sensor networks rely on the cumulative resources derived from a large-scale deployment of sensors (Estrin et al., 1999). This distributed processing is harnessed by giving each sensor only local responsibilities for the larger task. A base station performs more complex processing after receiving information from an appropriate subset of the sensor nodes. This distribution of workload helps to maintain a low-cost system by minimizing the resources of individual sensors.

Besides the obvious disadvantage of limited processing power on the design of applications and protocols, some less obvious disadvantages exist. First, less processing of the sensor readings can be done. For example, collaboration among neighboring sensors and local data processing is restricted to simple situations that are not data-intensive or computationally intensive. Security protocols, which are discussed later, often derive higher levels of security by using longer keys, more complex algorithms, or both. Symmetric and asymmetric security protocols are discussed in Section 13.2, where trade-offs between these two options are explained in greater detail. A review of Section 13.2 with consideration for resource-constrained sensor-node capabilities suggests which are most suitable for wireless sensor networks. Security of the application, although an important component for certain sensor applications, cannot consume resources such that the sensor is unable to achieve its primary purpose. Hence reduced computation implies lower levels of security.

As a final example, consider data compression and FEC. Since the overhead of wireless communication consumes a significant percentage of the sensor node's power, reducing the size of the data packets and the number of retransmissions could reduce the power consumption of the sensor node substantially. However, care must be taken that aggressive data compression and error-correcting codes do not consume excessive power. The appropriate use of these techniques in wireless sensor networks requires additional investigation before determining their optimal use.

9.1.5 Storage constraints

In addition to the limited computing power, sensor nodes also have very limited storage. Typical sensors have no permanent storage devices such as hard disks. A few kilobytes of nonvolatile memory may be available, and additional kilobytes of volatile memory usually are available. The operating system and application code must run in this extremely limited area. An example of such an operating system is TinyOS (Hill et al., 2000) which runs on a Mica sensor and requires only about 3500 bytes. The application running on the sensor nodes is compiled with the OS into a single program. All variables, values, and temporary workspace must fit within the remaining space.

This requires each application be optimized for space utilization and places significant bounds on the complexity of the application. Data derived from sensors and the application requirements dictate some specific data-processing algorithms must be programmed into the sensor. Additional application requirements determine the processing of the data and the amount of storage required to perform the necessary computation. A similar requirement exists for embedded systems in general, and experience has shown that it is possible to develop programs in this environment that accomplish the intended tasks (Panda et al., 2001).

The extremely limited storage capacity of each sensor node presents additional challenges. Although caching of data, such as results from neighboring nodes, is useful for removing redundant communication from the system (Intanagonwiwat, Govindan, and Estrin, 2000), the opportunities for and benefits of caching data are restricted by the small storage capacity. Buffering of local measurements also reduces redundant communication, but the performance advantages similarly are diminished by the storage restrictions. Often there exists a trade-off between these memory requirements of an application and the execution time of that application. The lack of sufficient space on the sensor may require the use of less efficient algorithms. Consider the storage requirements of security protocols, where space requirements of large encryption keys may not be available. As a result, security is weakened by the lack of available memory on the sensor.

9.1.6 Limited input and output options

Sensor nodes have very limited input and output options owing to their small form factor. Even for devices as large as PDAs, a standard keyboard is not feasible. Sensor nodes are significantly smaller. Input and output peripherals will be almost nonexistent on a sensor node. There is no output display and no keyboard, mouse, or stylus for input. The output peripherals consist of at most a few light-emitting diodes (LEDs) and a speaker. LEDs could be programmed in novel ways to flash patterns, although at the cost of some power consumption and program code overhead. Consequently, LEDs are limited to conveying simple messages, such as indicating that the sensor is turned on, that it is functioning correctly, or that it is currently transmitting or receiving data. Speaker output is limited to at most frequency, duration, and volume level, which constrains the amount of information transmittable to the user.

Configuring and trouble-shooting a sensor node is made more challenging by the lack of I/O peripherals. Not only does diagnosis of individual sensors present a problem, but evaluating sensor network performance and finding software problems also are more difficult. Consider attempting to debug a protocol by monitoring the execution of a sensor. Without access to a display or the ability to obtain the measured sensor readings stored in memory, debugging the sensor becomes much more difficult. Furthermore, if the only option for obtaining information from a sensor node is through wireless communication, diagnosis of the system may interfere with operation of sensor network protocols, which may impede isolation of the problem.

Because direct human-to-sensor-node interaction is extremely limited, simulators and emulators are the primary mechanisms for debugging. These tools are effective only to the extent that they correctly model the operating environment of the sensors. For example, assumptions about the wireless channel characteristics, the responses of the sensors, and the energy capacity and energy drain of the sensor node affect the accuracy of the simulation results. Optimistic assumptions or simply unanticipated situations lead to unforeseen problems with sensor nodes. These discoveries are made only after deploying the sensors. Similarly, if the sensor deployment occurs in a test bed rather than a real application, some problems still may not be detected. The problem of tracking down the specification, design, or implementation mistakes is substantially more challenging after a large-scale deployment.

Sensor configuration also becomes a more difficult task because downloading even simple instructions likely is performed via radio. Acknowledgments from the sensor may be required to determine that the sensor has been configured properly. For large sensor networks, limited feedback options may hamper the effort to configure the sensors to collaborate according to the application demands. For scalability, automatic configuration of the network is most desirable, but preconfiguration and subsequent calibration of some sensor information still may be required before initial deployment. Depending on the region of deployment, such as a hostile or remote region, sensor calibration may be performed among the sensors in the network (Bychkovskiy et al., 2003). The limited user interface makes these tasks more challenging.

9.2 Security

Sensor networks are deployed for a wide range of purposes. Some of these sensors require absolutely no security. A noncritical application, such as a sensor that simply recognizes that someone has picked up an object and responds, may not need any security. On the other hand, sensors deployed for military applications require stringent security mechanisms. Other types of sensor networks have more relaxed security requirements. For example, a location tracking and management system that tracks employee movements or parts in an industrial research center may contain information useful to industrial spies. Similarly, biomedical sensors may contain personal and privileged information on patients that should not be accessible to the public (Schwiebert et al., 2001). Privacy laws may even dictate that the existence of the device should be concealed from those who should not have access to this information, such as potential employers. For example, an implanted glucose monitor could reveal to an unauthorized person that the wearer of the sensor has diabetes.

In the preceding examples, the sensor data need to be available, but only to those with a legitimate need for this information. Arbitrary access could lead to unintended problems. For many applications it is difficult to determine in advance the potential misuses of sensor data, implying that either security mechanisms exist as a precautionary measure or needed security mechanisms may be omitted because of an oversight when considering the security requirements of the application.

Security requirements of wireless networks are discussed in Chaps. 12 and 13, but it is still worthwhile to consider the particular limitations of sensor nodes that complicate the application of security protocols to sensor networks.

9.2.1 Small keys

As mentioned earlier, wireless sensor nodes have significantly less storage space than other wireless computers. Since a significant amount of this modest space is dedicated to the program code and the data processing, storing large keys is not practical. For example, a 1024-bit

RSA[1] key would take up a significant fraction of the memory on a sensor. Since the level of security increases with larger keys, smaller keys reduce the security dramatically. If an adversary has access to significantly more powerful computational resources than the sensor node possesses, the adversary can break the security through brute force in a relatively short amount of time. Because security concerns must remain secondary to the main tasks of the sensor node, the security protocols must be adapted to the existing constraints and operate within the available space. Therefore, security protocols must be adopted that can provide sufficient levels of security with these smaller keys.

9.2.2 Limited computation

Another issue that has been mentioned before is the limited computing power of the sensor nodes. In general, security protocols perform additional computations to increase the level of security. In other words, more complicated algorithms offer improved security, but this requires that sufficient computing power is available. Security protocols exist that can operate on more constrained devices, but not at the same level of performance. The power available to the sensor nodes is also limited, and extensive computations for security purposes limit the lifetime of the sensor node or reduce the energy available for other sensor tasks. Hence one of the challenges for wireless sensor networks is to provide security that meets or exceeds the requirements of the application without consuming too many computing resources.

9.2.3 Changing network membership

Over the lifetime of a sensor network, the active membership of the sensor network varies. This variation may arise because of sensor nodes powering their radios off to conserve energy (Chen et al., 2001; Xu, Heidemann, and Estrin, 2001). Sensor-node failures and sensors that die owing to depletion of their energy also change the membership of the network. Security protocols that rely on sharing keys between neighboring nodes need to continue to operate even though the neighbors of a sensor node may change frequently during the lifetime of the sensor network. In addition, routing and other distributed tasks may rely on authentication of nodes. Storing keys and related data for a large number of neighboring sensors often is impractical. On the other hand, a single routing node or even a small subset of nodes cannot be assumed without limiting the flexibility of the network protocols.

---

[1]RSA is named after its inventors: Ronald L. Rivest, Adi Shamir, and Leonard Adleman.

Using a single key for the entire network is an attractive option for resolving this problem. However, this approach introduces the drawback that a single compromised sensor node is sufficient to allow an adversary to decrypt any message in the network, as well as interfere with the operation of any other sensor in the network. Sharing keys between individual pairs of neighboring sensors increases security but leads to problems with key distribution and key management. Handling changing network membership requires a key distribution and management scheme that is both scalable and resilient to adversarial attacks.

9.2.4 Arbitrary topology

Sensor networks are deployed in different ways. For example, sensors in biomedical applications are likely to be implanted in specific locations; the neighbors of each sensor are predetermined. For applications such as transportation monitoring and management, sensors are deployed with less precision but still with a well-defined distribution. On the other hand, sensors deployed for large-scale tracking and monitoring operations are likely to be deployed in an arbitrary fashion. Examples of such applications are military surveillance, forest fire detection, and animal tracking. For these applications, sensors are strewn from low-flying aircraft or vehicles or by hand. Not only is the exact position of each sensor somewhat arbitrary, but also the density of nodes varies over the region, and which nodes are neighbors of each sensor is unpredictable before network deployment.

Because both the number and identity of neighboring sensors are not known prior to network deployment, preconfiguration of security keys based on sensor IDs is not possible. Instead, keys that must be shared between neighboring sensors must be distributed and determined after the sensors have been deployed. This means that each sensor must have the ability to assign keys that are cryptographically secure based on communication with neighboring sensors. Relying on communication with a base station or some other central authority would not be scalable. For this reason, local collaboration among neighboring sensors must be sufficient to establish the keys.

Care must be taken with this approach to prevent malicious nodes from adding themselves to the network. In other words, an adversary should not be able to deploy a malicious node that can establish secure communication with existing sensor nodes. Sensor nodes must not authenticate nodes that do not belong to the sensor network. Periodic deployment of additional sensors is beneficial for some applications, especially when sensors are relatively inexpensive and do not have a lifetime that is sufficient for a given application. For example, a military surveillance application may require the occasional redeployment of additional sensors in order to maintain adequate network coverage.

Security protocols for wireless sensor networks must be robust to node failures and the subsequent addition of nodes (Jamshaid and Schwiebert, 2004). At the same time, some mechanism must be available that prevents sensors that do not belong in the sensor network from masquerading as legitimate participants in the sensor application. To prevent this, security protocols must have some technique that is difficult to compromise and is sufficiently powerful to block adversarial nodes from joining the network. At the same time, these protocols must satisfy the space and computational limits.

9.3 Mobility

Ad hoc networks are not necessarily mobile. The ad hoc nature of the network arises simply from the lack of fixed infrastructure. Of course, the absence of a fixed network infrastructure leads directly to the conclusion that mobility is a reasonable feature for these devices. A user carrying such a device will not recognize any practical reason for having to remain stationary while using the device. For certain applications, such as animal tracking or sensors embedded in clothing and the human body, the sensors necessarily are mobile. For PDAs and other small portable wireless devices, mobility is a required property because the user transports this device as he or she moves about.

For wireless sensors, mobility is not required for many applications. For example, sensors embedded in buildings, roads, and bridges are stationary. Sensors deployed for large-scale monitoring and surveillance applications, such as forest fire detection, are less likely to be mobile. Certainly, users of the network will not redeploy such nodes frequently because there are simply too many nodes to make this possible with a reasonable amount of effort. It has been proposed, however, that sensors may be equipped with limited mobility to allow for automated repositioning of these sensors after the initial deployment (Zou and Chakrabarty, 2003). The main objective for providing mobility for these nodes is to allow repositioning of sensor nodes to create a more uniform distribution that enhances the ability of the sensor network to cover the area of interest adequately. Sensors that consume a modest amount of power for mobility could reposition themselves periodically to overcome uneven node distributions and uneven coverage that arises as nodes die. This repositioning could be useful for efficient and reliable routing and delivery of packets (Grossglauser and Tse, 2002).

Mobility of ad hoc nodes presents a number of additional challenges to deploying and sustaining energy-efficient ad hoc networks. Although some energy may be used in mobility if the node is providing the movement on its own, as opposed to being transported by a human or an animal, the overhead is unavoidable because of the effects that mobility has on the wireless communication protocols. In other words, the overhead occurs because mobility is useful to the user.

9.3.1 Mobility requirements

A mobile ad hoc node may need to move in order to remain with the person, vehicle, or animal that carries the sensor. Mobility also may occur because of application needs. For example, nodes may move to enhance their coverage of an area, to achieve or maintain an even distribution of nodes, or to react to changing application requirements. As a node moves, connectivity with a different set of nodes in the network may be required. In addition, as a node moves, this modifies the connectivity of nodes that either had a direct link to this node in the past or that obtain a direct link in the future.

Some applications must support mobility for the convenience of the end user. Consider the case of a user with mobile devices embedded in a vehicle, his clothing, his body, or carried on his person. The user would like to continue using the device even though he is moving about. Although this presents challenges to the underlying network protocols, the reality is that users do not adopt technology that inconveniences them too much. Hence it is most advantageous to provide users with easy-to-use devices instead of imposing limitations that seem arbitrary and unreasonable to an end user who is not knowledgeable about the underlying technology.

Other applications may use mobility to assist the application. In these cases, mobility is provided to enhance the effectiveness of the sensor network in achieving the objectives of the underlying application. For example, mobility improves wireless network capacity by allowing nodes to buffer messages until delivery to the destination is feasible using only a modest amount of wireless bandwidth (Grossglauser and Tse, 2002). In other cases, mobility may be required for the initial positioning of sensor nodes or to reposition those sensor nodes (Zou and Chakrabarty, 2003). Application objectives determines the movement of the sensors; whether the current arrangement of sensors satisfies these objectives, and whether repositioning of some sensors can improve the sensor network's ability to maintain the application objectives. Assuming that sensors can move without a major consumption of power, mobility may be an energy-efficient and cost-effective method of improving sensor performance. For example, if a significant number of sensor nodes have died, moving the remaining sensors into a new configuration may be adequate for application needs while eliminating the cost of redeploying some additional sensors. However, protocols for determining how often and where to move individual nodes need to be developed, along with techniques for determining whether or not this redeployment fulfills the application objectives. In fact, this decision-making process may consume more power than actual movement of the sensors would.

9.3.2 Loss of connectivity

Loss of connectivity owing to mobility is possible during any sufficiently long data transfer. The time until connectivity is lost depends on the range of the devices, the initial distance between the pair of devices, and the speed at which the devices move away from each other.

Although connectivity is lost, there may be a need to reestablish a connection between the two endpoints of the communication. For example, a person using a PDA to check e-mail would expect to continue reading his e-mail even as he walks or drives around. The initial point of connectivity to the mail server may be far away from the place where the user finishes reading his e-mail. Similarly, a sensor in a car may need to complete a data transfer even as the car moves beyond the range of the current point of connectivity.

Loss of connectivity between a pair of nodes could occur because of the movement of any of the nodes along the path between these two nodes. If the two endpoints are stationary, but some other node along the path moves out of radio range, another path must be created to allow communication to continue. While this new path is being formed, packets must be buffered or dropped. In either case, delay is introduced into the network. In some cases this delay may not be significant, but for real-time transmission of sensor readings or other types of data, the delay may interfere with correct operation of the protocol. Simply flooding each packet from the source to the destination can circumvent such problems. Unfortunately, flooding is not an efficient strategy unless nodes are moving too quickly to maintain routing paths because nodes must make a large number of redundant data transmissions when using flooding. Since wireless communication overhead is a significant component of the power usage of a sensor node, flooding the network with packets consumes too much energy. Techniques for dealing with broken paths in an energy-efficient and robust manner are necessary for time-critical data.

Connectivity may be lost not only between a pair of nodes but also with the rest of the network. If the network becomes partitioned, then separate parts of the network cannot share data until the network becomes connected again. If a significant period of time passes between a node disconnecting from the network and the node reconnecting with the network, resynchronizing the node with the rest of the network is difficult. In some cases, reestablishing security between the node and the rest of the network may be complicated if the node is separated from the rest of the network for a long period of time. In general, we cannot avoid prolonged disconnection at the protocol level, so the only option is to design protocols that are flexible enough to deal with this situation. In some sense, prolonged disconnection from the network is not significantly different from a transient fault that keeps a node from communicating with the rest of the network, assuming that the node still retains the state information it possessed at the time of disconnection.

9.3.3 Data loss

As mentioned earlier in this chapter, transmission errors are much higher with wireless communication than over fiber optic or other wired links. In addition to the higher data loss rate owing to the wireless medium, movement of nodes leads to data loss. As described earlier, packets are lost owing to the movement of any node along the path from the source to the destination or the path from the destination back to the source if acknowledgment packets or other data are returned to the source. In some cases, the packets lost due to wireless channel errors can be recovered by locally retransmitting the packet (Parsa and Garcia-Luna-Aceves, 1994). However, when movement of the nodes causes the connection to be broken, retransmitting the packet does not solve the problem. Since the receiver is not within the range of the transmitter, subsequent attempts at sending the same packet will fail unless the transmitter or the receiver move close enough to be within range of the other node. Instead, an alternative path must be found in order to deliver the message.

Depending on the importance of the data that are lost, the packet may be retransmitted or dropped. Retransmission of packets is appropriate for packets with a high priority, but packets that have a low priority should be dropped. In some cases, packets that arrive shortly after the lost packet contain redundant information, which may allow this packet simply to be dropped. Real-time data may be dropped simply because the delay involved in resending the packet may exceed the deadline for delivering the packet. When mobility breaks a connection, the delay to reestablish the path is likely to be much higher than the delay that arises when data are lost owing to channel errors. The mechanisms used to rebuild the path must be sensitive to the requirements of the application. In some cases, aggressive techniques must be used to find a suitable alternative. In other cases, more conservative approaches that allow energy savings may be more desirable. The trade-offs between increasing the delay in delivering packets and the overhead of reconstructing paths quickly must be balanced based on the application's quality of service (QoS) requirements.

9.3.4 Group communication

Since the physical location of a mobile node changes, the neighbors of that node also change. In addition, mobile neighbors may move away from a node, so even if a particular node is stationary, there is no guarantee that this node will maintain the same set of neighbors. If enough mobile nodes move a sufficient distance, almost all nodes experience a continual fluctuation in neighbors. These changes in neighborhood often are unpredictable, so the mobility results in a pair of nodes no longer being neighbors at any time during the exchange of messages. The protocol must be robust enough to handle this in a clean way.

Mobility presents additional challenges for group communication because the structure of the group could constantly change. For example, assume that the group is represented as a multicast tree. As nodes move about, the structure of the tree changes, which could modify the parent and children of each node. Forwarding messages to ensure that the messages are transmitted to every node in the group could be difficult, even without any strong requirements for reliability. When a node receives a packet, the node may not be sure that its current parent has received the same packet, and the node may not know whether the packet needs to be transmitted to the children. Resending every packet to all the neighbors could result in a great deal of redundant communication, draining the power and wasting the bandwidth. Furthermore, a long period of time could be required before all the nodes stop sending this particular redundant packet. This leads to delay in delivering other packets. In cases where reliable communication is required, acknowledgments may be transmitted up the tree. Delivery of these acknowledgment packets could suffer from the same problems as the original packet delivery did.

Besides the difficulty of maintaining the logical group owing to mobility, the group also could become disconnected for periods of time. For example, if an ad hoc node moves beyond the communication range of other nodes in the network, there is no way to transmit the information to this particular node. Although the information still could be buffered, there may be limited buffer space on the other nodes. Besides, there is no guarantee that the disconnected node will return to the group any time soon or even at all. It is possible that this node has failed, depleted its power, been turned off, or otherwise become permanently separated from the network. Instead of buffering messages for a node that will not return, the node simply should be removed from the group. Buffering is no longer required in this case, but determining when to remove a node from the group is not trivial.

In certain situations, reliable protocols need to determine that all members of a group have received a packet before initiating the transmission of addition packets. Thus a single node disconnecting from the network on a temporary basis could stop or substantially reduce the flow of packets among members of the network. This is obviously not desirable for any long period of time, so some alternatives should be available when nodes are disconnected from the network for some period of time.

9.3.5 Maintaining consistent views

A set of nodes needs to maintain consistency for several reasons. For example, hop-by-hop routing of messages can lead to loops in the routing tables if nodes do not have consistent information. As loops arise, they need to be recognized and removed, essentially by achieving some local consistency. This fixes the routing paths so that packets can make progress. Another reason that consistency is needed in a network is to reduce the number of retransmitted packets. Each time a packet is transmitted, the transmitter and all receivers consume power. In addition, wireless bandwidth is a relatively scarce commodity in most cases; unnecessary retransmissions place an additional burden on this resource. Since packets often are transmitted over multiple hops, the retransmission of a single packet could lead to a significant amount of extra traffic. This extra traffic can be reduced when nodes have common knowledge of the packet status. In other words, when the transmitter of the packet learns that the receiver has obtained the packet, no retransmission is attempted. Consistent information between the sender and the receiver allows the suppression of these retransmissions.

Maintaining a consistent view in a wireless ad hoc network is difficult because acknowledgments and other mechanisms for determining whether or not a retransmission is required work less effectively with higher packet loss rates and wireless channel interference. The mobility of nodes makes this maintenance even more difficult because a transmission that occurred originally over a single hop may require a retransmission over several hops. This means that a new path must be found between the source and destination, and the nodes along the path consume additional power to accomplish this retransmission. Nodes can use flooding to transmit data packets when a consistent view of routing information does not exist, but flooding also increases power consumption.

Mobility also increases the difficulty of maintaining consistent information among nodes because mobility changes the network topology. In effect, this information changes by virtue of the mobility and needs to be propagated because of mobility. In practice, all nodes may not need to be notified of every change in network topology. At any given time in a large network, each node is communicating with only a small subset of the nodes in the network. (The only real exception to this situation is a broadcast or other group operation.) However, since the distance between nodes could be several hops, movement of any nodes along a path could require the rebuilding of a valid path. This leads to communication delay and either temporary buffering of packets or dropping of packets. Later we will consider protocols that address these problems by not maintaining explicit paths between the source and destination (Niculescu and Nath, 2003). Although this is an attractive option, it cannot be used in all situations.

For high rates of mobility or when nodes are at the edge of their transmission range, this mobility may interrupt the delivery of a packet. The transmitter of the packet may not hear a retransmission request from the receiver, so the transmitter is uncertain about whether the packet was lost or not. Another problem is the successful delivery of the packet but the inability to return an acknowledgment because the transmitting node has moved beyond the transmission range of the receiver. Either of these two situations leads to inconsistent information on the status of packets—whether they have been received or not. Lack of consistency can lead to retransmissions or additional delays in receiving these packets. Thus mobility adds an additional dimension, beyond lossy wireless channels, to situations in which inconsistent information arises. Protocols for resolving lost packets need to be robust to the mobility effects that can take place during data and acknowledgment transmission and reception.

9.4 Summary

In this chapter we discussed several of the unique challenges for using networks of wireless sensors. Besides describing the challenges, we compared the characteristics and capabilities of existing sensors with the properties of traditional networking environments. Wireless sensor networks require different solutions because of the differing characteristics of both the wireless medium and the sensor nodes that make up the network. As we have discussed, protocols for wireless sensor networks are more useful if they incorporate these special features into their design. By understanding these unique aspects of wireless sensor networks, the protocols presented in Chap. 10 become more understandable.

9.5 References

Akyildiz, I. F., W. Su, Y. Sankarasubramaniam, and E. Cayirci, "Wireless Sensor Networks: A Survey," *Computer Networks* 38(4):393, 2002.

Blazevic, L., L. Buttyan, S. Capkun, S. Giordano, J. Hubaux, and J. Le Boudec, "Self-Organization in Mobile Ad Hoc Networks: The Approach of Terminodes," *IEEE Communications Magazine* 39(6):166, 2001.

Bychkovskiy, V., S. Megerian, D. Estrin, and M. Potkonjak, "A Collaborative Approach to In-Place Sensor Calibration," in *Proceedings of the 2nd International Workshop on*

Information Processing in Sensor Networks (IPSN '03). Lecture Notes in Computer Science 2634:301, 2003.

Chen, B., K. Jamieson, H. Balakrishnan, and R. Morris, "Span: An Energy Efficient Coordination Algorithm for Topology Maintenance in Ad Hoc Wireless Networks," in *International Conference on Mobile Computing and Networking (MobiCom)*; Rome, Italy, ACM, 2001, p. 221.

Doherty, L., B. A. Warneke, B. E. Boser, and K. S. J. Pister, "Energy and Performance Considerations for Smart Dust," *International Journal of Parallel Distributed Systems and Networks* 4(3):121, 2001.

Dulman, S., L. V. Hoesel, T. Nieberg, and P. Havinga, "Collaborative Communication Protocols for Wireless Sensor Networks," in *European Research on Middleware and Architectures for Complex and Embedded Systems Workshop*, Pisa, Italy; IEEE Computer Society Press, 2003.

Estrin, D., R. Govindan, J. Heidemann, and S. Kumar, "Next Generation Challenges: Scalable Coordination in Sensor Networks," in *International Conference on Mobile Computing and Networking (MobiCom)*; Seattle, WA, ACM, 1999, p. 263.

Grossglauser, M., and D. Tse, "Mobility Increases the Capacity of Ad Hoc Wireless Networks," *IEEE/ACM Transactions on Networking* 10(4):477, 2002.

Hill, J. L., and D. E. Culler, "Mica: A Wireless Platform for Deeply Embedded Networks," *IEEE Micro* 2(6):12, 2002.

Hill, J. L., R. Szewczyk, A. Woo, S. Hollar, D. Culler, and K. Pister, "System Architecture Directions for Networked Sensors," in *Proceedings of the 9th International Conference on Architectural Support for Programming Languages and Operating Systems*; Cambridge, MA, ACM, 2000.

Intanagonwiwat, C., R. Govindan, and D. Estrin, "Directed Diffusion: A Scalable and Robust Communication Paradigm for Sensor Networks," in *International Conference on Mobile Computing and Networking (MobiCom)*; Boston, MA, ACM, 2000, p. 56.

Jacobson, V., "Congestion Avoidance and Control," *ACM SIGCOMM Computer Communication Review* 25(1):157, 1995.

Jamshaid, K., and L. Schwiebert, "SEKEN (Secure and Efficient Key Exchange for Sensor Networks)," in *IEEE Performance, Computing, and Communications Conference (IPCCC)*; Phoenix, AZ, IEEE Computer Society, 2004, p. 415.

Johnson, D. B., "Routing in Ad Hoc Networks of Mobile Hosts," in *Proceedings of the IEEE Workshop on Mobile Computing Systems and Applications*; Santa Cruz, CA, IEEE Computer Society, 1994, p. 158.

Lang, T., and D. Floreani, "Performance Evaluation of Different TCP Error Detection and Congestion Control Strategies Over a Wireless Link," *ACM SIGMETRICS Performance Evaluation Review* 28(3):30, 2000.

McFarland, B., and M. Wong, "The Family Dynamics of 802.11," *Queue* 1(3):28, 2003.

Mockapetris, P. V., and K. J. Dunlap, "Development of the Domain Name System," *ACM SIGCOMM Computer Communication Review* 25(1):112, 1995.

Niculescu, D., and B. Nath, "Trajectory Based Forwarding and Its Applications," in *International Conference on Mobile Computing and Networking (MobiCom)*; San Diego, CA, ACM, 2003, p. 260.

Panda, P. R., F. Catthoor, N. D. Dutt, K. Danckaert, E. Brockmeyer, C. Kulkarni, A. Vandercappelle, and P. G. Kjeldsberg, "Data and Memory Optimization Techniques for Embedded Systems," *ACM Transactions on Design Automation of Electronic Systems (TODAES)* 6(2):149, 2001.

Parsa, C., and J. J. Garcia-Luna-Aceves, "Improving TCP Performance Over Wireless Networks at the Link Layer," *Mobile Networks and Applications* 5(1):57, 2000.

Perkins, C. E., and P. Bhagwat, "Highly Dynamic Destination-Sequenced Distance-Vector Routing (DSDV) for Mobile Computers," in *Proceedings of the Conference on Communications Architectures, Protocols and Applications (SIGCOMM)*; London, UK, ACM, 1994, p. 234.

Powers, R. A., "Batteries for Low Power Electronics," *Proceedings of the IEEE* 83(4):687, 1995.

Rekhter, Y., and T. Li, *A Border Gateway Protocol 4 (BGP-4)*, Internet Engineering Task Force, RFC 1771, March 1995.

Scaglione, A., and S. Servetto, "On the Interdependence of Routing and Data Compression in MultiHop Sensor Networks," in *International Conference on Mobile Computing and Networking (MobiCom)*; Atlanta, GA, ACM, 2002, p. 140.

Schwiebert, L., S. K. S. Gupta, J. Weinmann, et al., "Research Challenges in Wireless Networks of Biomedical Sensors," in *International Conference on Mobile Computing and Networking (MobiCom)*; Rome, Italy, ACM, 2001, p. 151.

Shenck, N. S., and J. A. Paradiso, "Energy Scavenging with Shoe-Mounted Piezoelectrics," *IEEE Micro* 21(3):30, 2001.

Welsh, E., W. Fish, and J. P. Frantz, "Gnomes: A Testbed for Low Power Heterogeneous Wireless Sensor Networks," in *International Symposium on Circuits and Systems*. Vol. 4. 2003, p. IV-836. Held in Bangkok, Thailand. Published by IEEE Press.

Xu, Y., J. Heidemann, and D. Estrin, "Geography-Informed Energy Conservation for Ad Hoc Routing," in *International Conference on Mobile Computing and Networking (MobiCom)*; Rome, Italy, ACM, 2001, p. 70.

Ye, W., J. Heidemann, and D. Estrin, "An Energy-Efficient MAC Protocol for Wireless Sensor Networks," in *Proceedings of the 21st International Annual Joint Conference of the IEEE Computer and Communications Societies (INFOCOM)*, Vol. 3; New York, NY, IEEE Computer Society, 2002. p. 1567.

Zou, Y., and K. Chakrabarty, "Sensor Deployment and Target Localization Based on Virtual Forces," in *Proceedings of the 22nd International Annual Joint Conference of the IEEE Computer and Communications Societies (INFOCOM)*; San Francisco, CA, IEEE Computer Society, 2003, p. 293.

10

Protocols

In Chap. 9 we described the many challenges of ad hoc networking, including mobile ad hoc networking. Most of these topics remain areas of active research, partly because the topics are new and partly because the characteristics of wireless networks are changing so rapidly that existing techniques may not be ideal or even suitable for future networks. As research progresses in these challenging problems, new approaches will be developed, and better protocols will be designed. Many of the core strategies, however, seem to have been identified. Therefore, in this chapter we take a closer look at the protocols that can be used to address these challenges.

10.1 Autoconfiguration

Ad hoc networks, sensor networks in particular, benefit from being self-configurable. The initial deployment of sensors in an area could be rather large. Allowing these sensors to configure themselves into a functioning network has a number of advantages. First, human intervention may be impractical or impossible. For example, sensors may be deployed in an environment that is unsafe for humans, whether at a military location, a remote geographic location, or an industrial setting that is not suitable for humans. Second, large sensor networks require significant labor to configure each sensor individually. Preconfiguring the network becomes impractical if sensors are deployed randomly or in an ad hoc fashion. Preconfiguration is also challenging if sensors that have exhausted their battery supply or have failed are replaced periodically with new sensors. Automatically configuring nodes in an ad hoc network achieves good results if the overhead of configuring nodes is modest and the algorithms for automatically configuring the nodes lead to good solutions. In this section we will describe a number of possible uses for self-configuration in an ad hoc network. Based on this discussion, it will become clear what the network requires for configuration, as well as the specific functions of this autoconfiguration. Furthermore, this discussion helps to clarify the advantages of using automatic configuration as opposed to manual configuration.

10.1.1 Neighborhood discovery

Neighborhood information in an ad hoc network informs each node about the surrounding set of nodes in the network (Conta and Deering, 1998; Deering and Hinden, 1998). The use of this neighborhood information varies with the application; however, nodes require some knowledge of their neighbors for collaboration and coordination of efforts. For example, if a sensor observes an event, the node may wish to check with neighboring nodes to determine whether or not other sensor nodes also have observed the same event. Multiple sensors that detect the same event increase the level of confidence in the observation. These readings can be combined using a soft decision parley algorithm, such as the one proposed in Van Dyck (2002). In addition, a single composite message could be returned to the base station rather than separate messages from each sensor (Heinzelman, Chandrakasan, and Balakrishnan, 2002; Intanagonwiwat, Govindan, and Estrin, 2000; Lindsey, Raghavendra, and Sivalingam, 2002). Furthermore, when multihop routing is used, the network may need neighborhood information for routing packets.

Neighborhood discovery may choose to allow only bidirectional links or also include unidirectional links. Unidirectional links occur when node A can communicate directly with node B, although node B has no direct connection to node A. Unidirectional links could arise because of differences in transmission power among nodes (Prakash, 2001) or as a result of obstacles. For example, a node at a higher elevation may be able to transmit to a lower node, but not vice versa unless the node at a lower elevation transmits with significantly more power. Allowing unidirectional links may increase the network capacity because there are more wireless communication links, but bidirectional links allow a node to receive acknowledgments or listen for retransmission of packets. Listening for retransmission allows passive acknowledgments because if the receiver forwards the packet, it obviously received the packet.

Detection of unidirectional links is nontrivial because the node able to transmit on this link will have difficulty determining if the receiver can acquire the message. Assistance from other nodes is required to determine this. In addition, link availability may be variable, and the sender has no direct method to measure the quality of the link at a given time. For this reason, many protocols restrict themselves to only bidirectional links.

Detecting neighboring nodes is less difficult with a bidirectional link because both nodes can send and receive on the same link. Each node simply sends a probe message. Each node that receives a probe message responds to the sender with an acknowledgment. After a few rounds, every node knows all the neighboring nodes that share a link with this node, assuming that the error rate on the links is relatively low. When a mobile node moves into a new area, it detects other nodes in the area by promiscuously listening to messages intended for other nodes (Pei and Gerla, 2001). After the mobile node determines some of its neighbors, it initiates neighbor discovery to identify remaining neighbors and notify these nodes of its presence in the neighborhood.

10.1.2 Topology discovery

Topology information in an ad hoc network consists of the locations of nodes relative to each other, including the density of nodes in a particular region. The topology of the network also could include the number of nodes in the network and the size of the network—in other words, the physical dimensions of the sensor network and the number of nodes in this area. Based on this information, an average network density can be determined, from which the regions of high and low sensor-node density can be extrapolated. Topology information also can include the number of neighbors each node in the network has, as well as the average number of neighbors per node. This information is useful in deciding how to organize the wireless network.

The most obvious reason for conducting topology discovery is to build routing paths. A wireless sensor network may be deployed over a large area with a need to communicate with a base station at some specific location. In other cases, the base station may be mobile (Ye et al., 2002), but nodes still need topology information in order to route packets to the base station. Knowledge of the topology also provides an insight into the distribution of sensor nodes in the area of interest. This information then can be used to evaluate the coverage or exposure of the sensor network (Meguerdichian et al., 2001a, 2001b). The topology information also can be used to schedule nodes for sleep cycles to prolong the lifetime of the network, as well as to select coordinators, cluster heads, or other group leaders to enable hierarchical collaboration in the sensor network.

Over the life of the network, the network topology changes as sensor nodes stop working because of faults or lack of power. Monitoring these changes allows the network to figure out when and where to deploy additional sensors to maintain a suitable quality of service (QoS) level. Determining this information manually could take a significant amount of time and effort, so discovering the topology information and maintaining this information have practical advantages in making the ad hoc network cost-effective.

Topology discovery can be built on neighborhood discovery. In general, individual nodes do not need to be aware of the entire network topology. However, accumulating this information at a base station or retaining this information in some distributed fashion leads to better protocol designs. For example, knowledge of the average number of neighbors each node has can be used to select which nodes should be coordinators of areas of interest within a sensor network. These coordinators then could be combined in a hierarchical manner to provide robust and scalable support for communication across the sensor network.

Routing of data may not rely on generic network topology information because orientation toward the base station or any particular source of a request may be sufficient for routing the information to this location (Niculescu and Nath, 2003; Salhieh and Schwiebert, 2004). However, holes in the sensor network or other pathological situations such as obstacles are examples of topologic information that could be used to supplement the base routing protocol—either for efficiency or for robustness.

10.1.3 Medium access control schedule construction

Wireless bandwidth is a relatively scarce resource in wireless sensor networks, so this bandwidth must be used efficiently by making sure that as many packets as possible are delivered correctly. Because of the inherent broadcast nature and energy required for wireless communication, collisions must be avoided. To avoid collisions, each node cannot transmit a message in any given area that interferes with a message from another node. Spatial reuse of the wireless bandwidth becomes possible by limiting the transmission range of a node and by coordinating communication across neighboring regions. In this way, wireless bandwidth usage is improved. Protocols for avoiding collisions can be divided into essentially two categories: scheduling protocols and collision-resolution protocols.

Scheduling protocols assign some of the bandwidth to each node (or at least each node that wishes to transmit a message) (Lu, Bharghavan, and Srkant, 1997). Static assignments give a fixed amount of bandwidth to each node without regard to its transmission requirements. Dynamic assignments instead allocate bandwidth to nodes according to their scheduling schemes and their current communication needs. There are different ways this bandwidth can be divided among the nodes. In addition to spatial reuse realized through limiting the transmission range, the bandwidth could be divided by frequency or time. Frequency division medium access (FDMA) assigns one of a number of frequencies to each node so that multiple nodes can transmit simultaneously without interfering with each other (Wei and Cioffi, 2002). Time division multiple access (TDMA) creates multiple time slots and assigns one or more of these time slots to each node so that the same frequency can be shared among multiple nodes without incurring any interference (Sohrabi et al., 2000).

One advantage of TDMA in wireless sensor networks is that a sensor node can turn off its radio transmitter and receiver until it is ready to send or receive a message. Since having the receiver powered up uses energy even when it receives no messages, turning the receiver off for a significant number of communication slots saves a dramatic amount of power. Power usage of various wireless transceivers in transmit, receive, and idle states can be found in Chen, Sivalingam, and Agrawal (1999). However, since wireless sensors will send data only sporadically, the power consumption could be very high for leaving the transceiver in an idle state compared with powering down the transceiver (Rabaey et al., 2000). For this reason, TDMA-type scheduling protocols have been considered the most practical approach for wireless sensor networks (Pottie and Kaiser, 2000). On the other hand, TDMA schedules require sufficient synchronization among nodes to ensure that all nodes conform to the schedule (Shih et al., 2001).

To build a TDMA schedule, nodes must have neighborhood information. Once the number of neighbors has been determined, the number and size of time slots can be determined. For consistency across the network, the size of time slots can be fixed based on the expected (or maximum) packet size. The maximum number of neighbors or the average number of neighbors also can determine the number of time slots, both of which also require topology information. Once the local information is determined, some coordination is required among neighboring regions because the neighbors of a given node vary somewhat from the neighbors of its own neighbors. Coordination across adjacent neighborhoods prevents collisions and allows efficient bandwidth use. Dynamic TDMA schedules (Dyson and Hass, 1999), which allocate slots only to nodes with data to transmit, result in more efficient bandwidth usage because slots are not idle owing to being assigned to nodes with no data to transmit. In addition, nodes with a larger volume of data can obtain more than one slot to transmit data more quickly. Dynamic TDMA schedules are accomplished by having a short reservation period consisting of a number of minislots—each corresponding to one packet transmission time slot. A node indicates its desire to transmit during the current interval by randomly selecting one of these minislots and attempting to reserve it. If a node succeeds in obtaining a particular slot, it sends its packet during that time slot. In this way, nodes without data to send do not waste a preassigned time interval, allowing bandwidth to be used more efficiently.

Another option is to use an approach taken in protocols such as S-MAC (Ye, Heidemann, and Estrin, 2002) and a proposed variant of Aloha that uses preamble sampling (El-Hoiydi, 2002). These low-power MAC protocols have been proposed for use in wireless sensor networks to avoid the energy and system overhead of maintaining synchronization and creating a schedule. Instead, these protocols turn the radio off when idle to conserve power and turn the radio on periodically to listen for transmission requests. When a node wants to send a packet, it transmits a start symbol for a long enough period of time that the receiver wakes up before the transmitting node sends the packet. Although this avoids the overhead of scheduling time slots, there is the possibility that the transmissions wake up many nodes. This increases the power consumption as some nodes that could remain idle assume that the message is destined for them. These nodes can turn themselves off again by processing the packet header and determining that the message is not being sent to them. Even so, this processing overhead drains energy from these sensor nodes.

Besides TDMA and FDMA scheduling, Code Division Multiple Access (CDMA) has been proposed for wireless communication (Hu, 1993). CDMA signals are particularly difficult to detect by unintended receivers, making them ideal for military applications. We do not consider CDMA further here because the processing overhead is significantly higher, which prevents their application in sensor networks given current technology limitations.

The second type of wireless channel scheduling is collision avoidance. An example of collision avoidance is IEEE 802.11 (ANSI, 1999), which can use request-to-send (RTS) packets followed by clear-to-send (CTS) packets to reserve a channel. After the reservation is obtained, the packet can be transmitted. For low-power wireless devices, listening for RTS packets can be a significant energy drain on these devices. In addition, collisions still can occur because of the hidden terminal problem (Tobagi and Kleinrock, 1975), as well as other causes of interference. The hidden terminal problem arises because a node receiving a message from one node also may be in the communication range of another node sending an RTS packet. Interference from these two messages prevents the node from receiving the packet from either node. An example of the hidden terminal problem is shown in Fig. 10.1, where the transmission range of each node is shown as a circle centered at that node. In this figure, node A is sending a message to node B. Node C is out of the range of node A, so node C does not know that node B is currently receiving a data packet from node A. Therefore, node C sends an RTS to node B, interfering with (corrupting) the packet from node A to node B.

A related problem is the exposed terminal problem (Bharghavan et al., 1994), where a node cannot send a message because it hears another node

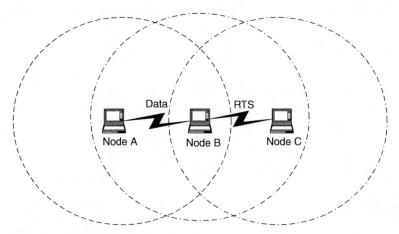

Figure 10.1 Hidden terminal problem.

transmitting, even though the recipients of the two messages would not experience any interference. Although the exposed terminal problem does not cause network interference, it does reduce the efficiency of the wireless channel. Figure 10.2 depicts a simple example of the exposed terminal problem. The network consists of four nodes. Node B is sending a message to node A, but node C is also in the transmission range of node B. Since node C overhears the transmission from node B to node A, node C will not initiate a transmission to node D, even though these two transmissions would not interfere with each other because node A cannot hear node C and node D cannot hear node B.

Scalability concerns also have been raised about collision-avoidance protocols because there is no mechanism for ensuring fairness in packet

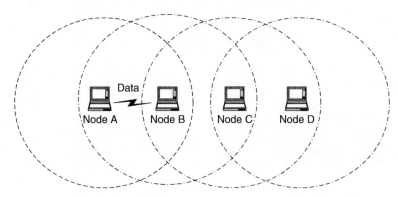

Figure 10.2 Exposed terminal problem.

transmissions (Xu and Saadawi, 2002). This occurs when a small subset of the nodes seize the wireless channel on a frequent basis, preventing other nodes from transmitting data in a timely manner. For wireless sensor nodes, which have limited storage available for buffering packets, a significant delay in transmitting packets could result in dropped packets. For time-critical data, the delay in obtaining the wireless channel also could prevent nodes from meeting their application requirements. For these reasons, collision-avoidance protocols similar to IEEE 802.11 are not well-suited for wireless sensor networks for many applications.

10.1.4 Security protocol configuration

Security protocols for wireless ad hoc networks are required for many applications. Military applications have an obvious need for security, but other types of sensors, such as those carrying medical or personal information, need adequate security as well. In some cases, less obvious needs of security also may exist, such as a desire for added privacy by not disclosing one's identity to surrounding sensor nodes. Data encryption can be used to protect the confidentiality of data. Encryption in wireless networks is discussed in Section 13.2.

Besides the need to encrypt or authenticate data produced by sensors, control information such as routing updates also may require authentication to prevent a malicious node from keeping data from being routed properly in the network (Zhou and Haas, 1999). As discussed earlier, sensor nodes have limited power, computing resources, and storage; symmetric key encryption is a more likely choice for these networks. A discussion of symmetric versus asymmetric key encryption and decryption also is found in Section 13.2.

The different demands for security, such as encryption and authentication, are addressed in Chaps. 12 and 13; in this section we consider the configuration issues for establishing security among various ad hoc wireless nodes. We are interested in the key distribution problem, which deals with providing keys to each of the nodes in the network (Jamshaid and Schwiebert, 2004). Several options exist for key distribution, each with its own advantages and disadvantages.

The simplest key distribution protocol is to preassign keys to each of the sensor nodes. This is an economical and efficient solution when each sensor node needs security only between itself and at most a few base stations. The base station information can be provided when the sensor network is established, and the security information can be embedded into each sensor node prior to deployment. Whether these keys are used for encryption of messages sent to the base station, decryption of messages received from the base station, or authentication of packets, the necessary information is available prior to deploying the network.

However, when encryption or authentication is required among the sensor nodes, preassigning keys is a much less attractive option. First, keys are likely to be required between each pair of neighboring sensor nodes that wish to communicate directly with each other. Depending on the method of deployment, it may be difficult to predetermine which nodes will be neighbors in the assembled network. Even if the sensors are positioned manually, keeping track of the location of each sensor and ensuring that each is placed correctly could be difficult. Determining later that the protocol is not functioning properly because a sensor node was placed in the wrong location, identifying which sensor should be moved, and repositioning this sensor would be a time-consuming and labor-intensive process. If an automated deployment strategy is used, such as dropping a large number of sensors from an airplane, repositioning the sensors is quite complicated. In addition, as sensor nodes exhaust their power or otherwise fail, neighbors of a sensor could change, which would have to be anticipated in advance in order to preassign keys to the sensor nodes (Jamshaid and Schwiebert, 2004). Furthermore, later deployment of additional sensors will be difficult to integrate into the existing framework of remaining sensors.

For applications in which secure communication between sensor nodes is required, predeployed keys are not practical. Since control information, such as routing, likely requires authentication for networks with a reasonable level of security, most large wireless sensor networks require mechanisms for authenticating neighboring sensors. For this reason, other key establishment and key distribution protocols are required.

As of the writing of this book, key distribution in wireless sensor networks is a topic that has attracted limited research interest; no general solution has been proposed. It seems reasonable to include the participation of the base station or some other more powerful node in establishing and distributing keys. The base station also can perform some initial authentication to verify that particular nodes are not impostors but rather legitimate members of the wireless ad hoc network. However, scalable, secure, and robust protocols for achieving key distribution have yet to appear. As wireless sensor networks become more widespread, it is reasonable to assume that such protocols will be designed.

10.2 Energy-Efficient Communication

As mentioned earlier, wireless communication is the largest single consumer of power in an ad hoc node (Pottie and Kaiser, 2000). Controlling the communication cost is fundamental to achieving a long operating lifetime for the wireless node. Hardware designs that lead to more efficient radio transceiver designs help to some extent. The need to transmit with sufficient power, as well as to receive the signal and distinguish the signal from noise, limits the impact hardware designs alone can make on the energy overhead of communication. Even though future hardware designs may reduce the power drain of the radio significantly, complementary software protocols are required to maximize the energy efficiency of the communication. In this section we look at a number of protocols that have been proposed for reducing the power consumed by the radio. As we discuss each protocol, it becomes clear that some of these solutions are orthogonal to others, allowing them to be combined to achieve still further efficiencies than using either in isolation.

Although this section focuses on reducing energy consumption of wireless radio communications, other mechanisms for wireless communication have been proposed, some of which consume orders of magnitude less power than radio communications. For example, as part of the smart dust project (Warneke et al., 2001), the researchers have proposed using a laser beam, reflected by the sensor nodes using a novel mirror design, to transmit a binary signal. This is an intriguing idea, although not suitable for all wireless communication because it requires clear line of sight as well as pinpoint accuracy in aiming the laser. However, for some applications, this is an excellent method, which also suggests that other novel low-power approaches to energy-efficient communication may yet be proposed.

10.2.1 Multihop routing

Energy consumed to transmit a packet on a wireless channel increases substantially with the distance a packet is sent (Shih et al., 2001). Ensuring that a packet is transmitted with a sufficient signal-to-noise (SNR) ratio requires that the energy used for communication increases at a rate of at least the square of the distance between the sender and the receiver. For sensor nodes placed close to the ground, the energy requirements increase with the distance raised to the fourth power owing to ground reflections that occur with the small antenna heights used for wireless sensors (Pottie and Kaiser, 2000). With polynomials of this order, even a modest increase in distance leads to a significant increase in the power expended for communication. Instead of transmitting each packet directly from the sensor node to the base station, nodes can send messages on a hop-by-hop basis. In many cases, hop-by-hop transmissions reduce the total energy consumption.

Multihop communication also has the advantage of reducing interference in the network. When a wireless node sends a message, no other node within the transmission range can receive another packet. For this reason, reducing the signal strength of the transmitter decreases the transmission range, which increases the number of sensors that can transmit at the same time without interfering with each other. This is known as *spatial reuse of the bandwidth*. Nodes at different locations can transmit at the same time, in the same way that nodes in the same region can transmit at the same time using different frequencies. Thus multihop communication increases both concurrency and overall network bandwidth. On the other hand, using multihop communication means that, on average, more transmissions must be performed to deliver a packet from the source to the destination. Consequently, the benefits of spatial reuse must exceed the additional transmissions required to deliver the average packet. If spatial reuse doubles the number of simultaneous transmissions possible and the destination is twice the previous distance, then spatial reuse through multihop communication is not effective. Fortunately, this situation is unlikely to occur because spatial reuse increases the diameter of the network linearly, whereas there is a quadratic increase in bandwidth reuse. The reason for this discrepancy is that the diameter of the transmission determines the distance, which in turn determines the number of hops needed to reach the destination, whereas the amount of interference caused by a transmission depends on the circle formed by the transmission radius, which determines the level of potential spatial reuse. Figure 10.3 depicts how spatial reuse in a wireless sensor network allows multiple sensors to transmit a message at the same time without interference. Because the transmission ranges of nodes A and B do not intersect, the neighbors of these two nodes can receive their respective transmissions at the same time without interference.

To provide multihop communication, coordination between the sender and receiver is accomplished along each hop in the end-to-end transmission. This requires that the transceiver be active at both nodes and that other nodes within the transmission range not use the wireless channel. In addition, multihop transmission introduces some delay into the network, so depending on the application requirements, the number of hops may need to be controlled to give a timely response. Coordinating schedules adds some overhead among neighboring nodes to ensure successful transmission without interfering with other current or pending transmissions. For a large sensor network, however, this overhead is trivial compared with the overhead in scheduling direct transmissions to a base station or some other centralized sink. Tight time synchronization may be needed among sensors to avoid collisions. Since the cost of keeping sensor nodes synchronized is relatively high, maintaining a schedule is more difficult than simply creating the schedule.

In reality, the energy required to transmit a packet depends on some constant cost to power the transmitter circuitry plus an additional factor based on the packet size and distance. If the distance is small, the energy to power up the electronics may be the predominant factor (Shih et al., 2001). For this reason, dividing a dense network into arbitrarily

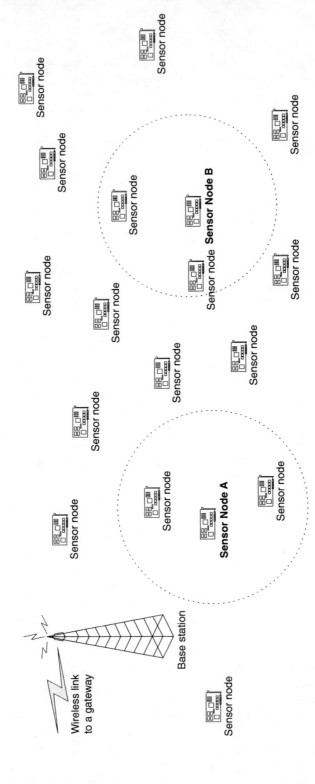

Figure 10.3 Illustration of spatial reuse.

small hops is likely to consume more power than using fewer longer hops once the distance between hops becomes small enough. Depending on the energy-consumption characteristics of the transceiver circuitry, an ideal distance and number of hops can be determined. Even when an ideal distance from the standpoint of energy consumption has been determined, shorter hops may be chosen for two reasons. First, spatial reuse of bandwidth improves with smaller transmission distances, so shorter hops may be selected to increase the network bandwidth effectively. Second, using a larger transmission distance tends to increase the number of neighbors each node has. This leads to an increase in the amount of information maintained by each node, as well as additional overhead in scheduling the channel. For both of these reasons, the best choice for hop distance may not match the ideal distance as computed by the energy consumption of the transceivers.

10.2.2 Communication scheduling

As discussed in the preceding section, scheduling wireless communication among sensor nodes offers a number of opportunities for saving power. Scheduling nodes to operate on different frequencies or at different times reduces the number of collisions. Collisions consume additional energy because transmitted packets are not received correctly and hence need retransmissions. Because wireless bandwidth is limited, retransmitting packets can reduce the available bandwidth to less than the amount required for other pending messages. Collisions that lead to retransmissions also delay the delivery of packets, which may be a drawback for time-sensitive or real-time applications.

TDMA allows nodes to power down the transceiver when no packet is transmitted or received by this node. Of course, such scheduling relies on synchronization among the sensor nodes, which adds some communication overhead (Elson and Estrin , 2001). Even if tight synchronization is required, energy savings from communication scheduling must exceed the overhead of maintaining the schedule.

Table 10.1 presents the power usage of two wireless radios employed on typical mobile devices. As the table shows, the wireless radio uses nearly as much power to receive a message as to transmit a message.

TABLE 10.1   Average Current for Two IEEE 802.11b Cards

Chipset	Sleep (mA)	Idle (mA)	Receive (mA)	Transmit (mA)
ORiNOCO PC Gold	12	161	190	280
Cisco AIR-PCM350	9	216	260	375

SOURCE: Shih et al., 2001.

Reservation
slots

Figure 10.4 Static TDMA schedule.

In addition, a radio uses about the same amount of power whether currently receiving a message or idle waiting for a message to arrive. Significant energy savings are realized only when the radio is turned off or put in reduced energy state such as a sleep mode.

TDMA works by preceding each set of time slots with a number of minislots, each long enough for a request to be made to transmit a packet in the corresponding frame. Each sensor node listens during this slot-reservation window to see if any packets are destined for it. The sensor node also randomly selects one of the minislots and attempts to reserve the corresponding slot whenever it has a packet to transmit. In general, the rate of packets produced by a sensor node may be quite modest, so a fixed schedule may be suitable. For example, if there are 10 wireless nodes in the same region, each node could be assigned permanently one of 10 slots. Figure 10.4 shows how the allocation could be organized in such a case, with a short reservation period followed by each sensor receiving one of the longer 500-ms transmission time slots. The nodes then simply listen during the reservation period to determine if another sensor node wishes to send a packet to this sensor during the corresponding time slot. Based on this information, each sensor node can determine when to turn its wireless radio on and when to turn it off. This minimizes the time the wireless radio consumes power, leading to increased energy efficiency.

10.2.3   Duplicate message suppression

To further reduce power consumption for communication, transmitters can send fewer packets. One way of accomplishing this is for sensor nodes to transmit less frequently and thereby compress multiple sensor readings into a single packet. When sensor readings do not change, there may be no need to report the same information repeatedly. Periodic transmission may be sufficient to inform neighboring sensors, as well as the base station, that the node is still operational and that its readings have not changed. Intervals for communication are based on the application requirements, but it is reasonable to expect that this relatively straightforward energy savings can be realized for most applications.

Besides suppressing duplicate data from a single sensor, multiple sensors with identical readings can combine their data into a single message

(Intanagonwiwat, Govindan, and Estrin, 2000; Krishnamachari, Estrin, and Wicker, 2002). Since multihop communication will be used for most large sensor networks, local communication among neighboring sensors to achieve a consensus, followed by the transmission of a single combined packet over a number of hops, leads to a large energy savings for sensors that normally function as routers on the path to the base station.

Nodes can combine duplicate messages using a number of approaches. The first is simply to have neighboring nodes communicate with each other to determine if other sensors also have detected the same event. In the likely event that a similar reading is obtained by neighboring sensors, these readings could be combined into a single message that is forwarded to the base station. The result is that a single packet is forwarded from a local set of nodes to the base station (Kumar, 2003).

The second approach is to perform duplicate message detection along the path from the sensor nodes to the base station (Intanagonwiwat, Govindan, and Estrin, 2000). Returning a large number of redundant data to the base station, known as the *broadcast storm problem,* can overwhelm the base station (Tseng et al., 2002). Instead, the base station receives only a single message, which reduces the overhead for processing at the base station, as well as avoiding the flooding of the base station with a large number of duplicate packets. Besides preventing the base station from being overwhelmed with an excessive number of messages that provide no additional useful information, the sensor nodes close to the base station also receive less traffic, thus extending their lifetime.

Suppressing duplicate packets enroute to the base station requires handling two problems. The first is detection of duplicate packets. One typical mechanism for correctly detecting duplicate packets is to buffer or cache packets for a short period of time and then to compare each incoming packet with this buffered information. Packets with duplicate information are discarded. Another option is to retain packets for some period of time before transmitting them along the next hop. The receiving node compares additional packets received during this period of time with this packet, and only one matching packet needs to be forwarded. One advantage of this technique is that a counter could be attached with the packet that is ultimately sent, allowing the base station as well as intermediate nodes to determine the number of sensor nodes that detected the same event. This may be useful in determining the validity of the information, provided that the counter is assigned and incremented properly. If multiple sensors detect the same event, the level of confidence that the event occurred can be increased. If only a single node detects this event, the level of confidence is lower because a faulty sensor or sensor reading could be responsible for generating the event.

The second requirement for duplicate packet detection is to determine if two packets actually represent the same event. For example, consider the situation in which the sensor application is tracking animals in some region of interest. If it detects two animals in different areas, these two events should not be combined into a single event but rather should be considered separate events. Likewise, events that occur at slightly different times should not be regarded as the same event but should be recognized as distinct events. Consequently, determination of duplicate events means that the system must have some synchronization to detect events that occur at the same time (Elson and Estrin, 2001), as well as location information to determine where the events occurred. Without this information, duplicate packet detection may not obtain sufficient accuracy to make duplicate packet suppression a feasible option. However, if these problems are resolved with sufficient accuracy, the power savings could be significant, especially for intermediate sensor nodes responsible for forwarding data to the base station. On the other hand, redundant data sent to the base station could be critical data for certain applications, such as military target tracking.

10.2.4 Message aggregation

Duplicate message suppression is one method for reducing the volume of information that is transmitted through the sensor network. This is essentially loss-free data compression, especially if a counter is included that indicates how many sensors generated matching packets.

Message aggregation is often a lossy data-compression technique. Essentially, readings from multiple sensors are combined into a single packet (Heinzelman, Chandrakasan, and Balakrishnan, 2002; Lindsey, Raghavendra, and Sivalingam, 2002). In some cases, these sensors have produced the same data, but in other cases, the results may not be the same. Combining messages that are the same is equivalent to duplicate message suppression, but some messages with different information also could be combined. This often leads to the loss of some information.

Modest message aggregation results in combining the data from multiple packets. This leads to a modest improvement in total data transmission because only a single message header is required. The disadvantages of this approach are that variable packet sizes are more difficult to buffer and process, and the probability of losing a packet owing to communication errors increases with packet size. Although methods exist for resolving this problem, the overhead is likely to be greater than the savings obtained from removing a few headers.

More aggressive message aggregation, such as PEGASIS (Lindsey, Raghavendra, and Sivalingam, 2002), compresses data from multiple sensors into something that extracts the most relevant information from the data without transmitting all the raw data to the base station. For example, the readings of multiple sensors could be combined into some general statistical information, such as the count, average, standard deviation, minimum, and maximum readings. The base station then could use this composite information to reach a decision about whether or not the sensor readings are valid. One method for accomplishing this aggregation is to perform local processing of the data and then to send the result to the base station. In essence, sensors trade processing power for transmission power. If raw sensor readings were transmitted over several sensors to reach the destination, the energy consumption could be significant. A dramatic reduction in the number of messages or the size of messages saves enough energy to compensate for the additional overhead of processing the packets locally. The required processing must be modest because the storage space and processing capabilities on a sensor node are quite limited. It should be possible to determine the ratio of data compression and energy consumption of the processing performed at the sensor node for different applications. Based on this evaluation, one can determine whether or not an overall energy savings will be achieved.

Since processing may be done at a single sensor node and the resulting energy savings are spread across many sensor nodes, the network may achieve a total energy savings even if an individual sensor node has a net loss of energy from performing data aggregation. In this case, a sensible approach is to rotate the data-aggregation responsibility among sensor nodes so that the power of one particular sensor node is not depleted rapidly (Heinzelman, Chandrakasan, and Balakrishnan, 2002; Lindsey, Raghavendra, and Sivalingam, 2002). Some sensor nodes also may be positioned in more effective locations for accomplishing this aggregation. For example, in a randomly distributed sensor network, some sensors have more neighbors than others. Sensors with more neighbors are better situated for receiving data from surrounding sensors and aggregating these packets into a single packet. Furthermore, sensor data readings often are event-triggered, so the location of the event determines which sensors may participate in the message-aggregation step. For example, an animal-tracking application generates sensor readings where the animals are located, which probably is not equally distributed across the network. Opportunities for data aggregation are best at the source of the sensor readings, so sensors in the neighborhood of the event should perform the message aggregation. DFuse presents distributed algorithms and techniques for aggregating messages (Kumar et al., 2003). These realities limit the effectiveness of distributing the processing among remote sensor nodes that are not associated or responsible for the event.

A second method of performing data aggregation is to perform incremental aggregation as the data move through the network (Intanagonwiwat, Govindan, and Estrin, 2000). In this case, messages are combined along the path from the sensor nodes to the base station. As with duplicate-packet suppression, the feasibility of this approach depends on the time interval in which the packets arrive, as well as the ability of the algorithm to distinguish unrelated sensor readings from sensor readings that can be combined. The extent to which distinct readings must be differentiated, as well as the mechanisms for aggregating multiple readings, is application-dependent. This helps to determine when this approach is suitable for a sensor network task.

A third method proposed for aggregating messages is to employ a clustering approach. A specific sensor is designated as the cluster leader or cluster head, and this node is responsible for combining readings from sensors in the cluster (Heinzelman, Chandrakasan, and Balakrishnan, 2002). We discuss this concept of clustering later in this chapter.

In summary, there are multiple methods of aggregating messages. Each has its own unique advantages and disadvantages. For some applications, data aggregation may not be desirable because the loss of any information may hamper performing the sensing task correctly. It also may be the case that the data processing is too complex to be performed by a sensor node. For many applications, however, aggregating packets can result in significant energy savings.

10.2.5 Dual-radio scheduling

As discussed earlier, a radio consumes similar amounts of power whether transmitting, receiving, or idle. In order to transmit messages across a sensor network, the receiver needs to be on while the transmitter sends the message. If there were a way to determine when a transmitter was ready to send a message, the radio could be turned off or placed in an ultra-low-power deep-sleep mode until some other sensor node is ready to send a packet to this sensor. Unfortunately, the only method of communicating between most sensors is by transmitting a message between them. This presents a dilemma—a message must be transmitted to indicate that communication is desired, but a message cannot be received if the radio is not in the receive mode. Although this problem may appear to be insurmountable, two different solutions have been proposed.

The first is to equip a sensor with two radios, each of which operates on a different frequency (Shih et al., 2001). The data channel radio provides a wireless channel for communicating data among sensor nodes. The other radio, which uses a control channel, is an extremely low-power radio used to send control information to wake up the receiver. The protocol works by transmitting a brief message from the transmitter to the receiver using the control channel radio. Receipt of this message causes the receiver to power up its data channel radio. An acknowledgment could be returned over the control channel to inform the transmitter that the message has been received or to indicate that the receiver has powered up its radio. Once the transmitting sensor is reasonably sure that the receiver's radio is turned on, the packet can be transmitted. Since the control channel radio consumes much less power than the data channel radio, the control channel radio can be left on constantly. Leaving the data channel powered down until needed can realize a significant power savings. However, this approach introduces some delay into the transmission process. Therefore, for large networks with time-sensitive data, this method may not be suitable. In addition, the added cost and complexity of providing two radios on a single sensor instead of one radio may be impractical for low-cost sensor applications. For most types of sensors and sensor applications, however, this approach could result in significant energy savings, provided that an extremely low-power control channel radio can be built at a cost that does not exceed the sensor's cost target.

A second approach to solving this problem is to use two identical radios that operate on different frequencies (Singh and Ragavendra, 1998). This has the advantage of reducing engineering costs because a single radio design is needed. The control channel radio consumes the same amount of power as the data channel, so leaving the control channel radio on all the time does not solve the problem. Instead, the control channel radio is turned on periodically to see if a message needs to be transmitted to this sensor on the data channel. The control channel stays on for a brief period of time to listen for requests. On receiving such a request, the data channel is turned on, and the sensor node sends an acknowledgment to indicate that it is ready to receive the message. When a node wishes to transmit a packet, the control channel sends a signal on a continuous basis until enough time has passed for the receiver to wake up. Assuming that each sensor node adheres to this frequency of turning on the control channel radio, both sensor nodes eventually synchronize, and data transmission occurs. Two different frequencies are used so that sensor nodes using the control channel do not interfere with current data transmissions. This also makes it much easier for a sensor node listening on the control channel to determine that a node is making a request to send a packet. The chief drawback to this approach is the added delay that occurs in the network. If a packet must be transmitted over many hops, there could be considerable delay in sending a packet from the source to the destination. For small to medium-sized sensor networks, however, the hop-by-hop delay in synchronizing the two sensors does not typically lead to a significant delay.

10.2.6 Sleep-mode scheduling

The density of a network depends on the number of nodes in the region occupied by the sensor nodes, as well as the transmission range of the sensors. As the transmission range of a sensor increases, the number of direct neighbors increases. This effectively increases the density of the network. Similarly, distributing more sensors in the network increases the average number of neighbors, which also increases network density. With a random deployment, the density of the network varies across the region of deployment. Some areas will have more sensors, and some will have fewer. As nodes die or additional nodes are added to the sensor network, the density also changes. With mobile nodes, the density in a particular section of the network is not constant but varies according to the relative movement of sensors into and out of each area.

When sensor networks are dense, all nodes along a path do not need their radios on in order for packets to be transmitted from the source to the destination. Because the network is dense, multiple sensors could receive the same packet and forward this packet along the path to its destination. Energy can be saved in sufficiently dense networks by having only some radios turned on (Chen et al., 2001). Through local coordination among sensors in a neighborhood, routing connectivity with the base station or elsewhere can be maintained using a subset of these nodes. To extend the lifetime of individual sensors, nodes can be scheduled to take turns serving as routers so that power is not depleted at specific nodes while other nodes retain a large quantity of power. Typically, the selection of a specific sensor as the current router depends on the energy of the surrounding sensors relative to the power remaining at this sensor, along with the location of each sensor to ensure connectivity.

10.2.7 Clustering

Clustering is a hierarchical approach to support routing and data aggregation in an ad hoc wireless network (Steenstrup, Beranek, and Newman, 2000). The wireless network is divided into regions, in which each region forms a cluster with a single node, the cluster head, designated as the leader of the cluster. Figure 10.5 shows an example of a sensor network divided into two clusters. A dotted line shows the division between the two clusters. Nodes in each cluster send a message to the cluster head, which then combines these data and transfers the information directly to the base station.

In a larger sensor network, cluster heads may communicate on a hop-by-hop basis with the base station, but in a small sensor network such as the one shown in Fig. 10.5, direct communication between the base station and the cluster head may be reasonable. A number of algorithms have been proposed for clustering sensors, but we focus first on the

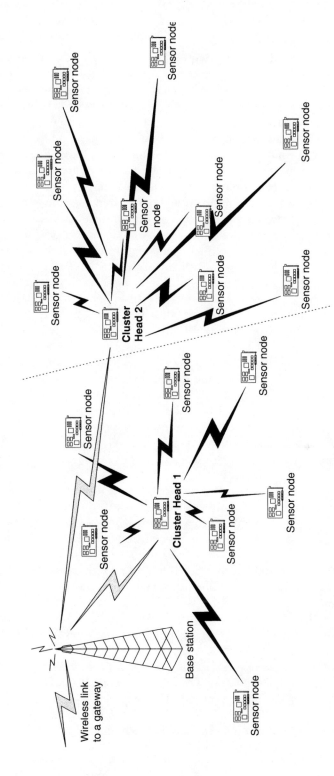

Figure 10.5 Cluster heads in a clustered wireless sensor network.

basic ideas of clustering. Later we take a closer look at techniques for building clusters.

The cluster head is responsible for coordinating activities among the sensors within the cluster. For example, the cluster head is responsible for routing packets from among nodes in its cluster or to nodes outside the cluster. This function also puts the cluster head in an ideal position for performing duplicate packet suppression or message aggregation. Since each event that occurs is likely to be sensed by neighboring sensors and relayed to the cluster head at approximately the same time, the cluster head can readily determine if more than one sensor observed the same event. It also may be able to combine multiple sensor readings into a single aggregate packet that it relays to the base station.

When events occur at the boundary of two or more clusters, each cluster head is unable to suppress duplicate information from all sensors observing the event and may have difficulty aggregating data because only a fraction of the sensor readings are from sensors in this cluster (Kochhal, Schwiebert, and Gupta, 2003). Thus clustering is an attractive option for squelching redundant data or compressing similar sensor readings, but it may not provide a complete solution. Cooperation among cluster heads of adjacent clusters may be necessary to maximize data aggregation or duplicate message detection. For this reason, the clusters may need to be chosen carefully to maximize the usefulness of clustering. In addition, care must be taken in clustering the ad hoc nodes to ensure that there are not a large number of orphaned nodes that subsequently must merge into relatively small clusters. A large number of small clusters means more nodes than necessary serve as cluster heads and expend additional energy.

A better approach divides the network into a number of equal clusters, in which each cluster contains roughly the same number of nodes. However, this could be difficult to achieve without a great deal of overhead. Depending on the amount of information available about the network topology, determining the ideal number of clusters also may be difficult. In practice, the best clustering for a given network topology is difficult to determine because an optimal clustering for random graphs, as well as many regular graphs, is an NP-complete problem (Garey and Johnson, 1979). In addition, mobility adds further complication to this process because a good clustering may become very inferior later. Clustering is likely to be a relatively expensive operation in terms of not only the energy consumed in clustering the sensors but also the time involved in building the cluster. Thus frequent creation of new clusters is an unattractive option that should be avoided if possible.

Clustering can be based on a simple two-level hierarchy, or cluster heads can be combined into metaclusters leading to a hierarchy of levels that culminate in a single node or a few nodes at the highest level.

Building a multilevel clustering hierarchy may be advantageous when routing information travels across multiple clusters. Nodes on the border between two adjacent clusters can serve as gateways for forwarding data from one cluster head to another. In essence, cluster heads form a routing backbone (Chen et al., 2001) for transmitting information among clusters, as well as providing a scalable mechanism for transmitting sensor readings from individual sensors to a base station.

10.3 Mobility Requirements

Mobility of the ad hoc wireless network nodes introduces additional complexity into the protocol designs. Decisions that are optimal at one point in time may be significantly suboptimal in the future because nodes have moved to new locations. In addition to the resulting loss of quality, routing protocols need to be modified to work properly despite mobility. Otherwise, messages sent between a source and a destination may not be delivered successfully. Similarly, clustering nodes into a hierarchy is complicated by node mobility, especially if the cluster heads move beyond the range of other nodes in the cluster. This could require frequent selection of new cluster heads and may make clustering of nodes impractical when high rates of mobility occur.

10.3.1 Movement detection

In order for a protocol to address mobility, there must be some mechanism for detecting mobility. Besides determining if a node is mobile, relative mobility is also important. For instance, if all the nodes in the ad hoc network move together, the mobility is not an issue. Consider a number of ad hoc device users seated on a train. All these devices may be moving at a high rate of speed, but as long as the passengers remain seated and the devices stay with these passengers, the ad hoc network does not detect any movement. Of course, connections with a base station not located on the train would be subject to the effects of mobility, but within the ad hoc network, no node mobility is detected.

As this example indicates, detecting mobility is more complicated than determining that a node has changed location. Use of the Global Positioning System (GPS) is sufficient to determine that a node is moving, but the relative movement of nodes is required to make mobility a factor in the protocols. In other words, detection of movement is a necessary condition for ad hoc network node movement but not a sufficient condition. Instead, nodes need some mechanism for detecting that they are moving toward each other or away from each other.

One technique for measuring relative movement of nodes is the received signal strength indicator (RSSI) (Niculescu and Nath, 2001). Measuring the received signal and monitoring the changes in the signal strength determine the RSSI. Ideally, the value of the RSSI is proportional to the distance between the sender and receiver. Hence the signal strength increases as a node moves toward the sender and decreases as a node moves away from the receiver. An advantage of this approach is that any node in the transmission range can determine its relative position from the sender, so nodes other than the receiver can update their positions. The primary disadvantage of using RSSI to determine locations is the weak correlation between the RSSI and the actual location. Multipath interference, as well as obstacles that block or absorb part of the signal strength, can give an inaccurate estimate of location (Patwari and Hero, 2003). Although RSSI may not give a good location, it gives a useful indication on the quality of a current connection between two nodes. This information may be of more practical value than the relative location of the node. For example, if one node is moving to a position where there is an obstacle between the two nodes, the relative distance is less important than the inability to communicate between the two nodes.

10.3.2   Patterns of movement

In an ad hoc mobile network, nodes move in an arbitrary pattern depending on the movement of the user or object that carries the node. For this reason, it is difficult to determine what mobility patterns arise in practice. Determination of this pattern is valuable so that reasonable simulation studies of wireless protocols can be performed. It is reasonable to expect that the performance of various algorithms for ad hoc networks will differ depending on the mobility pattern. A common movement model is the random waypoint model (Bettstetter, Hartenstein, and Perez-Costa, 2002). In this model, nodes are stationary for a length of time randomly chosen from some time distribution. Nodes then move in a randomly chosen direction for a random distance. On moving this distance, nodes then pause again for a random length of time. The random waypoint model has been used in many different papers with slight variations. Although easy to use, a random movement model may not reflect realistic movement patterns of mobile nodes in typical ad hoc networks. Movement patterns based on real user mobility patterns allow more accurate analysis of protocols.

Certain protocols may work well for random movement patterns, but real users may not follow a random movement pattern. For example, there may be locations where users tend to congregate, such as a coffeeshop. Users may move to the coffeeshop and away from the coffeeshop but not to random buildings near the coffeeshop. Similarly, within the coffeeshop, users may move to specific locations, such as counters or printers, more frequently than random locations throughout the shop.

Now that experimental mobile wireless networks have been deployed in campus settings and other reasonably large configurations, some measurements of real-world mobility patterns can be obtained (Kotz and Essien, 2002). Whether or not mobile ad hoc networks exhibit similar mobility patterns is not clear, but these studies provide data that are helpful in developing simulation models for studying mobile ad hoc wireless networks (Jardosh et al., 2003).

10.3.3 Changing group dynamics

Mobility breaks communication links between neighboring nodes. In addition, any grouping of sensor nodes also can change because of mobility. For instance, clustering usually strives to build clusters where members of the cluster are relatively close to the cluster head and to distribute the cluster heads so that they are geographically dispersed. When we introduce mobility into the equation, these choices lead to temporary solutions. Even if a good clustering is made initially, the quality of the clustering can degrade after an extended period of mobility. Of course, there is also the possibility that clustering improves after mobility, but protocols still need to handle any potential degradation introduced by mobility.

Changing group dynamics can cause problems for routing messages because delivery of a message relies on the existence of a path from the source to the destination. The routing path may require rebuilding if the source, destination, or any node along the path moves outside the range of any neighboring node on the routing path. On a more positive side, however, it has been shown that the capacity of a wireless network can be improved by mobility because nodes theoretically could buffer packets until the source and destination nodes move close to each other (Grossglauser and Tse, 2002). This introduces significant latency into the network, which requires a correspondingly large amount of buffer space. On the other hand, this result does illustrate the idea that mobility in an ad hoc wireless network can be exploited to improve performance in some cases.

Group communication often relies on the construction of a logical tree on the underlying network links, which means that specific nodes on the tree have the responsibility of forwarding packets from their parent to their immediate children or leaf nodes. When a node moves, restoring communication so that paths exist between the multicast source and the leaf nodes becomes quite challenging. For example, in Fig. 10.6a, one node in the multicast tree, node B, has three children that must receive the packets, and node B moves out of range of its parent; then either a node that can transmit to all three children must be found, or more than one parent node must be found to replace node B. In many cases,

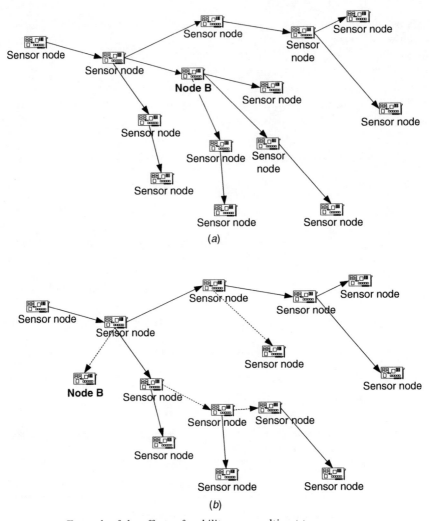

Figure 10.6 Example of the effects of mobility on a multicast tree.

more than one node is required to replace the forwarding previously performed by node B. This situation is shown in Fig. 10.6b, in which the three children that are orphaned each need to find a new parent. Node B is able to keep the same parent, but this is not true in all cases. This leads to additional bandwidth contention, as well as increasing the chance of future disconnections, because more nodes perform the forwarding functions.

Mobility also may cause one of the leaf nodes to move beyond the transmission range of the other nodes. Thus the network may become disconnected. If this disconnection is temporary, it should be possible to buffer packets for a short period of time and deliver them when this node is no longer disconnected. If the multicast group requires ordering that needs an acknowledgment from each intended recipient, the progress of the multicast group will be delayed until the separated leaf node is reconnected to the multicast tree. Other options are to remove a node when it becomes disconnected or to relax the consistency model so that progress can be made even when one or more group members are disconnected.

10.3.4   Resynchronization

For group communication, all members of the group may need to maintain consistent information. In other words, group members must receive information generated by each member in order to have the same view of the system state. A database application with data distributed among the various mobile ad hoc nodes must ensure that each node receives all transactions. When a mobile node moves away from the rest of the network, notification of these transactions is not possible until this node moves back into range. Once this node returns to the network, the database contents must be synchronized with the changes that have occurred while this node was separated from the other nodes.

During this time, the separated mobile node also may wish to perform its own transactions on the data. If these transactions are not prevented because of being disconnected from the network, then these changes are made only to the local copy of the data. Once this node reconnects with the network, changes to the database made by this node need to be synchronized with any changes that have been made to these data by other nodes in the group. This synchronization can be complicated by the fact that when this node reconnects with the rest of the network, other nodes have become disconnected. Depending on the size of the group and the mobility patterns of the group members, the frequency with which all the group members are connected to the network simultaneously may be quite small. Most of the time, at least one group member may be separated from the network. In this case, synchronizing the data among the group members may be quite difficult because only partial information about the updates is available at any given point in time. In addition, a lengthy period of time may have passed between when a node separated from the network and when the same node rejoins the network. This could require buffering a large number of transactions so that updates can be made. In cases in which updates are inconsistent, some decision must be made on how to resolve this.

There are standard techniques from distributed databases that can be applied to this problem (Bernstein and Goodman, 1981), including conservative schemes that do not allow disconnected nodes to make updates and optimistic schemes that do allow updates. In the case of conservative schemes, data inconsistencies are not possible, but nodes may be prevented from modifying data for long periods of time, thus decreasing the efficiency and usability of an application drastically. Optimistic approaches allow transactions to be performed even when the group members are disconnected from each other. This improves the efficiency of the application but requires that conflicting updates be reconciled. In some cases, this can be automated using certain rules. In other cases, human involvement may be needed to resolve these inconsistencies.

Although this problem has been studied for distributed databases, the limited resources available to the mobile nodes make certain options impractical, such as buffering of a significant number of changes. Less resource-intensive solutions are required for mobile ad hoc nodes. Even though such complications can arise without mobility, mobility can lead to extended periods of time when communication among group members is not possible. Unreliable communications and node faults along with the inherent problems associated with node mobility make these problems more challenging and complicated.

10.4   Summary

Most of these topics we examined in Chap. 10 remain areas of active research. The main reason is that the characteristics and applications of wireless sensor networks are changing so rapidly that existing techniques may not be ideal or even suitable for future networks. However, certain system requirements, such as scalability, will remain. For this reason, many of the core strategies seem to have been identified. In Chap. 10, we took a careful look at a number of protocols that have been proposed. Knowledge of these protocols provides an excellent introduction for the protocols we examine in Chap. 11.

10.5   References

ANSI/IEEE Standard 802.11, *Wireless LAN Medium Access Control (MAC) Sublayer*, 1999, and ISO/IEC 8802-11:1999, *Physical Layer Specifications*, 1999.

Bernstein, P. A., and N. Goodman, "Concurrency Control in Distributed Database Systems," *ACM Computing Surveys* 13(2):185, 1981.

Bettstetter, C., H. Hartenstein, and X. Perez-Costa, "Stochastic Properties of the Random Waypoint Mobility Model: Epoch Length, Direction Distribution, and Cell Change Rate," in *International Workshop on Modeling Analysis and Simulation of Wireless and Mobile Systems*; Atlanta, GA, ACM, 2002, p. 7.

Bharghavan, V., A. Demers, S. Shenker, and L. Zhang, "MACAW: A Media Access Protocol for Wireless LANs," in *Proceedings of the Conference on Applications, Technologies, Architectures, and Protocols for Computer Communication (SIGCOMM)*; London, UK, ACM. 1994, p. 212.

Chen, B., K. Jamieson, H. Balakrishnan, and R. Morris, "Span: An Energy Efficient Coordination Algorithm for Topology Maintenance in Ad Hoc Wireless Networks," in *International Conference on Mobile Computing and Networking (MobiCom)*; Rome, Italy, ACM, 2001, p. 221.

Chen, J.-C., K. Sivalingam, and P. Agrawal, "Performance Comparison of Battery Power Consumption in Wireless Multiple Access Protocols," *Wireless Networks* 5(6):445, 1999.

Conta, A., and S. Deering, *Internet Control Message Protocol (ICMPv6) for the Internet Protocol Version 6 (IPv6) Specification.* Internet Engineering Task Force, RFC 2463, December 1998.

Deering, S., and R. Hinden, *Internet Protocol, Version 6 (IPv6) Specification.* Internet Engineering Task Force, RFC 2460, December 1998.

Dyson, D., and Z. Hass, "A Dynamic Packet Reservation Multiple Access Scheme for Wireless ATM," *Mobile Networks and Applications* 4(2):87, 1999.

El-Hoiydi, A., "Aloha with Preamble Sampling for Sporadic Traffic in Ad Hoc Wireless Sensor Networks," in *Proceedings of IEEE International Conference on Communications*, Vol. 5; New York, NY, IEEE Communications Society, 2002, p. 3418.

Elson, J., and D. Estrin, "Time Synchronization for Wireless Sensor Networks," in *International Parallel and Distributed Processing Systems (IPDPS) Workshop on Parallel and Distributed Computing Issues in Wireless Networks and Mobile Computing*; San Francisco, CA, IEEE Computer Society, 2001, p. 1965.

Estrin, D., R. Govindan, J. Heidemann, and S. Kumar, "Next Generation Challenges: Scalable Coordination in Sensor Networks," in *International Conference on Mobile Computing and Networking (MobiCom)*; Seattle, WA, ACM, 1999, p. 263.

Garey, M., and D. Johnson, *Computers and Intractability: A Guide to the Theory of NP-Completeness.* New York: W. H. Freeman, 1979.

Grossglauser, M., and D. Tse, "Mobility Increases the Capacity of Ad Hoc Wireless Networks," *IEEE/ACM Transactions on Networking* 10(4):477, 2002.

Heinzelman, W., A. Chandrakasan, and H. Balakrishnan, "Energy-Efficient Communication Protocol for Wireless Microsensor Networks," *IEEE Transactions on Wireless Communication* 1(4):660, 2002.

Hu, L., "Distributed Code Assignments for CDMA Packet Radio Networks," *IEEE/ACM Transactions on Networking* 1(6)668, 1993.

Intanagonwiwat, C., R. Govindan, and D. Estrin, "Directed Diffusion: A Scalable and Robust Communication Paradigm for Sensor Networks," in *International Conference on Mobile Computing and Networking (MobiCom)*; Boston, MA, ACM, 2000, p. 56.

Jamshaid, K., and L. Schwiebert, "SEKEN (Secure and Efficient Key Exchange for Sensor Networks)," in *IEEE Performance Computing and Communications Conference (IPCCC)*; Phoenix, AZ, IEEE Computer Society, 2004, p. 415.

Jardosh, A., E. M. Belding-Royer, K. C. Almeroth, and S. Suri, "Towards Realistic Mobility Models for Mobile Ad Hoc Networks," in *International Conference on Mobile Computing and Networking (MobiCom)*; San Diego, CA, ACM, 2003, p. 217.

Kochhal, M., L. Schwiebert, and S. K. S. Gupta, "Role-Based Hierarchical Self Organization for Ad hoc Wireless Sensor Networks," in *ACM International Workshop on Wireless Sensor Networks and Applications;* San Diego, CA, ACM, 2003, p. 98.

Kotz, D., and K. Essien, "Analysis of a Campus-Wide Wireless Network," in *International Conference on Mobile Computing and Networking (MobiCom)*; Atlanta, GA, ACM, 2002, p. 107.

Krishnamachari, B., D. Estrin, and S. Wicker, "The Impact of Data Aggregation in Wireless Sensor Networks," in *ICDCS International Workshop of Distributed Event Based Systems (DEBS)*; Vienna, Austria, IEEE Computer Society, 2002, p. 575.

Kumar, M., "A Consensus Protocol for Wireless Sensor Networks," M.S. thesis, Wayne State University. August 2003.

Kumar, R., M. Wolenetz, B. Agarwalla, J. Shin, P. Hutto, A. Paul, and U. Ramachandran, "DFuse: A Framework for Distributed Data Fusion," in *ACM Conference on Embedded Networked Sensor Systems (SenSys)*; Los Angeles, CA, ACM, 2003, p. 114.

Lindsey, S., C. Raghavendra, and K. M. Sivalingam, "Data Gathering Algorithms in Sensor Networks Using Energy Metrics," *IEEE Transactions on Parallel and Distributed Systems* 13(9):924, 2002.

Lu, S., V. Bharghavan, and R. Srkant, "Fair Scheduling in Wireless Packet Networks," *ACM SIGCOMM Computer Communication Review* 27(4):63, 1997.

Meguerdichian, S., F. Koushanfar, M. Potkonjak, and M. B. Srivastava, "Coverage Problems in Wireless Ad-Hoc Sensor Networks," in *Proceedings of the 20th International Annual Joint Conference of the IEEE Computer and Communications Societies INFOCOM*; Anchorage, Alaska, IEEE Computer Society, 2001a, p. 1380.

Meguerdichian, S., F. Koushanfar, G. Qu, and M. Potkonjak, "Exposure in Wireless Ad-Hoc Sensor Networks," in *International Conference on Mobile Computing and Networking (MobiCOM)*; Rome, Italy, ACM, 2001b, p. 139.

Niculescu, D., and B. Nath, "Trajectory Based Forwarding and Its Applications," in *International Conference on Mobile Computing and Networking (MobiCom)*; San Diego, CA, ACM, 2003, p. 260.

Niculescu, D., and B. Nath, "Ad Hoc Positioning System (APS)," in *IEEE Global Telecommunications Conference (Globecom) 2001*, Vol. 5; San Antonio, TX, IEEE Communications Society, 2001, p. 2926.

Patwari, N., and A. Hero, "Using Proximity and Quantized RSS for Sensor Localization in Wireless Networks," in *ACM International Workshop on Wireless Sensor Networks and Applications;* San Diego, CA, ACM, 2003, p. 20.

Pei, G., and M. Gerla, "Mobility Management for Hierarchical Wireless Networks," *Mobile Networks and Applications* 6(4):331, 2001.

Pottie, G. J., and W. J. Kaiser, "Wireless Integrated Network Sensors," *Communications of the ACM* 43(5):51, 2000.

Prakash, R., "A Routing Algorithm for Wireless Ad Hoc Networks with Unidirectional Links," *Wireless Networks* 7(6):617, 2001.

Rabaey, J., M. Ammer, J. da Silva, D. Patel, and S. Roundy, "PicoRadio Supports Ad Hoc Ultra-Low Power Wireless Networking," *IEEE Computer Magazine* 33(7):42, 2000.

Salhieh, A., and L. Schwiebert, "Evaluation of Cartesian-Based Routing Metrics for Wireless Sensor Networks," in *Communication Networks and Distributed Systems Modeling and Simulation (CNDS)*; San Diego, CA, The Society for Modeling and Simulation International, 2004.

Shih, E., P. Bahl, and M. J. Sinclair, "Wake on Wireless: An Event Driven Energy Saving Strategy for Battery Operated Devices," in *International Conference on Mobile Computing and Networking (MobiCom)*; Atlanta, GA, ACM, 2002, p. 160.

Shih, E., S. Cho, N. Ickes, R. Min, A. Sinha, A. Wang, and A. Chandrakasan, "Physical Layer Driven Protocol and Algorithm Design for Energy-Efficient Wireless Sensor Networks," in *International Conference on Mobile Computing and Networking (MobiCom)*; Rome, Italy, ACM, 2001, p. 272.

Singh, S., and C. S. Ragavendra, "PAMAS—Power Aware Multi-Access Protocol with Signaling for Ad Hoc Networks," *ACM SIGCOMM Computer Communication Review* 28(3):5, 1998.

Sohrabi, K., J. Gao, V. Ailawadhi, and G. Pottie, "Protocols for Self-Organization of a Wireless Sensor Network," *IEEE Personal Communications* 7(5):16, 2000.

Steenstrup, M., "Cluster-Based Networks," in *Ad Hoc Networking,* Charles E. Perkins (ed.). Reading, MA: Addison-Wesley, 2000, p. 75.

Tobagi, F., and L. Kleinrock, "Packet Switching in Radio Channels: II. The Hidden Terminal Problem in Carrier Sense Multiple-Access and the Busy-Tone Solution," *IEEE Transactions on Communications* 23(12):1417, 1975.

Tseng, Y.-C., S.-Y. Ni, Y.-S. Chen, and J.-P. Sheu, "The Broadcast Storm Problem in a Mobile Ad Hoc Network," *Wireless Networks* 8(2–3):153, 2002.

Van Dyck, R. E., "Detection Performance in Self-Organized Wireless Sensor Networks," in *IEEE International Symposium on Information Theory;* Lausanne, Switzerland, IEEE, 2002, p. 13.

Warneke, B., M. Last, B. Leibowitz, and K. S. J. Pister, "Smart Dust: Communicating with a Cubic-Millimeter Computer," *IEEE Computer Magazine* 34(1):44, 2001.

Wei, Y., and J. M. Cioffi, "FDMA Capacity of Gaussian Multiple-Access Channels with ISI," *IEEE Transactions on Communications* 50(1):102, 2002.

Xu, S., and T. Saadawi, "Revealing the Problems with 802.11 Medium Access Control Protocol in Multi-Hop Wireless Ad Hoc Networks," *Computer Networks* 38(4):531, 2002.

Ye, F., H. Luo, J. Cheng, S. Lu, and L. Zhang, "A Two-Tier Data Dissemination Model for Large-Scale Wireless Sensor Networks," in *International Conference on Mobile Computing and Networking (MobiCom)*; Atlanta, GA, ACM, 2002, p. 148.

Ye, W., J. Heidemann, and D. Estrin, "An Energy-Efficient MAC Protocol for Wireless Sensor Networks," in *Proceedings of the 21st International Annual Joint Conference of the IEEE Computer and Communications Societies (INFOCOM)*, Vol. 3; New York, NY, IEEE Computer Society, 2002, p. 1567.

Zhou, L., and Z. J. Haas, "Securing Ad Hoc Networks," *IEEE Networks* 13(6):24, 1999.

11

Approaches and Solutions

In previous chapters we have considered a number of challenges to using ad hoc networks, in particular wireless sensor networks. As we discussed earlier, large-scale deployment of wireless sensor networks has become feasible only recently. Until recently, the cost of these nodes was too high, the size of sensor nodes was too large, and the capabilities for sensing and computation were too limited for many applications. Because wireless sensor networks became broadly applicable only a short time ago, the protocols for using sensor nodes and networking them together efficiently are relatively new. Although the protocols will continue to improve and new ideas will be developed, we can recognize a number of general approaches. By reviewing these approaches, the essential ideas that have been developed to date can be understood and evaluated. Future protocols will improve on these protocols, but this progress will benefit from knowledge of the existing protocols.

11.1 Deployment and Configuration

Although configurations vary with the application requirements, general deployments of wireless sensors consist of a large number of inexpensive sensors, each with a limited operating lifetime. Initial placement of the sensors must be done in a cost-effective way, or the advantages of using low-cost sensors are lost in the overhead associated with labor-intensive installation. Periodic replacement of faulty or energy-depleted sensors must be performed to preserve the ability of the sensor networks to meet the application requirements. This too must be accomplished without excessive costs. These considerations have motivated the protocols designed for the initial deployment and configuration, as well as subsequent redeployments to replenish the networks.

11.1.1 Random deployment

To enable limited human interaction in the placement of sensors, random deployment has been suggested. Random deployment occurs because sensors are strewn about the area of interest rather than being placed individually at specific locations. Examples include dropping sensors from an airplane to facilitate military applications that require monitoring enemy troops or a strategically important location (Estrin et al., 1999). In this case, the cost of human-intensive deployment of sensors is not the only consideration; the danger associated with military operations also may prevent manual placement of the sensors. Sensors dropped from the air land in an arbitrary arrangement. The sensor coverage is likely to be less even than manual placement, which may require that extra sensors be used to increase the chance of having sufficient coverage throughout the sensing area. Given the locations of the sensors and the sensing range of each sensor, the coverage of the area of interest can be computed (Meguerdichian et al., 2001). A worst-case analysis could be conducted to figure out how many extra sensors to redeploy.

Another alternative is the use of mobile sensors that can move to better positions after deployment (Zou and Chakrabarty, 2003). The movement can be based on the location of each sensor, the number of sensors in the network, and their ideal positions. This may require on-site coordination among sensor nodes because obstacles or uneven terrain can affect the sensing range and thus their ideal placements.

11.1.2 Scalability

Whether random deployment is employed or some other technique is adopted for positioning the sensors, scalability becomes a concern when many sensors are used. As the number of sensors grows, manual deployment usually grows superlinearly with the size of the network. As the number of nodes increases, additional personnel may be needed to configure the sensors. Furthermore, interaction among neighboring sensors may introduce additional overhead so that increasing the network density leads to a superlinear increase in the deployment overhead.

Because the primary advantages of sensor networks are based on the low cost of the sensors, a large number of sensor nodes will be used to monitor the environment near the point of interest. For this reason, scalability is important in deployment and configuration, as well as the actual sensing functional requirements, such as routing and event detection. To make the configuration process scalable, the process must be at least partially automated. The first step in configuring the network is to organize the nodes into a network. We discuss protocols for this self-organization process in the following section.

11.1.3 Self-organization

For sensor networks of any reasonably large size, some organization into a localized hierarchical structure is useful. Self-organization is the process of nodes in the network imposing some simple scalable organization on the network. Thus self-organization makes the process less labor-intensive, lowering costs and time for deploying the network.

Self-organization can create a number of structures needed for other higher-layer protocols. For example, nodes can self-organize to assign a unique address to each sensor node (Chevallay, Van Dyck, and Hall, 2002), or address size (in bits) can be reduced to reduce the packet overhead (Schurgers, Kulkarni, and Srivastava, 2002). Coordination among sensors is required to ensure that local addresses are unique, which is useful for messages exchanged among neighboring nodes; however, when messages need to be sent to destinations that are farther away, a unique address is required.

Self-organization also has been proposed as a way of creating clusters using the LEACH protocol (Heinzelman, Chandrakasan, and Balakrishnan, 2002). In the simplest case, clusters provide a two-level hierarchy—each sensor forwards readings to the cluster leader (cluster head) responsible for this sensor. The cluster head then aggregates data from multiple sensors and forwards these composite data to the base station.

LEACH proposes a number of different techniques for choosing cluster heads. In the most basic case, cluster heads select themselves using a random number based on the number of sensors in the network, the number of sensors that should serve as cluster heads at any point in time, and how long it has been since this sensor was a cluster head. Based on this information, some randomly chosen nodes become cluster heads. Nodes that do not become cluster heads listen for signals from neighboring cluster heads. A sensor joins the cluster of the cluster head with the strongest received signal. Because cluster heads are chosen randomly, the performance of the network could be poor if the cluster heads do not have a roughly uniform distribution. For example, if all cluster heads end up on one-half of the sensor area, the nodes on the other side of the region will have to expend much more energy to reach a cluster head. Even if the clusters exhibit a better distribution, there is a reasonable possibility that the number of nodes per cluster will not be uniform. For this reason, other options, such as allowing the base station to select the cluster heads or having some other fixed rotation among the sensor nodes may perform better (Heinzelman, 2000).

11.1.4 Security protocol configuration

Security protocols for wireless sensor networks also require configuration (Carman, Kruss, and Matt, 2000). In most cases, security protocols require some authentication and coordination among neighboring sensors. To prevent the compromise of one sensor from compromising the entire sensor network, some pairwise key exchange is required. Unless the positioning of the sensors is predetermined, the identification of neighbors of each sensor node cannot be determined until deployment. Protocols that require preassignment of keys may not work in situations where each pair of neighboring nodes needs a unique key. Because of the potentially large number of sensors being deployed and the limited storage space of each sensor, assigning a key for every potential pair of sensors is not practical. The preassigned keys may be given to sensors that end up not being neighbors. The same problem cannot occur if the sensors are placed manually, but this assumes that no mistakes are made in placing the sensors. If a sensor is misplaced, then these keys cannot be used because communication occurs between neighboring nodes, and the absence of keys between physically neighboring nodes prevents secure communication. Tracking down such erroneous placements could be labor-intensive, leading to substantial delays in the deployment and significantly increasing the cost associated with the sensor deployment. Deployments of additional sensors to replace sensors that are no longer functional also is complicated if these newly deployed sensors must work with some of the sensors that have been deployed already.

Instead, secure communication between neighboring nodes should be established based on some configuration performed after the sensors have been deployed. This requires some key-establishment protocol. To prevent impostors from infiltrating the network, a fixed key that is shared with the base station may be required, although care must be taken to ensure that compromising any sensor node does not result in the entire network being compromised.

Sensor nodes generally communicate with only a subset of the other sensor nodes. For example, communication between a sensor node and the cluster head is needed with a hierarchical organization of the sensor networks. Similarly, each cluster head may need to communicate with a subset of the other cluster heads. Communication with neighboring nodes generally is sufficient for a flat organization. Thus pairwise security keys are needed among only a few of the possible pairs of sensor nodes, but the selection of these pairs cannot be made before deployment of the network unless the nodes are placed carefully, as explained earlier.

The SEKEN protocol (Jamshaid and Schwiebert, 2004) provides a secure protocol for the pairwise key setup in wireless sensor nodes. The protocol assumes that each sensor node is provided with a private *device identifier* that has been stored in the base station. In addition, the public key of the base station is loaded into each sensor node, which eliminates the need for a trusted certification authority. Each node, starting with the node nearest the base station, forwards a packet to the base station with its device identifier and a timestamp encrypted using the base station's public key. The base station allocates a key for each sensor node, with sensor nodes that already have been assigned a key forwarding request from another node, e.g., B1, to the base station after appending their own encrypted ID so that the base station can gain some understanding of the network topology. Once more than one node has been assigned a security key, the base station also provides keys that can be shared with the neighboring node that initially forwarded the message for B1. The protocol also handles the addition of more nodes or the removal of nodes. Although the SEKEN protocol supports only a linear array of sensor nodes, extensions to a generic two-dimensional sensor network are possible.

11.1.5 Reconfiguration/redeployment

The lifetime of different sensors within the same network could vary dramatically. The most energy-consuming aspect of sensor operation is communication. Because protocols may place additional demands on some subset of the sensors, such as the sensors near the base station or the sensors far from the base station, redeployment of some sensors may be required to meet the application requirements. Even if most of the sensors are still operational, segments of the sensing field may have inadequate coverage because of the death of sensors in that region.

Redeployment of sensors requires that these newly deployed sensors be incorporated into the existing network. Security keys, locally assigned addresses, and other application-specific parameters that are determined through self-organization need to be supplied to these newly introduced sensors. The existing self-organization network protocols may require a reconfiguration protocol to add new sensor nodes in an already organized network. For example, when using random cluster head selection such as LEACH, it may be necessary to update all the sensors in the network when additional sensors are added. Otherwise, an inadequate or wrong number of cluster heads may be chosen because the number of sensors in the network has changed, and nodes no longer have an accurate estimate of the energy remaining in other nodes.

11.1.6 Location determination

For many sensor applications, the physical location of the sensors is very important. For example, consider an application where the sensors monitor some large area for events, such as a forest fire detection system in a large national park. If an event occurs—a fire starts—the information must be relayed from the sensors that detected the fire to the base station so that the park rangers can be informed. A crucial piece of information is the location of the sensors that detected the fire. Otherwise, the park rangers have no idea where the fire started.

Location information is also useful for routing protocols, where the distance and direction from the source to the destination can be useful for finding routing paths. Location information also can simplify the selection of cluster heads and clusters, as well as other approaches for organizing local groups of sensors. Location information also can be used to perform sleep scheduling or to construct time division multiple access (TDMA) communication schedules. Although knowledge of a node's neighbors is sufficient for constructing these schedules, position information can make the process more efficient because nodes can compute their own neighbors more easily, as well as their two-hop neighbors, without exchanging several rounds of messages.

Determining the location of individual sensors is not difficult in theory; simply equip each sensor with Global Positioning System (GPS) capabilities, and then each sensor can record and store its location. For stationary sensors, this is particularly attractive because the GPS readings are not required after startup. In practice, however, the cost of equipping each sensor with GPS may make the sensors too expensive. In addition, the added power and size requirements for a GPS receiver also may make these sensors unsuitable for certain applications. Eventually, the cost of GPS receivers will decline, but it will be some time before use of GPS on typical sensors becomes affordable.

One option for addressing this problem is to provide GPS (or manually configured location information) to a subset of the sensor nodes. These sensors then function as beacons or anchors that can be used by other sensors in the network to determine their coordinates (Savvides, Han, and Srivastava, 2001). By listening to several beacons, a node can determine an approximate location. For instance, the distance can be estimated by using the received signal strength (Niculescu and Nath , 2001), by measuring the delay between sonic and radiofrequency (RF) transmissions (Girod et al., 2002; Savvides, Han, and Srivastava, 2001), or simply by considering the transmission range and location of each beacon (Bulusu, Heidemann, and Estrin, 2000; Savarese, Rabaey, and Beutel, 2001). Given the distance between a sensor and at least three beacons, the position of the sensor can be approximated. Once a node has obtained an approximate location, that node can serve as a beacon for other sensors. In Fig. 11.1, for example, node A can be used as a beacon for location information for node B once node A approximates its location. In this way, each sensor obtains an initial estimate of its location using only a few sensors that are equipped with GPS. The initial location estimate is refined further using an iterative approach where each sensor uses additional information from neighboring sensors to obtain a better approximation (Patwari and Hero, 2003; Savarese,

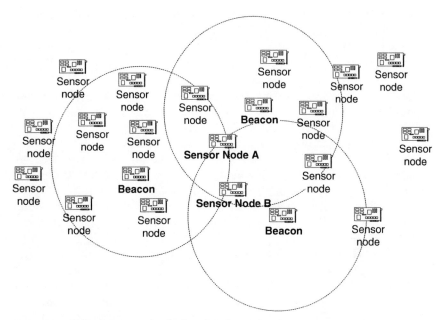

Figure 11.1 Using beacon nodes for location determination.

Rabaey, and Beutel, 2001). After a series of refinement steps, the sensors achieve a location that has sufficient accuracy for the application. In the forest fire monitoring application, for example, a position that is accurate within, say, 10 m should be satisfactory. Being notified that a fire has started at the sensor location, the smoke and flames should be visible even if the recorded location of the sensor is off by a few meters.

Another option is to produce a local coordinate system (Bulusu, Heidemann, and Estrin, 2000; Capkun and Hubaux, 2001; Shang et al., 2003). This coordinate system can be used to assign a relative location for each sensor. The coordinate system has its own orientation, although the distances among sensors should be preserved as close to the actual distances as possible. When a sensor detects an event, the relative location within the sensor network needs to be translated into an absolute location that the application can understand. As long as the base station has its absolute physical position, as well as its location based on the relative coordinate system, a translation is possible. Once the base station is incorporated into the local coordinate system, finding a sensor's location is straightforward. The simplest way to resolve local coordinate systems is to align them. This can be accomplished easily by providing a compass with each sensor node or at least a sufficient number of sensor nodes (Niculescu and Nath, 2003b). With the addition of a compass, the orientation of the device with respect to North can be determined. Using this information to construct the coordinate axes allows the coordinate systems to be aligned. In the absence of a compass to provide a common orientation to the sensor nodes, converging coordinate systems is significantly more difficult.

Another drawback to local coordinate systems is the overhead involved in building and maintaining the coordinates. Especially when nodes are mobile but also as nodes die and more sensors are introduced into the network, updates in the coordinate system may need to propagate throughout the sensor network. Local positioning systems (Capkun and Hubaux, 2001) have been proposed as a technique for defining local coordinate systems without the participation of all the nodes in the network. This idea has been extended to allow the construction of a coordinate system maintained only by the sensors involved in a particular communication (Niculescu and Nath, 2003b). In other words, all nodes on the path from the source to the destination, or on the paths to multiple destinations in the case of group communication, need to share a coordinate system so that messages can be transmitted between the source and the destination(s). Nodes not involved in the communication can be unaware of this coordinate system. This limits the number of nodes that must participate in the construction of the local coordinate system. This may be advantageous when the nodes are mobile because fewer node locations need to be updated, and unneeded information is not maintained. The drawbacks are that a number of local coordinate systems may need to be constructed at any given point in time, there is a time delay associated with building these coordinate systems, and changes in network topology can require that paths be reassembled and additional nodes incorporated into the coordinate system.

The main advantage of using GPS to determine the position of sensors is the flexibility in translating the location into a value that is meaningful to users and network components outside the wireless sensor network. Although having only a few beacon nodes equipped with GPS receivers lowers the cost of deployment, there is still added cost and complexity. In addition, GPS cannot be used inside buildings and other places where GPS signals cannot be received. The main advantage of constructing a local coordinate system is that the limitations, cost, and complexity of using GPS receivers with the sensors are eliminated. The main drawbacks of this approach are that the location determined for each sensor may be less accurate, and constructing the local coordinate system often requires participation of the entire network.

11.2 Routing

Routing is the fundamental component of the network layer protocols. Routing allows the multiple devices that make up the network to exchange information. For this reason, routing in wireless sensor networks has been well studied by the networking research community. A number of protocols have been proposed for routing in wireless sensor networks, and additional protocols continue to be proposed. Each has its own advantages and disadvantages. As with other topics discussed in this chapter, routing in wireless sensor networks is a topic of ongoing research, but many of the core approaches have been identified. Further progress on these different protocols will determine which, if any, of these protocols ends up becoming predominant.

Routing in wireless sensor networks differs from routing in other types of networks for two reasons: (1) the nodes have limited resources for gathering routing information and maintaining routing tables, and (2) the traffic patterns that are generated are likely to be different. The traffic patterns differ because the sensor readings are a primary source of traffic. Some applications accumulate readings peridically, but for many applications, the primary traffic occurs when events are sensed. The sensing and reporting of events are likely to be somewhat random because events are likely to occur in an unpredictable and arbitrary pattern. Once a sensor detects an event, however, tracking that event in the network becomes more predictable. For example, the event may proceed from one location to another along a trajectory that can be predicted with some accuracy (Goel and Imielinski, 2001). Another reason that traffic patterns in wireless sensor networks differ is that the vast majority of the traffic has a base station as either a source or a destination of the message. Communication between two sensors is much less common, with the exception of neighboring sensors that require local collaboration.

When a node senses an event of interest to the application, which is indicated by a sensor reading that is outside the normal operating range, the node must transmit this event to the base station. Of course, some local processing certainly is possible, and this processing may result in some messages being suppressed or combined with the readings of nearby sensors. However, we ignore these considerations in this section. Instead, we focus on the protocols used to transfer sensor readings from the source to the destination once a decision has been made to send this message.

The protocols that we discuss in this section generally assume one of two models. Either all the sensors provide periodic feedback, or the primary source of traffic is from responses from sensors that detect events. The best routing algorithm depends on these traffic characteristics. In some applications, both types of traffic may be required, which warrants using a hybrid approach or using more than one approach.

11.2.1 Event-driven routing

Because sensors interact with the physical world, the data that are routed in the network are based on the readings of the sensors. Rather than having the base station query particular nodes for their current readings, it has been suggested that having the base station query the network for sensor readings with particular characteristics is a more useful model for wireless sensor networks (Intanagonwiwat, Govindan, and Estrin, 2000). For example, the base station might send a request that all nodes that have detected a temperature above 120°F report their readings. This is useful for detecting a fire. The request is flooded throughout the network, and the nodes that respond are those with this reading. In other words, rather than requesting sensor readings from a particular sensor node, we request a particular type of sensor reading from any sensor node that possesses such a reading. This type of network interaction is consistent with the kinds of communications we anticipate from a typical sensor network application.

11.2.2 Periodic sensor readings

Another reasonable type of communication that a sensor network may generate is a periodic response from all the sensors. On a regular basis, all the sensors transmit their sensor readings to the base station (Heinzelman, Chandrakasan, and Balakrishnan, 2002). This gives the base station a snapshot of the entire region covered by the sensors. It also might be useful for diagnosing sensor faults. For example, when a particular sensor has a reading that is not shared by any of the neighboring sensors, it is reasonable to suspect that that sensor is reporting incorrect information because it is either faulty or malicious.

An example application where such periodic readings may need to be accumulated is the temperature monitoring of a large building. If you have ever worked or lived in a large building with multiple floors that has a central heating or air-conditioning system, then you know that the temperature in the building can vary dramatically from room to room. There may be rooms that are uncomfortably cold, whereas other rooms in the same building are unbearably hot. The problem is that the heat distribution is not controlled efficiently so either the heating or air conditioning is not distributed effectively. In some cases, this may result from having too few thermostats or having them placed with a distribution that is less than ideal such as near a computer monitor. An alternative approach to this problem uses a large number of wireless sensors to monitor the temperature in various locations of the building (Kintner-Meyer and Brambley, 2002). For instance, placing sensors in each room, with multiple sensors in a large room, will allow for the accumulation of more data. These data can be used to modify the heating or cooling flow in the building. Not only does this increase the comfort of the building occupants, but also achieves a greater energy efficiency.

It is well known that a system cannot be controlled at a rate faster than the feedback of changes can be obtained. A real-world example is climate control in a car. If the temperature is too hot or too cold, you adjust the output of the heater (or air conditioner). After the adjustment, however, you need to wait until the temperature changes as a result of the change in settings. Otherwise, you simply continue to change things without knowing if you are getting close to your goal—you may simply oscillate between a car that is too hot and too cold. For this reason, some feedback time is required in this system. Sensors measure the temperature and send it to the control center (base station). Based on these measurements, the temperature or distribution of heat is modified. After waiting enough time to ensure that the modifications have affected the system, the sensors take new measurements. Further adjustments then are made. This ongoing cycle can be achieved best by getting periodic readings from all the sensors.

There are two mechanisms for conducting periodic readings. In the first method, each sensor sends its message individually to the base station; in the other method, nodes have a hierarchical structure where nodes transmit their messages to a higher-level node that combines these messages into a single packet that is forwarded higher up the hierarchy. In the case of a two-level hierarchy, for instance, nodes send to their cluster head, which forwards the message directly to the base station. Each of these approaches has advantages and disadvantages, so we consider both approaches. Briefly, sending separate messages to the base station can result in more energy consumption, but combining sensor readings into a single packet can lead to loss of information.

We first consider the idea of sending separate messages. One approach is to send a message directly to the base station from each sensor (Heinzelman, Chandrakasan, and Balakrishnan, 2002). There are two disadvantages to this approach. First, contention for bandwidth becomes acute and severely limits the scalability of this approach. Second, nodes far from the base station expend significantly more energy to transmit their messages. Therefore, the nodes far from the base station deplete their energy much more quickly. On the other hand, if sensors use multihop communication to transmit messages to the base station, then sensors close to the base station must consume a large percentage of their power forwarding messages from other sensors to the base station. This causes the sensors near the base station to run out of energy much sooner than the other nodes in the network.

To avoid either of these situations, nodes can use multihop transmission only when nodes closer to the base station have enough power and memory to perform data forwarding. The base station can send signals to synchronize nodes and create a transmission schedule for the sensor nodes. Even with modest overhead to accumulate the remaining energy levels at the neighbors, the performance of the protocol may better balance energy usage among the sensor nodes. The nodes in the network dissipate their energy at approximately the same rate whether they are close to the base station or far from the base station. This approach is best when the sensor readings cannot be aggregated into a single packet or a few packets. When a high percentage of data aggregation is possible, it is better to aggregate packets at a neighboring node and forward only a single packet or a small number of packets.

When the sensor readings can be aggregated or compressed into a very small fraction of the original data volume, combining or clustering data readings into a single packet and then forwarding only that single packet can save significant amounts of energy. This observation has been incorporated into the LEACH (Heinzelman, Chandrakasan, and Balakrishnan, 2002) and PEGASIS (Lindsey, Raghavendra, and Sivalingam, 2002) protocols.

The simplest version of the LEACH protocol randomly selects a number of the sensor nodes to function as cluster heads, each of which advertises its availability. Each cluster head performs its duties for some period of time, after which different sensors take over the role of cluster head. Since the responsibilities of the cluster head lead to additional energy consumption, periodically changing the cluster heads balances the energy consumption more or less evenly among the sensors throughout the network.

Experiments suggest that having approximately 5 percent of the nodes function as cluster heads at any time gives good performance. Each sensor node that is not a cluster head listens to these advertisements and selects the closest cluster head. Once each cluster head has determined the membership of its cluster, a schedule is created for the transmissions from the sensor nodes in the cluster to the cluster head. On receiving messages from the sensors in this cluster, the cluster head sends a single packet to the base station that combines the readings of all the sensors. The assumption is that each sensor is significantly closer to its cluster head than to the base station. Because the cluster head sends only a single packet a relatively long distance, the amount of energy consumed in communicating with the base station is reduced significantly. This leads to a substantial increase in the lifetime of the sensor nodes.

In LEACH, each sensor chooses the closest cluster head, and the cluster heads are chosen randomly, so there is no assurance as to the distribution of clusters. Since a number of nodes may be assigned to the same cluster, the data may need to be aggregated into a very small fraction of the originally transmitted information. Depending on the application, this level of compression, which is likely to be lossy compression, may or may not be suitable. In some cases, there is a natural trade-off between reducing the power consumption for transmitting information to the base station and obtaining high-quality information at the base station.

Randomly selecting the cluster heads depends on each sensor having some idea of its probability of becoming the cluster head. Each sensor then independently makes a decision on whether or not to become a cluster head. This approach implies that achieving the desired number of cluster heads may not always occur. In some instances, more than an optimal number of sensors may decide to become cluster heads. In other cases, fewer than optimal may. In addition, there is no guarantee that the cluster heads are distributed evenly in a geographic sense. If a poor distribution occurs, some sensors may need to transmit their values a long distance to reach the nearest cluster. In extreme cases, some sensors may send further to reach their chosen cluster head than sending directly to the base station. For these reasons, other options for selecting the cluster heads have been proposed (Heinzelman, 2000). One example is allowing the base station to preselect the cluster heads. This may require sensors to inform the base station periodically of their remaining energy so that sensors can be chosen in a way that extends the life of the entire network as long as possible.

A second approach to aggregating data is the PEGASIS protocol (Lindsey, Raghavendra, and Sivalingam, 2002). In this protocol, a chain of sensors forms starting from sensors far from the base station and ending with the sensors close to the base station. The sensor readings are aggregated on a hop-by-hop basis as they travel through each sensor node. Figure 11.2 shows a possible chain for this sensor network. The messages are aggregated hop by hop until a single packet is delivered to the base station. The chain is created so that each link of the chain is between two nearby sensors to the extent possible. The problem of finding the optimal chain is NP-complete, so the PEGASIS protocol uses a heuristic that produces a reasonably good chain. Since the data are aggregated over short hops, there are no long-distance transmissions from a cluster head to a base station. This reduces the overall energy usage; there are essentially no messages that travel a long distance, provided that there are sensor nodes near the base station. In addition, no cluster heads need to be chosen, which means that the energy requirements placed on each sensor node are roughly equivalent. However, there are two drawbacks to this approach. The first is that a significant delay can occur in transmitting the sensor readings to the base station. Since a sensor must receive the data, process the data, combine these data with the local readings, and then transmit the aggregated data to the next sensor, delay occurs at each sensor in the chain. Since each sensor is included in the chain, the delay grows linearly with the number of sensors in the network.

The delay could become very significant for large sensor networks, and this limits scalability. For many applications, sensor readings must be obtained within some short time after the readings are obtained. When

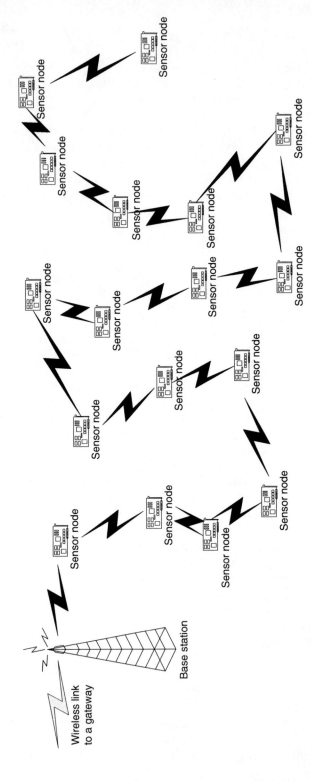

Figure 11.2 Sensor chain for data aggregation.

monitoring environmental conditions, timely feedback may be required to react to sensor readings. For instance, the temperature measurements in a high-rise office building may not be delivered for a considerable length of time if they must travel from sensor to sensor throughout every room in the building.

The modified version of the PEGASIS protocol addresses the issue of excessive delay (Lindsey, Raghavendra, and Sivalingam, 2002). In essence, a tree is created so that a hierarchy of nodes is created, and the nodes forward messages up the hierarchy. This results in more energy consumption but reduces the delay from linear time to $O(\log n)$ time for n sensors. In fact, the energy-delay product is better with a tree hierarchy because the time decreases much more compared with the increased energy consumption. Any sensor that is not a leaf node in this tree has additional resource demands because it must receive multiple messages and aggregate those messages. However, nodes at higher levels of the hierarchy perform roughly the same amount of work as those at lower layers of the hierarchy, assuming that each parent in the tree has the same number of children. By rotating the responsibility to serve as a parent node, the energy consumption among sensor nodes can be maintained at approximately an equivalent level.

The second drawback is that a very high level of data aggregation is required. Otherwise, the packet size grows significantly. If the sensor readings are all aggregated into a single packet the same size as the original, then either the original packet must be padded to be unnecessarily large, which wastes energy, or the data that eventually reach the base station are a small fraction of the data generated by the sensors. For example, with 100 sensors in the wireless network, the base station receives only about 1 percent of the data produced by the sensors. For some applications, such as counting the number of sensors with a particular reading, this is sufficient. However, for many applications, where the unique reading of each sensor is required, this level of aggregation is not acceptable. On the other hand, if the packet size is permitted to grow, then the energy savings decrease rapidly, and sensors near the base station consume a disproportionately large amount of energy. In fact, unless significant aggregation rates can be obtained, the advantages of hop-by-hop aggregation are limited.

With periodic transmission of the sensor readings from all the sensors in the network, the best choice of protocol depends on the data-aggregation rate. At least some modest data aggregation is always possible because the packet headers from two different messages can be replaced with a single header. Compression or aggregation of the packet data, however, depends on the application requirements and the type of sensor readings obtained. If the application can tolerate some loss of data, then more aggressive compression of the data is possible, which can result in significant energy savings for communication. On the other hand, if only a limited amount of aggregation is possible for a given application, protocols such as LEACH and PEGASIS do not deliver a significant reduction in energy consumption. In extreme cases, where essentially no data aggregation is possible, LEACH and PEGASIS probably consume more energy than protocols that do not rely on data aggregation.

11.2.3 Diffusion routing

As mentioned earlier, routing in wireless sensor networks is either periodic feedback of readings from all the sensors in the network or feedback from sensors that have readings that match particular query values. In the preceding section we considered protocols that support the periodic feedback of readings from the sensors. In this section we consider protocols for routing messages from sensors that have readings that match the requirements of a request from the base station or some other node (Intanagonwiwat, Govindan, and Estrin, 2000).

For purposes of illustration, we assume that the request comes from the base station. Although other network nodes could generate the request, a base station or other gateway node that connects the sensor node with the outside world or a larger network is the most likely source of the queries. The process is initiated by distributing a request from the base station to all the nodes in the network. This generally is accomplished by controlled flooding, except when the request is hard-coded in the sensor application.

Controlled flooding consumes some resources but is the only way to get the request to all the sensor nodes. There are different ways of controlling the flooding. For example, the base station simply could send a message with enough power that it reaches all the sensor nodes. The message also could be sent on a hop-by-hop basis from one sensor node to another. Each sensor would forward a particular request once so that only a limited number of messages are sent. A third option is to forward messages along particular predefined paths or directions so that all the nodes receive the message (Niculescu and Nath, 2003a).

Once the request has been disseminated, the sensors with readings that match the request transmit their readings back to the base station. As an example, we consider an animal-tracking application. Suppose that we are conducting a study of how particular animals move about a certain area. Sensors capable of detecting these animals are deployed in the area of interest. The base station sends a query asking that any sensor that currently detects an animal report the relevant information, such as location and number of animals sensed. The query might include parameters such as the frequency of responses and the length of time the animals should be tracked. The problem then becomes one of routing the information from these sensors back to the base station in an energy-efficient method.

Directed diffusion (Intanagonwiwat, Govindan, and Estrin, 2000) solves this problem by initially sending messages from the sensor nodes to the base station in a reverse flooding operation. Figure 11.3 shows the initial transmission of the query from the base station to the sensor nodes. Messages move toward the base station from each sensor with corresponding readings. Nodes that receive the message and are along a path to the destination forward this message toward the base station. Sensor nodes that receive multiple copies of the same message suppress forwarding. In effect, messages are funneled from individual sensor nodes to the base station.

Figure 11.4 depicts an initial pattern of packet routing from the sensors that detected the event to the base station. Of course, this way of delivering messages to the base station is not particularly energy efficient. To improve the energy efficiency, it is necessary to suppress some of the sensors forwarding messages toward the base station.

Figure 11.5 shows one possible set of routing paths that could remain after some duplicate path suppression. The directed diffusion protocol uses both positive and negative feedback to either encourage or discourage sensor nodes from forwarding messages toward the base station. This feedback can be based on, for example, the delay in receiving data. In this case, a sensor that receives the same message from multiple nodes sends positive feedback to the first and negative feedback to the others. Another option is to have a node send with less frequency unless it receives positive reinforcement. In effect, the forwarding of the message times out after some number of packets are forwarded. This is commonly referred to as *soft state* because the forwarding does not continue unless a message is received periodically to retain this state. This feedback propagates throughout the network to suppress multiple transmissions. Eventually, messages use a single path from the source to the base station.

Maintaining a single path from each sensor to the base station, perhaps with some data aggregation along this path when similar packets merge from two separate sensors, is the most energy-efficient way of delivering messages to the base station. On the other hand, sensor nodes are less reliable than other network components because the power source is self-contained, and these nodes are relatively inexpensive. For this reason, the probability of delivering a message from a distant sensor to the base station increases if multiple paths are supported. This is especially true when the paths are disjointed or essentially disjointed so that there are no common nodes on these paths.

The trade-off between energy efficiency and the high probability of delivering packets depends on the application's desired QoS. The best

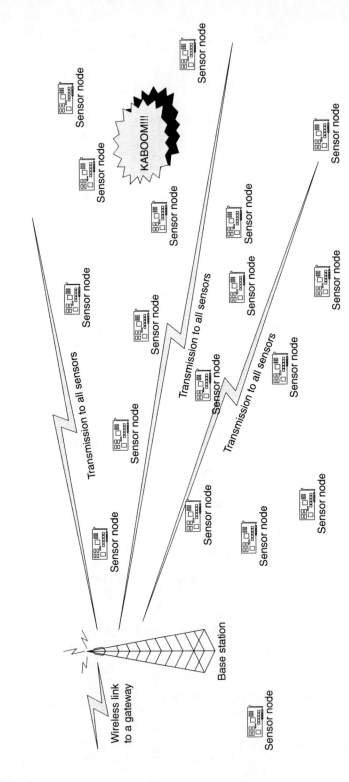

Figure 11.3 Directed diffusion request dissemination.

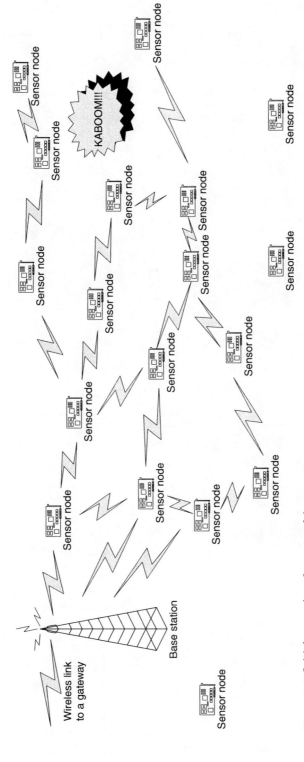

Figure 11.4 Initial processing of an event-driven query.

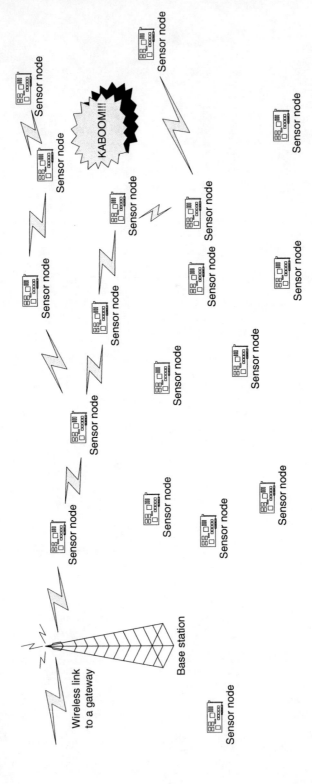

Figure 11.5 Directed diffusion routing paths after redundant path suppression.

choice of path redundancy also varies based on the characteristics of the sensors, including the mean time to failure and the sleep pattern.

11.2.4 Directional routing

Routing in large sensor networks requires a scalable mechanism for delivering messages. Routing tables are not a good option because sensors have limited storage space, and consuming a large amount of this space for storing routing information is not realistic. Furthermore, centralized routing is not a viable option because contacting a central node in order to determine routing paths is not energy efficient either. Besides, this could add significantly to the routing delay. Although it may be possible to maintain routing paths to a few base stations or other central nodes, general-purpose routing is not scalable if a path for every possible destination needs to be stored and maintained. One method for enabling scalable routing is to have the routing paths implicit in the network. In other words, the routing path is effectively encoded in the network.

For a dense wireless sensor network or mobile ad hoc network, we can assume that, on average, there is sufficient sensor coverage so that a sensor is available in any direction in which we need to route. For example, there could be additional nodes that are approximately North, South, East, and West of the current sensor so that messages can be forwarded in some direction toward the destination. For this type of network, the location of the destination relative to any node in the network is sufficient to determine the orientation in which to route the message. Given the orientation in which the message must be sent, the next sensor that can be used to forward the message can be determined. Figure 11.6 depicts a scenario in which a message is forwarded on a direct line from the sensor node to the base station or as close to a straight line as possible given the positions of the sensor nodes. Each hop of the path is determined on a greedy basis. Although the path in this case follows very closely with the direct path from the sensor to the base station, this is not possible in all cases.

Various techniques have been proposed for enabling this sort of routing, including Cartesian routing (Finn, 1987), Trajectory-based forwarding (Niculescu and Nath, 2003a), and directional source aware routing (Salhieh et al., 2001). Although each of these protocols differs in their particulars, the general idea is similar—routing messages based on the geographic location of each sensor as well as the destination.

Cartesian routing creates a straight line from the source of the message to the destination. The message is then routed on a hop-by-hop basis through sensors that are on this path. This is particularly useful for mobile nodes because maintaining long routing paths when intermediate nodes are moving on the path is difficult. These nodes may

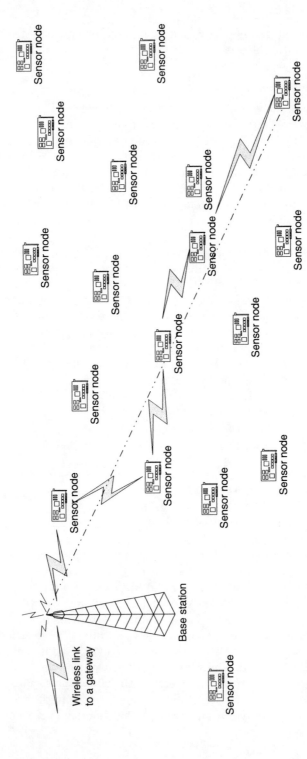

Figure 11.6 Simple example of geographic forwarding.

move out of range of the neighboring nodes on the path, thus preventing packets from being forwarded until an alternative path is built when a traditional routing protocol is used. However, no routing path is maintained with Cartesian routing. Instead, when the message arrives at one node, it simply finds the best current node along the path toward the destination and forwards to this node. The node chosen may vary with each packet depending on the mobility patterns and the speed with which nodes move. As long as there is adequate density in the network, there should be a node available at each hop along the path that can forward the message on toward its destination.

With the directional source aware protocol (Salhieh et al., 2001), routing again is done based on the current location and the location of the destination. In order to avoid the need for creating location information for each of the sensors, the sensors are assumed to be distributed in a regular topology. For example, a two-dimensional grid can place the sensor nodes. Although random deployment has some advantages, a regular placement of each sensor leads to a more uniform coverage of the region. In addition, a random deployment of sensors can be augmented with mobile sensors to move these sensors into a more regular distribution. In this case, nodes can perform routing easily as long as each node knows its relative location within the sensor network. Fixed sensors are assumed. Although some initial repositioning of the nodes does not compromise the protocol, support for mobility within the network has not been considered by this protocol. Research on the directional source aware routing protocol (Salhieh and Schwiebert, 2004) has focused on selecting paths for delivering messages between sensor nodes, as well as delivering messages from the sensors to a base station that maintains the lifetime of the sensor network for as long as possible.

Trajectory-based forwarding (Niculescu and Nath, 2003a) is a significant generalization of Cartesian routing. Trajectory-based forwarding uses the geographic location of the ad hoc nodes to determine the route, which avoids having to store routing tables or make use of other centralized routing facilities. Rather than sending along a direct path from the source to the destination, a message can follow any path specified with a parametric equation. For instance, the trajectory could be a sine wave that originates at the source and terminates at the destination (Niculescu and Nath, 2003a). At each hop in the path, nodes forward messages to the most appropriate node toward the destination. The determination of the most appropriate node is made in a number of different ways. For example, the node that is closest to the path of the trajectory, the node that provides the most progress toward the destination while maintaining some close association to the trajectory, or some node that represents a compromise between these two choices could be chosen.

Direction-based routing schemes rely on the position of the current node and the destination in order to deliver messages to the destination. Whether the route is a straight path from the source to the destination or some other path, the intermediate nodes required to deliver the message are not important, so no information is maintained about these nodes. The assumption of sufficient network density guarantees that either there is a suitable candidate node for forwarding the message toward the destination or one will appear shortly because of network mobility. In the absence of mobility, the path may terminate at some point before the destination is reached. An example is when the path is traveling due West, and there is a hole in the network that prevents a message from being forwarded West. In this case, the message must be misrouted—routed temporarily in an incorrect direction—so that the path can be restored later. The best path around a hole cannot be determined using only local information. This requires some additional knowledge of the perimeter of the hole to make the best decision on how to send the message. For example, routing the message North instead of West may result in a very long path, whereas routing the message South instead could lead to a much shorter path. Of course, the opposite situation is also possible, where routing North could be significantly shorter. Without knowledge of the extent of the hole, there is no mechanism for making the correct choice.

A protocol for finding the perimeter of a sensor network, as well as the perimeters of any holes in the network, has been proposed (Martincic and Schwiebert, 2004). The protocol constructs a perimeter at network initialization time by exchanging local information on which nodes are surrounded by neighboring nodes. Nodes that are surrounded by other nodes are not on the edge of the network and are not on the boundary of a hole. By exchanging this local information, the boundaries can be determined without requiring the exchange of global information on the positions of all the sensors. Later the network can make use of the perimeter information for routing, as well as collaboration among neighboring nodes. If there are mobile nodes in the network, then the holes and the boundaries of these holes, as well as the boundary of the entire network, can change. Thus the perimeter information must be updated to maintain valid data for use in routing around obstacles.

11.2.5 Group communication

Routing information from a group of sensors to the base station can be accomplished as individual transmissions from each of these sensors. However, it is more energy efficient to combine these readings into a single message if these sensors are close together and have obtained similar readings. In addition, a higher level of confidence can be placed on the sensor readings if multiple adjacent sensors obtain the same reading. Similarly, a lower level of confidence should be placed on a sensor reading if neighboring sensors that are functioning at the same time do not have the same observation.

In many cases it should be possible to determine that several sensors have generated the same readings using some local communication and local processing. A resource-constrained protocol for combining these sensor readings into a single value with some greater confidence in the correctness of the sensors' observations should allow this collaboration to be performed on a local basis rather than requiring global communication or the participation of many nodes. Ideally, each node should share information only with other sensors that overlap its sensing range. If the radio transmission range of each sensor is equivalent to at least twice its sensing range, a node can collaborate with all the sensors that overlap any portion of this sensor's sensing range using direct communication. Figure 11.7 shows the effect of having a radio transmission that is exactly twice the sensing range or slightly less than twice the sensing range. In the first case, sensor node A can communicate with node B. In the second case, the transmission range of node A is less, so although the sensing regions of node A and node B overlap, they cannot communicate with each other directly. The radio transmission range is shown with a solid circle, and the sensing range is shown with a dashed line. This allows each sensor to maintain its own local group of sensors and coordinate readings among these sensors.

A sensor reaches an agreement if a majority of its neighbors have obtained the same reading or one that is approximately the same (Kumar, 2003). If no other neighboring sensors share this same reading, a sensor can assume that there is some problem with its measurement. The measurement then could be retaken, or the sensor could perform other actions to recalibrate the sensor or reset components to try to recover from an error. Malicious sensors could avoid sending an acknowledgment of correct readings to a sensor node but could not prevent a consensus from being reached unless more than half the sensors were malicious or faulty. Sensors also could listen to the request for collaboration and refuse to forward messages from nodes that have not attempted to reach consensus or are transmitting bogus information. These procedures increase the confidence that the base station places in the sensor readings that are obtained from the network. The energy required for transmitting the same reading from the multiple sensors is also reduced significantly. Assuming that each sensor generates a message when it obtains an interesting sensor reading, reducing n sensor readings into a single message saves a significant amount of energy as the value of n increases.

Another option is to combine similar messages as they travel through the network (Intanagonwiwat, Govindan, and Estrin, 2000). As each

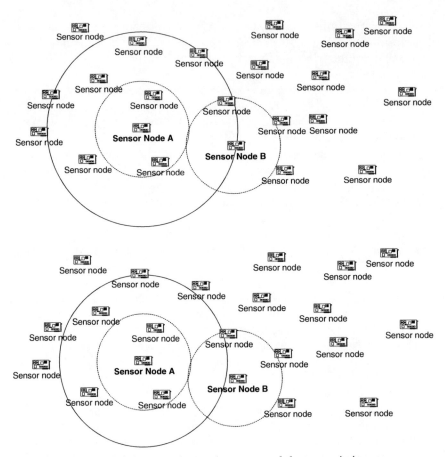

Figure 11.7 Relationship between the sensing range and the transmission range.

message approaches the base station, there are fewer remaining paths. Therefore, the probability increases that the message uses a node that has been or will be used for other equivalent messages from other sensors. To combine the messages as they travel through the network, it will be necessary to buffer some packets in order to wait for similar packets to arrive. In addition, some mechanism is required to determine that these packets are synchronized in the sense that the corresponding readings occurred at the same time. We discuss protocols for synchronizing sensor nodes in the next subsection.

Combining messages that record the same event can be done in the network between the sources and the base station whenever the paths used by these sources merge on the way to the base station. These messages can be merged only if they arrive at roughly the same time, or a significant amount of caching delay is introduced between arrival of the first message and arrival of subsequent matching messages. The opportunity to combine matching packets is likely to increase the closer a packet gets to the base station because the number of paths decreases closer to the base station as paths merge. Of course, the advantages of combining packets decreases the closer the packets are to the base station because there are fewer remaining hops. Thus the energy that can be saved by sending a single message instead of multiple messages decreases with fewer hops. The trade-off between the delay in transmitting the packets to the base station and the memory available for caching packets, the difficulty of matching equivalent responses, and the energy consumption for sending multiple messages determines whether this combining is practical. Combining these packets makes sense only when the application can tolerate significant delays and the sensors have sufficient resources for caching these messages. If the application needs to receive the sensor readings as quickly as possible, or if the sensor nodes cannot buffer packets for a substantial period of time, then combining messages within the network is not a realistic option.

11.2.6 Synchronization

Synchronization is an important element of many wireless sensor network applications. As mentioned elsewhere, data aggregation requires synchronization. Collaboration among sensors requires some synchronization. In addition, the base station needs accurate timestamps to perform the data aggregation. TDMA scheduling also requires local synchronization among sensor nodes. Although the synchronization does not always need to be extremely precise, some loose synchronization must be maintained for TDMA scheduling to work correctly.

Keeping all the sensors synchronized with each other would be an expensive operation, especially as the accuracy of the synchronization increases. Since sensors are expected to observe interesting information only occasionally, maintaining close synchronization at all times is not required. Furthermore, a great deal of power could be wasted maintaining synchrony among nodes that do not require it, such as nodes that do not communicate directly with each other. Instead, a protocol has been proposed that keeps a relatively loose level of synchrony among nodes until an event is sensed (Elson and Estrin, 2001). After sensing an event, the sensors establish a more accurate time synchronization, which can be used to determine whether or not events detected by multiple sensors occurred at the same time or not. Based on the locations of these sensors and whether or not the sensors observed an event at the same time, a determination can be made about whether a unique event has been observed by multiple sensors or different events are being reported.

By avoiding the overhead of closely synchronizing the sensors except when an event occurs, the energy requirements are reduced substantially. By synchronizing only when an event is registered, time synchronization is achieved when necessary. Another protocol, such as the Network Time Protocol (Mills, 1994). can be used to maintain loose synchronization among nodes so that closer synchronization can be established when needed. This loose synchronization allows nodes to maintain a schedule that can be used to contact neighboring nodes and establish the close synchronization required when events occur.

Because of the importance of time synchronization for many wireless sensor applications, other efficient synchronization protocols have been proposed. Another synchronization protocol that has been proposed is the Reference Broadcast Synchronization (RBS) protocol (Elson, Girod, and Estrin, 2002). The RBS protocol is a receiver-receiver-based protocol. Although multiple rounds of the RBS protocol can be used to obtain a better estimate, we describe just a single round. The protocol starts with a source sending a signal to two receivers. Each receiver records the time when this signal was received. Based on this, the two receivers determine their relative clock differences, assuming that both nodes take the same amount of time to process the signal after they receive it. There is clock drift among the sensor nodes because no two clocks run at exactly the same rate. Processing multiple signals over a period of time allows the clock drift to be estimated more closely. In addition, the difference in the clocks can be estimated more accurately with multiple samples. In fact, the RBS protocol can achieve synchronization accuracy of several microseconds.

A hierarchical approach can be used for synchronization in wireless sensor networks. This is the approach taken by the Timing-Sync Protocol for Sensor Networks (TPSN) (Ganeriwal, Kumar, and Srivastava, 2003). A root node, which could be a base station, initiates the synchronization process by sending a `level_discovery` protocol message to each neighbor. These nodes assign their level as level 1 and propagate the message. As the message radiates out from the root node, the levels are assigned based on the minimum number of hops needed to receive this message. Once each node has determined its level in the tree, each pair of nodes that wishes to synchronize sends a message to each other. For example, node A sends a message to node B, which then responds with another message back to node A. By exchanging the timestamps of when these packets were sent and received, the two clocks can be synchronized within a reasonably tight range. This protocol is an example of a sender-receiver synchronization protocol. The TPSN protocol is capable of synchronizing two neighboring nodes within several microseconds.

11.3 Fault Tolerance and Reliability

Ad hoc nodes, and especially mobile nodes, naturally are less reliable than traditional computing devices because communication links are transient. In addition, these nodes are expected to be inexpensive and often do not have the energy resources for long-term operation. For instance, the sensors may be deployed with battery power and replaced after some period of time when the batteries fail. It is unlikely that sensors designed for one-time use will be built with expensive components to prevent failure. It is more likely that the cost of sensor nodes will be modest enough that they can be replaced easily at a lower cost than designing them for long-term use. Sensor nodes also are more susceptible to failure because of their direct exposure to the environment. For these reasons, sensor applications and protocols must be designed with failures and fault recovery as basic assumptions. In other words, not only should the protocols and applications recover from faults, but they also should be designed to operate with the expectation that faults are part of the normal operating situation rather than being anomalous events that should be recovered from so that the application can return to normal operation. Protocols at all layers of the protocol stack must work together to keep faults from preventing the application from receiving correct information or otherwise performing the desired task. In this section we consider a number of protocols at various layers of the traditional protocol stack that could be used to make sensor applications and protocols provide the necessary level of fault tolerance and fault recovery. As can be seen in this discussion, these protocols can operate in conjunction with each other and often perform better together than separately. In other words, the sum is greater than the parts.

11.3.1 FEC and ARQ

Wireless communication is inherently unreliable compared with wired connections. Fiber optic cables, for example, have error rates that are negligible. For all practical purposes, transmission errors on fiber optic cables can be ignored in terms of network performance. Of course, these rare errors must be detected for reliable communications to be supported. However, the protocols can treat such errors as anomalies that should be corrected so that the system can return to the normal operating state as soon as possible. A classic example is the Transmission Control Protocol (TCP), which treats an error as network congestion (a lost packet) rather than a transmission error because errors on wired channels are so rare that handling these special cases separately would not make a significant difference in performance (Jacobson, 1995; Lang and Floreani, 2000). In fact, depending on the mechanism to handle errors, the system performance may decline because the overhead of using this mechanism when almost all packets do not require this mechanism could easily exceed the performance gains when this rare event arises.

Wireless connections, on the other hand, are much less reliable. Bit errors on wireless channels are a normal condition. For this reason, specific techniques are required to operate in the presence of errors. There are two general techniques for addressing bit errors at the physical layer (Lettieri, Schurgers, and Srivastava, 1999). One technique is Forward Error Correction (FEC), which uses extra bits in a packet to allow bit errors to be corrected when they arise during transmission. In general, the more error-correcting bits that are included, the more bit errors that can be corrected. On the other hand, including the number of error-correcting bits in a packet decreases the number of data bits that can be sent or increases the packet size. Automatic Repeat Request (ARQ) refers to a set of protocols used to request retransmission of packets that have errors. There are essentially three choices: (1) stop and wait, which does not send a packet until it receives a positive acknowledgment for the previous packet, (2) go-back-N, which resends all packets from the point at which an error was observed, and (3) selective repeat, which resends only the packets that have errors. Not resending packets that do not have an error saves wireless bandwidth but can require an arbitrarily large buffer. The choice between FEC and ARQ depends on a number of factors, including the error distribution on the wireless channel, whether bursty or uniform error patterns exist, and the probability of bits being erroneous. These two techniques can be used together as a hybrid approach or individually.

11.3.2   Agreement among sensor nodes (Reliability of measurements)

Wireless sensor nodes need to operate under a wide variety of environmental conditions. In addition, wireless sensor nodes are likely to be relatively inexpensive. For these reasons, the sensor readings may have a wide variability in accuracy (Elnahrawy and Nath, 2003). Faulty nodes are discussed in the next section, so in this section we will confine our discussion to sensor readings that are inaccurate because of limitations in the sensing materials. These could be viewed as transient faults, but we consider them to be simply the inherent nature of working with analog sensors; accuracy in a digital sense just is not possible. For this reason, there is an advantage in using redundant sensors—sensors with overlapping areas of sensing—to measure the target phenomenon. These sensors can then collaborate to improve the accuracy of the resulting readings. In other words, by reaching a consensus on interesting readings, sensors can more reliably transfer data to the base station. The other advantage of locally reaching consensus, of course, is that nodes expend less energy in the network in transmitting the sensor readings to the base station.

For wireless sensor networks that have been organized based on hierarchical clustering, there is a straightforward approach to performing consensus because the cluster head simply can combine the information received from each cluster member. The results will then be forwarded up the hierarchy to the base station. The drawback to this approach is that events of interest do not necessarily occur within the boundary of a single cluster. Thus the event may be sensed by nodes in multiple clusters, leading to the transfer of multiple packets, one from each cluster, back to the base station. Depending on the amount of redundancy near the event and how this redundancy is partitioned among adjacent clusters, the clusters may obtain different consensus sensor readings. This could result in the base station receiving multiple readings for the same event and then needing to resolve this discrepancy after receiving these packets. To combat this problem, a recent proposal has been to cluster the wireless sensor network based on the sensing capabilities of the sensors (Kochhal, Schwiebert, and Gupta, 2003). If this is done successfully, the boundaries between adjacent sensors should be better structured to reduce the number of clusters that need to participate in reaching consensus after an event is sensed. The essential idea is to refrain as far as possible from splitting sensors that are very close into separate clusters. Thus events that are near a sensor are likely to be observed primarily by sensors in the same cluster.

When the wireless sensor network is not organized into clusters, there are essentially two options—groups can be formed locally on an ad hoc basis as events occur (Kumar, 2003), or the groups can be formed ahead of time to build consensus when events occur (Gupta and Birman, 2002). An example can best illustrate the difference between these two options. If the groups are preformed, then we have the grouping shown in Fig. 11.8, where the sensors have been partitioned into a grid, and each sensor belongs to the group composed of the sensors in this grid. When an interesting sensor reading is obtained, the other sensors in the same group are consulted, and a consensus is generated with these sensors. The result is then forwarded to the base station.

Events that occur near a grid edge present challenges for reaching consensus. Either data must be exchanged between adjacent groups, or multiple partial results must be sent toward the base station for further processing. On the other hand, if the group is not formed ahead of time, then when an interesting sensor reading occurs, a node queries its neighbors, those with the same sensing range, for their current readings. Figure 11.9 demonstrates this possibility. An event occurs at some location within the sensor network. A group consisting of the sensors

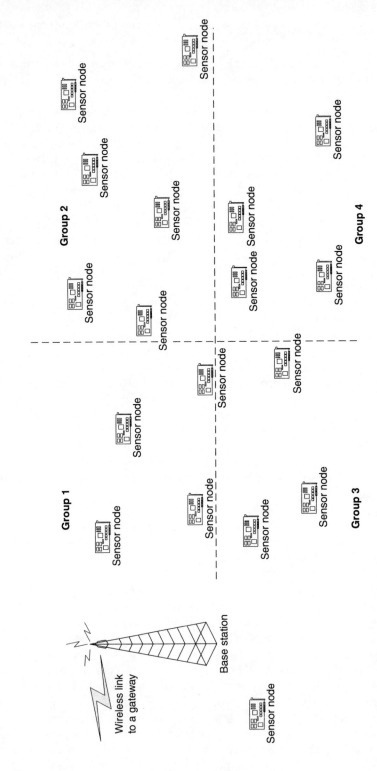

Figure 11.8 A priori grouping of wireless sensors.

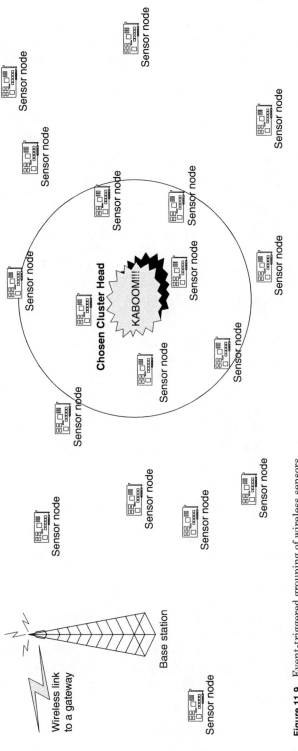

Figure 11.9 Event-triggered grouping of wireless sensors.

within the circle shown forms with one of these sensors serving as the cluster head. Although the problem of resolving events along grid edges is avoided, some additional delay may be experienced. The nodes then produce consensus by consulting these sensors, and the agreed-on reading is forwarded to the base station.

The main advantage of forming a group only after an event occurs is that the sensors that are able to observe this event reach consensus among themselves and forward this result to the base station. A second advantage is that the sensors do not need to maintain a great deal of state information. For example, there is no need to maintain a list of sensors within the group because the group forms only after the event occurs. The sensors then reach consensus with the neighbors, but these neighbors can change over time. For example, some sensors may be sleeping or have exhausted their power supplies. Thus a sensor with an interesting reading needs to coordinate only with active neighboring sensors. For wireless sensor networks that have frequent changes in active sensors, maintaining state information on the groups has limited usefulness. The primary disadvantage of this approach is that there is some delay in forming the consensus. Since sensors may not know their neighbors, the group formation must be conducted after an event is observed. Only after the group has been formed on the fly can the consensus-generation process begin. Although there is some added delay in generating consensus, forming the groups after events occur may enable a more accurate consensus value to be generated because all the nodes that could have observed the event are participants in the consensus-forming process.

11.3.3 Dealing with dead or faulty nodes

There are a number of strategies for dealing with dead or faulty nodes. Of course, these nodes cannot be used, but simply ignoring these nodes and operating the network without them may not be acceptable. For example, the dead or faulty nodes may leave holes in the sensor network coverage that prevent the wireless sensor network from performing the desired functions in at least part of the area of interest. Manually repairing or replacing the sensor nodes generally is not an attractive option unless the sensors have a relatively long life span or are costly. When the sensors have been deployed in a remote or inhospitable location, manual replacement may be impossible. Even when it is physically possible to repair or replace the sensors, this may not be a very cost-effective or scalable option.

Rather than replacing or repairing sensors, a better option might be to deploy extra sensors—in effect, creating an initial sensor network that is more dense than necessary. These extra sensors then can sleep until they are needed. For wireless sensors that have renewable energy sources, the sensors could recharge while sleeping. By rotating the active state and sleep state among the wireless sensor nodes, the overall lifetime of the sensor network could be extended in proportion to the additional sensors that are deployed. This has the added advantage of not leaving portions of the sensor network without adequate coverage from the time of failure until the time of repair.

Because faults are somewhat unpredictable, deploying additional sensors uniformly in the wireless sensor network may not be adequate. The failure or depletion of energy may not be uniform. For example, sensors closer to the base station may perform more communication and thus consume energy more rapidly. Even if the additional sensors are not deployed in a nonuniform pattern, such as placing more sensors close to the base station, the problem will not be solved completely. Other instances of sensor failure patterns may be unpredictable when the sensors are deployed. For example, the events concentrated in a particular region of the sensor network may cause these sensors to consume significantly more power than other sensors. Another example would be when most of the sensors in one region are destroyed. As an example, consider an elephant-tracking application. If a herd of elephants happens to stomp through a portion of the sensor network, destroying most of the sensors, the extra sensors that have been deployed will not prevent the sensor coverage from lapsing in this region. An avalanche, mudslide, or flash flood could produce a similar result. An additional localized deployment may be used to recover from such situations, but there is another option.

Mobile sensors could be used to recover from the scenarios such as those just described, where an unanticipated catastrophe occurs in one region of the sensor network (Zou and Chakrabarty, 2003). The extra sensors then could move to the depleted region to fill the hole. Locally coordinating this movement within the sensor network would be quite challenging, so participation of the base station or some other central node in determining when and where to move these extra sensors seems the most practical approach.

11.4 Energy Efficiency

As pointed out repeatedly in the past few chapters, limited energy is one of the primary distinguishing characteristics of wireless sensors. Because wireless communication consumes a relatively large percentage of the power, most of the solutions that have focused on improving energy efficiency have addressed reducing the number of messages generated. However, because wireless sensors have additional components and energy is such a precious commodity, it makes sense to optimize sensor design and operations across all functions.

11.4.1 Uniform power dissipation

One method of ensuring that the sensor network performs its functions as long as possible is to maximize the lifetime of the sensor network. Some authors have defined the network lifetime as the time until the first node dies (Chang and Tassiulas, 2000; Singh, Woo, and Raghavendra, 1998), but this is not a particularly useful metric for wireless sensor networks. Because a large wireless sensor network monitors a relatively large region and there generally is some overlap among the sensors, the death of a single sensor does not significantly reduce the performance of the typical wireless sensor network. In fact, if a single sensor failure did prevent the sensor network from fulfilling its intended task, then the wireless sensor network has no fault tolerance.

Instead, a more reasonable metric is when the sensor network can no longer perform its intended function. Although it is difficult to define this precisely for most applications, designing the protocols so that all the sensors die at roughly the same time or so that sensors die in random locations instead of in specific regions extends the lifetime of the entire sensor network as long as possible. Achieving uniform energy consumption rates is possible for sensor networks that provide periodic feedback to the base station, but for sensor networks that provide only event-driven feedback, organizing the protocols to balance the lifetime of all the sensors is probably impossible.

It is necessary to balance the energy consumption for wireless communication in order to achieve uniform power dissipation. This means that a balance must be found between sending messages on a hop-by-hop basis from the sensor nodes to the base station and sending them directly to the base station. Clustering seems like an attractive solution to this problem, provided that the cluster heads are rotated in a logical manner to balance the energy consumption among the sensors. However, one drawback to clustering is that some of the messages are sent away from the base station in order to reach the cluster head. The information from this sensor then must be sent from the cluster head to the base station, resulting in a larger total distance for the wireless communication. Since the energy used for a wireless communication increases with at least the square of the distance, over time this could become a significant inefficiency in the communication protocol. An approach that achieves a better balance between the two extremes of transmitting directly to the base station and using hop-by-hop communication without the inefficiencies of communicating with the cluster head would be attractive.

In practice, clustering works well when the messages are combined. In other words, when a relatively high percentage of data aggregation is permitted, sending to the cluster head is not likely to reduce the efficiency significantly. Protocols such as LEACH (Heinzelman, Chandrakasan, and Balakrishnan, 2002) and PEGASIS (Lindsey, Raghavendra, and

Sivalingam, 2002) are attractive options under these circumstances. On the other hand, the overhead in sending away from the base station in order to reach the cluster head can be costly when little data aggregation is possible. Instead, sending either directly to the base station or forwarding to the neighbor nearest to the base station, depending on the power available, could reduce the overall energy consumption and yield nearly uniform power dissipation. Although some overhead is involved in this approach because the remaining energy must be gathered from the neighbors, the energy consumption of selecting and rotating cluster heads may exceed this overhead. This is especially the case when nodes can gather this information by promiscuous monitoring of transmissions by neighboring nodes.

11.4.2 Sensor component power management

A second option for reducing energy consumption is to decrease the power consumption of a sensor node by powering down various components of a wireless sensor (Sinha and Chandrakasan, 2001). For example, power is consumed by a wireless transceiver whether a sensor is receiving a message or the transceiver is idle, although the power consumption in both cases is less than the energy consumed for transmitting a packet (Chen, Sivalingam, and Agrawal, 1999). However, since wireless sensors send data only sporadically, the power consumption could be very high for leaving the transceiver in an idle state compared with powering down the transceiver (Rabaey et al., 2000). Table 10.1 shows the energy consumed by a few wireless transceivers in each of these three modes.

If a sensor node is sleeping, turning off the wireless transceiver saves a significant fraction of the power. Powering down the sensors and the processor, except for a low-power watchdog process, can extend the lifetime of a wireless sensor for years. However, even when the sensor remains operational, powering down the transceiver can increase the life of the sensor significantly. Many wireless sensor nodes provide a range of power settings that determines which components are powered down and at what level each component is powered down.

Sensor networks should avoid protocols that require turning the transceiver on and off repeatedly over short intervals of time. Because additional energy is consumed in turning the wireless transceiver back on, more energy is consumed than a naive model of the energy consumption would predict (Sinha and Chandrakasan, 2001). Instead, the energy consumption of powering down the transceiver and then powering the transceiver back up should be included in the model. For transceivers that can be powered off for a significant fraction of the time, turning off the transceiver makes sense. However, if the transceiver must be on a significant fraction of the time, or if the intervals between being ready to transmit or receive a packet are short, then turning the sensor off actually might increase the total power consumed.

Some wireless transceivers have a number of power levels, which allow a relatively modest sleeping mode from which the sensor could be returned to full power with less energy. Choosing the best sleep mode for a given sensor is an optimization problem that can be difficult to solve in practice unless the sleeping times can be approximated with a high degree of accuracy and the energy consumption model is accurate. However, an optimal solution may not be necessary. Instead, an approximate solution that yields a significant percentage of the optimal energy savings could be adequate for many applications.

11.4.3 MAC layer protocols

Turning off the wireless transceiver or other sensor components is a physical-layer optimization for saving energy. However, there are other layers of the protocol stack where energy can be saved. For example, energy also can be saved at the MAC layer by scheduling the wireless transmissions. Two sources of overhead can be addressed at the MAC layer. The first is the communication for scheduling the wireless channel. The second is the energy wasted when packet collisions occur and packets must be retransmitted.

Wireless computer communication on typical wireless devices commonly uses the IEEE 802.11 protocol (ANSI, 1999). This protocol optionally sends Request-to-Send (RTS) and Clear-to-Send (CTS) packets to schedule the wireless channel. The other choice is to listen for the availability of the channel prior to transmitting a packet. If the channel is busy, then a node that wishes to send uses an exponential back off to request the wireless channel again at some point in the future. In order to reduce collisions, all nodes need to have their transceivers on so that they can detect the requests from other nodes. This means that even some nodes that are not planning to transmit or receive a message cannot turn off their transceivers or else delay transmission after turning their transceivers back on for some initial listening. In addition, the hidden terminal and exposed terminal problems mean that even with the RTS and CTS scheduling mechanisms, collisions still can occur. For these reasons, IEEE 802.11-type scheduling protocols are not popular for wireless sensor networks. Instead, a different protocol for wireless sensors may be more appropriate.

The most common approach for wireless sensor networks is to use a time division multiple access (TDMA) protocol (Pottie and Kaiser, 2000; Sohrabi et al., 2000). A TDMA protocol divides the wireless channel into different time segments. In each transmission phase, a number of reservation mini-slots are used to reserve each of the transmission slots. Sensors can indicate whether or not they wish to transmit a message during the scheduling time segment. If a sensor is successful in acquiring one of the time slots, then the message can be transmitted during this time interval. One advantage of using a TDMA protocol is that nodes that are not planning to send or receive a packet need to have their transceiver on only during the reservation time slot for the purpose of seeing if other sensors are sending a packet to them. The second advantage is that collisions are avoided, except for relatively small reservation packets, so that the wireless bandwidth can be used efficiently. Eliminating packet retransmissions and allowing inactive sensors to turn off their transceivers for most of the time can save significant power. However, TDMA protocols require that nodes be reasonably well synchronized in order to schedule transmissions during the correct time slot, to ensure that transmissions occur during the correct time slot, and to ensure that the receiver is prepared to receive the message when the transmitter starts the transmission. Since wireless sensor networks usually use hop-by-hop communication or need to schedule communication among only a relatively small set of cluster heads, synchronization on a local basis could be sufficient for achieving the necessary level of time synchrony among the senders and receivers.

11.4.4 Trade-offs between performance and energy efficiency

Energy efficiency also can be obtained at higher layers of the protocol stack, but this usually requires that the application accept more delay. One example is synchronizing the wireless sensors. Maintaining synchronous clocks is very useful when attempting to consolidate readings from multiple sensors—otherwise, it is difficult to determine whether two sensors observed the same event or two distinct events that occurred within a short period of time. However, keeping the sensors synchronized is not possible without a large number of messages and the resulting communication overhead. Instead, the sensors could maintain only loose synchronization and generate tight clock synchronization only after an interesting event occurs (Elson and Estrin , 2001). Although this is a feasible solution in many cases, some added delay is encountered. This additional delay could result in a performance penalty for some real-time applications but also offers significant energy savings.

Another approach to realizing energy savings in a wireless sensor network is to first aggregate data in the wireless sensor network and then to send only the aggregated sensor readings to the base station. This can result in significant energy savings for large sensor networks when the size of the aggregated sensor readings is comparable with the size of a single packet. However, aggregating packets can lead to the loss of information, which can have a negative impact on the performance of the sensor application. In addition, waiting for multiple packets to arrive so that they can be aggregated introduces delay into the transmission process. This delay can produce significant energy savings but does involve a reduction in the application performance for real-time sensor applications.

In essence, any energy-efficiency protocol, whether energy-efficient routing or application-layer protocols, involves some increased delay, loss of accuracy, or other performance penalty. Balancing the energy savings against the performance penalties and achieving the application requirements are one of the challenges in designing wireless sensor networking protocols.

11.5 Summary

In Chap. 11, we have reviewed many wireless sensor networking protocols that have been proposed in the past few years. From this review, we see the novel approaches the researchers have taken to meeting both the requirements and demands of a wireless sensor network. The protocols for using sensor nodes and networking them efficiently are relatively new. Although protocols for these problems will continue to improve and new ideas will be developed, we have seen many general approaches. Future research will improve on these protocols, but this progress will benefit from knowledge of the existing protocols. Therefore, the protocols presented in Chap. 11 are an attractive base upon which to build further research in wireless sensor networking.

11.6 References

ANSI/IEEE Standard 802.11, *Wireless LAN Medium Access Control (MAC) Sublayer* and ISO/IEC Standard 8802-11, *Physical Layer Specifications.* 1999.

Bulusu, N., J. Heidemann, and D. Estrin, "GPS-Less Low-Cost Outdoor Localization for Very Small Devices," *IEEE Personal Communications* 7(5):28, 2000.

Capkun, M., and J. Hubaux, "GPS-Free Positioning in Mobile Ad-Hoc Networks," in *Proceedings of the 34th Annual Hawaii International Conference on System Science*; Island of Maui, IEEE Computer Society, 2001, p. 3481.

Carman, D. W., P. S. Kruss, and B. J. Matt, "Constraints and Approaches for Distributed Sensor Network Security," in *NAI Labs Technical Report 00-010*, 2000.

Chang, J., and L. Tassiulas, "Energy Conserving Routing in Wireless Ad-Hoc Networks," in *Annual Joint Conference of the IEEE Computer and Communications Societies (INFOCOM)*, Vol. 1; Tel-Aviv, Israel, IEEE Computer Society, 2000, p. 22

Chen, J.-C., K. Sivalingam, and P. Agrawal, "Performance Comparison of Battery Power Consumption in Wireless Multiple Access Protocols," *Wireless Networks* 5(6):445, 1999.

Chevallay, C., R. E. Van Dyck, and T. A. Hall, "Self-Organization Protocols for Wireless Sensor Networks," in *36th Annual Conference on Information Sciences and Systems*; Princeton, NJ, Princeton University Press, 2002. (Published in CD)

Elnahrawy, E., and B. Nath, "Cleaning and Querying Noisy Sensors," in *ACM International Workshop on Wireless Sensor Networks and Applications*; San Diego, CA, ACM, 2003, p. 78.

Elson, J., and D. Estrin, "Time Synchronization for Wireless Sensor Networks," in *International Parallel and Distributed Processing Systems (IPDPS) Workshop on Parallel and Distributed Computing Issues in Wireless Networks and Mobile Computing*; San Francisco, CA, IEEE Computer Society, 2001, p. 1965.

Elson, J., L. Girod, and D. Estrin, "Fine-Grained Time Synchronization Using Reference Broadcasts," in *Proceedings of the Fifth Symposium on Operating Systems Design and Implementation (OSDI)*; Boston, MA, Usenix Association, 2002, p. 147.

Estrin, D., R. Govindan, J. Heidemann, and S. Kumar, "Next Generation Challenges: Scalable Coordination in Sensor Networks," in *International Conference on Mobile Computing and Networking (MobiCom)*; Seattle, WA, ACM, 1999, p. 263.

Finn, G., "Routing and Addressing Problems in Large Metropolitan-Scale Internetworks," Technical Report ISI Research Report ISI/RR-87-180, University of Southern California, Los Angeles, CA, March 1987.

Ganeriwal, S., R. Kumar, and M. Srivastava, "Timing-Sync Protocol for Sensor Networks," in *ACM Conference on Embedded Networked Sensor Systems (SenSys)*; Los Angeles, CA, ACM, 2003, p. 138.

Girod, L., V. Bychkovskiy, J. Elson, and D. Estrin, "Locating Tiny Sensors in Time and Space: A Case Study," in *IEEE International Conference on Computer Design: VLSI in Computers and Processors*; Freiburg, Germany, IEEE Computer Society, 2002, p. 214.

Goel, S., and T. Imielinski, "Prediction-Based Monitoring in Sensor Networks: Taking Lessons from MPEG," *Computer Communications* 31(5):82, 2001.

Gupta, I., and K. Birman, "Holistic Operations in Large-Scale Sensor Network Systems: A Probabilistic Peer-to-Peer Approach," in *International Workshop on Future Directions in Distributed Computing (FuDiCo)*. *Lecture Notes in Computer Science* 2584:180, 2003.

Heinzelman, W., "Application-Specific Protocol Architectures for Wireless Networks," Ph.D. thesis, Massachusetts Institute of Technology, June 2000.

Heinzelman, W., A. Chandrakasan, and H. Balakrishnan, "Energy-Efficient Communication Protocol for Wireless Microsensor Networks," *IEEE Transactions on Wireless Communication* 1(4):660, 2002.

Intanagonwiwat, C., R. Govindan, and D. Estrin, "Directed Diffusion: A Scalable and Robust Communication Paradigm for Sensor Networks," in *International Conference on Mobile Computing and Networking (MobiCom)*; Boston, MA, ACM, 2000, p. 56.

Jacobson, V., "Congestion Avoidance and Control," *ACM SIGCOMM Computer Communication Review* 25(1):157, 1995.

Jamshaid, K., and L. Schwiebert, "SEKEN (Secure and Efficient Key Exchange for Sensor Networks)," in *IEEE Performance Computing and Communications Conference (IPCCC)*; Phoenix, AZ, IEEE Computer Society, 2004, p. 415.

Kintner-Meyer, M., and M. R. Brambley, "Pros and Cons of Wireless," *ASHRAE (American Society of Heating, Refrigerating and Air-Conditioning Engineers) Journal* 44(11):54, 2002.

Kochhal, M., L. Schwiebert, and S. K. S. Gupta, "Role-Based Hierarchical Self-Organization for Ad Hoc Wireless Sensor Networks," in *ACM International Workshop on Wireless Sensor Networks and Applications*; San Diego, CA, ACM, 2003, p. 98.

Kumar, M., "A Consensus Protocol for Wireless Sensor Networks," M.S. thesis, Wayne State University, Detroit, MI, August 2003.

Lang, T., and D. Floreani, "Performance Evaluation of Different TCP Error Detection and Congestion Control Strategies over a Wireless Link," *ACM SIGMETRICS Performance Evaluation Review* 28(3):30, 2000.

Lettieri, P., C. Schurgers, and M. B. Srivastava, "Adaptive Link Layer Strategies for Energy Efficient Wireless Networking," *Wireless Networks* 5(5):339, 1999.

Lindsey, S., C. Raghavendra, and K. M. Sivalingam, "Data Gathering Algorithms in Sensor Networks Using Energy Metrics," *IEEE Transactions on Parallel and Distributed Systems* 13(9):924, 2002.

Martincic, F., and L. Schwiebert, "Distributed Perimeter Detection in Wireless Sensor Networks," in Networking Wireless Sensors Lab Technical Report WSU-CSC-NEWS/04-TR01, Detroit, MI, 2004.

Meguerdichian, S., F. Koushanfar, M. Potkonjak, and M. B. Srivastava, "Coverage Problems in Wireless Ad-Hoc Sensor Networks," in *Proceedings of the 20th International Annual Joint Conference of the IEEE Computer and Communications Societies INFOCOM*; Anchorage, Alaska, IEEE Computer Society, 2001, p. 1380.

Mills, D. L., "Internet Time Synchronization: The Network Time Protocol," in Z. Yang and T. A. Marsland (eds.), *Global States and Time in Distributed Systems*. New York, NY: IEEE Computer Society Press, 1994.

Niculescu, D., and B. Nath, "Ad Hoc Positioning System (APS)," in *IEEE Global Telecommunications Conference (Globecom) 2001*, Vol. 5; San Antonio, TX, IEEE Communications Society, 2001, p. 2926.

Niculescu, D., and B. Nath, "Trajectory Based Forwarding and Its Applications," in *International Conference on Mobile Computing and Networking (MobiCom)*; San Diego, CA, ACM, 2003a, p. 260.

Niculescu, D., and B. Nath, "Localized Positioning in Ad Hoc Networks," in *IEEE International Workshop on Sensor Network Protocols and Applications;* Anchorage, Alaska, IEEE, 2003b, p. 42.

Patwari, N., and A. Hero, "Using Proximity and Quantized RSS for Sensor Localization in Wireless Networks," in *ACM International Workshop on Wireless Sensor Networks and Applications*; San Diego, CA, ACM, 2003, p. 20.

Pottie, G. J., and W. J. Kaiser, "Wireless Integrated Network Sensors," *Communications of the ACM* 43(5):51, 2000.

Rabaey, J., M. Ammer, J. da Silva, D. Patel, and S. Roundy, "PicoRadio Supports Ad Hoc Ultra-Low Power Wireless Networking," *IEEE Computer Magazine* 33(7):42, 2000.

Salhieh, A., J. Weinmann, M. Kochhal, and L. Schwiebert, "Power Efficient Topologies for Wireless Sensor Networks," in *International Conference on Parallel Processing*; Valencia, Spain, IEEE Computer Society, 2001, p. 156.

Salhieh, A., and L. Schwiebert, "Evaluation of Cartesian-Based Routing Metrics for Wireless Sensor Networks," in *Communication Networks and Distributed Systems Modeling and Simulation (CNDS)*; San Diego, CA, The Society for Modeling and Simulation International, 2004.

Savarese, C., J. Rabaey, and J. Beutel, "Locationing in Distributed Ad-Hoc Wireless Sensor Networks," in *Proceedings of the International Conference on Acoustics, Speech, and Signal Processing (ICASSP)*, Vol. 4; Salt Lake City, Utah, IEEE, 2001, p. 2037.

Savvides, A., C.-C. Han, and M. B. Srivastava, "Dynamic Fine-Grained Localization in Ad-Hoc Networks of Sensors," in *International Conference on Mobile Computing and Networking (MobiCom)*; Rome, Italy, ACM, 2001, p. 166.

Schurgers, C., G. Kulkarni, and M. B. Srivastava, "Distributed On-Demand Address Assignment in Wireless Sensor Networks," *IEEE Transactions on Parallel and Distributed Systems* 13(10):1056, 2002.

Shang, Y., W. Ruml, Y. Zhang, and M. Fromherz, "Localization from Mere Connectivity," in *4th ACM International Symposium on Mobile Ad Hoc Networking and Computing*; Annapolis, Maryland, ACM, 2003, p. 201.

Singh, S., M. Woo, and C. Raghavendra, "Power-Aware Routing in Mobile Ad Hoc Networks," in *International Conference on Mobile Computing and Networking (MobiCom)*; Dallas, TX, ACM, 1998, p. 181.

Sinha, A., and A. Chandrakasan, "Dynamic Power Management in Wireless Sensor Networks," *IEEE Design and Test of Computers* 18(2):62, 2001.

Sohrabi, K., J. Gao, V. Ailawadhi, and G. Pottie, "Protocols for Self-Organization of a Wireless Sensor Network," *IEEE Personal Communications* 7(5):16, 2000.

Zou, Y., and K. Chakrabarty, "Sensor Deployment and Target Localization Based on Virtual Forces," in *Proceedings of the 22nd International Annual Joint Conference of the IEEE Computer and Communications Societies (INFOCOM)*; San Francisco, CA, IEEE Computer Society, 2003, p. 1293.

Chapter

12

Wireless Security

This chapter presents problems relating to computer security. We first discuss the traditional problems of security, and then we focus on how mobile and wireless systems introduce some additional problems and make some traditional problems more difficult. We conclude with a discussion of ad hoc networks and electronic commerce. Chapter 13 will describe how these problems are addressed.

12.1 Traditional Security Issues

In this section we present the traditional security criteria. While these apply to nonmobile computing as well, they are important for mobile computing not only because the same risks are present but also because certain properties of mobile computing *increase* some of these risks, such as eavesdropping in a wireless network. Below we introduce the major categories that describe security problems, specifically integrity, confidentiality, nonrepudiation, and availability.

12.1.1 Integrity

Integrity is the first aspect of providing security we describe. Integrity can refer to either system integrity or data integrity. A system provides integrity if it performs its intended function in an unimpaired manner, free from deliberate or inadvertent unauthorized manipulation of the system. Data maintains its integrity if the receiver of the data can verify that the data have not been modified; in addition, no one should be able to substitute fake data.

As an example of the need for data integrity, consider a company that has a payroll database that determines how much each employee is paid. If an unethical employee can gain access to the database and modify it, he can cause the company to write him paychecks of any amount. The company has a strong motivation to maintain integrity of this data. As another example, consider a law enforcement officer who seizes a disk from a suspect. The officer will need to show that the integrity of the disk's data was maintained if he wishes to submit it as evidence to a court.

12.1.2   Confidentiality

Confidentiality refers to data and is provided when only intended recipient(s) can read the data. Anyone other than the intended recipients either cannot retrieve the data because of access mechanism protections, or other means, such as encryption, protect the data even if they are stolen or intercepted.

In the preceding example, management of the company with the payroll database may not want every employee to know the CEO's salary. Only certain people should be able to have access to this information. Sometimes even the metadata need to have confidentiality preserved. The sender and recipients of an e-mail message may need to be protected as well as the contents of the message. For example, a person may not wish others to know that he subscribes to a newsletter for people who suffer from a particular disease.

12.1.3   Nonrepudiation

Nonrepudiation is a property of data and means that the sender should not be able to falsely deny (i.e., repudiate) sending the data. This property is important for electronic commerce because vendors do not want clients to be able to deny that they made purchases and thus must pay for any services or goods they received.

Currently, credit card companies lose billions of dollars to fraud, and those losses eventually wind up as charges to legitimate customers, as various charges to cover the "cost of doing business" for the company. Any online vendor would want nonrepudiation to prevent customers from claiming that they never made a purchase. Similarly, a law enforcement investigator who finds an incriminating e-mail message sent from a suspect must be able to prevent the suspect from denying that he sent the message, claiming it was a forgery by someone trying to frame him.

12.1.4   Availability

Availability is a property of systems where a third party with no access should not be able to block legitimate parties from using a resource.

Denial-of-service (DoS) attacks are fairly commonplace on the Internet. They can involve one site flooding another with traffic or one site sending a small stream of packets designed to exploit flaws in the operating system's software that take the site down (either crash or hang the operating system or disable any network communication to or from the site).

There have been numerous DoS attacks; notable ones include "syn flood," "smurf," "ping of death," and "teardrop." The "syn flood" attack (http://www.cert.org/advisories/CA-1996-21.html) creates many "half-open" Transmission Control Protocol (TCP) connections so that the target computer no longer accepts any new connections. The "smurf" attack (http://www.cert.org/advisories/CA-1998-01.html) sends an Internet Control Message Protocol (ICMP) packet to a broadcast address resulting in a large number of replies, flooding a local network. The "ping of death" attack (http://www.cert.org/advisories/CA-1996-26.html) crashes target machines by sending them ping packets larger than they can handle. The "teardrop" attack (http://www.cert.org/advisories/CA-1997-28.html) crashes machines that improperly handled fragmented TCP/IP packets with overlapping Internet protocol (IP) fragments.

The Computer Emergency Response Team Coordination Center (CERT/CC) presents information on various DOS attacks on their website (http://www.cert.org/tech_tips/denial_of_service.html). Similarly, SecurityFocus's bugtraq (http://www.securityfocus.com/archive/1) and SecuriTeam (http://www.securiteam.com/) are good forums for information on these problems. Many vulnerabilities are presented in McClure et al. (1999).

Distributed denial-of-service (DDoS) attacks are on the rise. Similar to a DoS attack, the DDoS attack sends a flood of traffic to disable a site. Unlike the DoS attack, though, the DDoS attack employs a large number (hundreds or more) of machines to participate in the attack. In a DoS attack, network engineers can configure firewalls and routers to block traffic from the offending machine. In a DDoS attack, blocking the malicious traffic becomes difficult because there are so many machines involved. The attacking machines typically are *zombies*, compromised machines unrelated to the attacker that run software for the attacker. By using zombies, attackers hide their trail because the zombies typically have all logging disabled and hide most of their connections to the attacker.

Any machine can be a zombie, from a business's machine, to a laboratory full of PCs, to a home user with a cable modem. Home users are popular targets because high-speed home Internet connections such as cable modems and digital subscriber line (DSL) connections allow their computers to be online continuously, many home computers are poorly protected, and few home users know how to determine if their machine is running some malicious programs in the background. Any individual flow from a DDoS attack may be unnoticeable—it is the aggregate flow, directed at a single site, that becomes devastating (Gibson, 2004).

Another aspect of attacks against availability is that the effects can either be temporary or permanent. Temporary attacks only prevent system availability while the attack occurs, whereas a permanent attack keeps the system down even after the attack ends. A permanent attack might crash the operating system (or the networking parts of the operating system) by sending a single malformed packet that keeps the system down until an operator reboots it. Alternatively, if the system runs on batteries, the attack may disable the system permanently by causing it to waste all its power on useless transmissions. In some cases the batteries are not intended to be replaced. See Section 13.2.2 for more details on resource-depletion/exhaustion attacks.

In January 2003, the Sapphire/Slammer SQL worm used a known buffer overflow vulnerability in Microsoft's SQL server to propagate itself across the Internet. It spread extremely rapidly, doubling the number of infected machines approximately every 8.5 seconds. After 3 minutes the propagation rate flattened out because of the adverse effect it had on the Internet by consuming nearly available bandwidth. Whether intended or not, it caused a massive DDoS attack on the Internet (Moore et al., 2003).

12.2   Mobile and Wireless Security Issues

Until now, we discussed general security issues that apply to all systems. We now focus on some of the problems that are either unique to or exacerbated by mobile computing, especially wireless systems. We will discuss detectability, limited resources, interception, and theft-of-service problems.

12.2.1   Detectability

One problem associated with wireless communication is *detectability*. Nonmobile users typically do not face this problem. In some circumstances, the mobile users do not want their wireless system to be detected, and this is part of the reason they are mobile. The best example of this is a military one, such as a soldier in the field not wanting the enemy to detect the presence of his mobile communication system and hence his physical location.

Even if strong encryption is used and the data cannot be deciphered, the mere presence of the signal can put the user at risk. If the enemy can detect the signal and locate its position, the device can be jammed by local radio frequency (RF) interference, the soldier can be captured by troops sent to that location, or he can be killed by remote weapons that target that location (e.g., bombs, artillery shells, etc.).

For maximum safety, the signal should not be able to be detected. It can change frequencies, try to blend in with background signals, use very directional antennas, or use the minimal power required for reception.

12.2.2 Resource depletion/exhaustion

Another problem unique to mobile systems is that the resources often are very limited. To keep the mobile unit small and lightweight, the designers often make compromises. The CPU speed may be an order of magnitude or more slower than that of conventional desktop machines. The network bandwidth may be similarly limited.

The biggest constraint on these systems is the battery. Often these systems run on internal batteries because AC power is not available owing to location (e.g., being outside) or because they are moving continually and would require a very long and impractical extension cord. The normal lifespan of the batteries in such devices may be measured in hours or weeks, but they are finite. Additionally, their lifespan is related to their activity, so the more a unit must compute and transmit, the shorter the time its battery lasts.

This leaves these devices open to resource-depletion and exhaustion attacks. The former involves an attack that shortens the lifespan of the battery, causing it to fail "naturally" at a later date but much sooner than it would normally. The latter involves an attack that consumes (and wastes) all the power in the battery, leaving the unit unable to function. In ad hoc networks, these attacks can cause key routing nodes in the network to fail, leaving parts of the network unreachable.

12.2.3 Physical intercept problems

One major difference between wired and wireless systems is the ease of physical intercept of the signal. In wireless systems, the signal is broadcast through the air, where any receiver can intercept it. This problem is related to the detectability problem because once the signal can be discerned, the data can be read. In general, the approaches to mitigate this problem involve directional antennas, low-power transmissions, and frequency-hopping/spread-spectrum technology at the physical layer and encryption techniques at higher layers. Later chapters will show how this problem is handled by specific networks, such as personal and local area networks.

12.2.4 Theft of service

A final problem we will discuss is theft of service. While this problem has plagued computer systems seemingly forever, wireless systems are particularly prone to it. Normally, a system requires a user name and password to gain access to it. However, it is common to install wireless local area network systems simply by taking them "out of the box" and plugging them into the network, and they "just work." Typically, by default, security settings are disabled, and if not, then the factory-set default passwords are commonly known. Unauthorized users simply need to be nearby, and their computers will connect automatically to the wireless network, get a dynamically assigned IP address, and put them on the Internet. It is so simple that users may not realize that they are on the *wrong* network. For example, a user may be sitting outside a coffee shop known to have a public wireless Internet connection. A business next door also may have a wireless network, and because of the user's physical location, she may have better reception to the next-door business's wireless network and connect to it, thinking that she is connecting to the coffee shop's network. Of course, this example assumes that the user has no malicious intent, nor any knowledge of what network she is using. Much effort has been put forth to document many of these networks, as we will see below.

12.2.5   War driving/walking/chalking

There is a whole class of *war* terms that originate from the term *war dialing*. Back in the 1980s, before the widespread popularity of the Internet, hackers and crackers would search for phone numbers with modems attached to them by using programs that would dial every number in an exchange and listen for the modem tones. The 1983 movie *War Games* featured the use of war dialing by the main character.

Since then, as wireless local area networks gained popularity faster than any awareness of any best practices in security, the new fad is to find wireless networks. *War driving* is the wireless equivalent of war dialing. The technique involves taking a computer with a wireless card running some detection software (http://www.netstumbler.com; http://kismetwireless.net; http://airsnort.schmoo.com; http://wardriving.com) and optionally a Global Positioning System (GPS) and driving around a city. The software detects the presence of wireless networks, and the GPS gives the location for later reference. There are sites that have composite maps built up of wireless access points and their protections, if any. Users can then look up the location of the nearest Internet connection. Of course, the legality of this is questionable and is beyond the scope of our discussion.

War walking and similar variants (e.g., war flying) reflect different modes of transportation. In this case, the term refers to scanning for wireless networks by using a lightweight computer (personal digital assistant or palmtop or small laptop) and walking around an area. A pedestrian often can get closer to the perimeter of a site than a vehicle and can get a more accurate picture of wireless networks that bleed out through the walls of buildings. On the other hand, war walking is slower than war driving and covers a much smaller area individually.

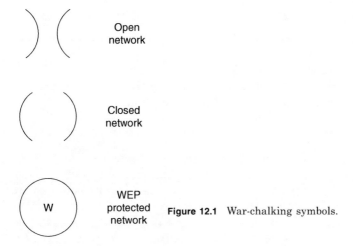

Figure 12.1 War-chalking symbols.

One other variant that started to become popular in 2002 is *war chalking*, which is the practice of marking the presence of wireless networks with chalk either on sidewalks or on the sides of buildings. The three symbols shown in Fig. 12.1 represent an open network, a closed network, and a Wired Equivalent Privacy (WEP) password protected network.

Between the war chalkers and the extensive databases to query on the locations of open wireless networks on the Internet, the presence of wireless networks must be considered public knowledge. Administrators cannot rely on any "security through obscurity" techniques to preserve the anonymity of their networks or, worse, their network's unprotected status.

12.3 Mobility

One of the essential characteristics of mobile computing is that the locations of the nodes change. Mobility provides many freedoms, but it also increases several security risks. Dynamically changing routes, potential lack of a trusted path, disconnected operation, and power limitations all increase the security risks.

12.4 Problems in Ad Hoc Networks

Single-hop and *multihop* are terms to describe infrastructure versus ad hoc–based networks. In the former, wireless stations communicate directly with the base station, whereas ad hoc networks must propagate messages from one wireless station to the next until they reach the destination or a boarder (typically the Internet). Ad hoc networks form on the fly, without a fixed infrastructure.

Data in ad hoc networks typically pass through several other ad hoc nodes. Typically, there is no guarantee as to the identity of these intermediate nodes, so "man in the middle" attacks can be used to copy or corrupt data in transit. Because nodes are mobile, the route between any two nodes is dynamic, even if the endpoints are stationary.

12.4.1 Routing

There are several security risks associated with routing. The first is *spoofing*, in which one node impersonates another. An example of this type of attack is "ARP spoofing", in which one machine responds to an Address Resolution Protocol (ARP) request meant for another machine. After doing so, traffic intended for the original machine is sent to the impersonator. Another attack "ARP cache poisoning," causes all traffic to pass through a malicious node that permits "man-in-the-middle" attacks in which the malicious node sees all traffic and can either silently copy it or actively corrupt it as it passes through.

A route between two nodes can be disabled by two malicious nodes that share a common segment along the route. In an ad hoc network, key routing nodes can be disabled via a resource-exhaustion attack in this manner.

12.4.2 Prekeying

We will cover encryption in Chap. 13, but one problem when using encryption or authentication is key management, which involves creating, sharing, storing, and revoking encryption keys. Public key encryption (described in Section 13.2) is one way to avoid needing a key exchange. If a symmetric key algorithm is used, then the two endpoints must agree on a key, either via a key-exchange protocol, such as IKE or Diffie-Hellman (Harkins and Carrel, 1998; Diffie and Hellman, 1976) or decide on a key a priori. Of course, if a fixed key is used, then there is no easy way to handle compromised keys, nor to change them periodically to avoid the risk of exposure through use. For some applications, because of either network or processor limitations or application constraints, a high-cost key-exchange protocol is not practical. For example, an ad hoc network of low-powered sensors that report weather data periodically, such as temperature and dew point, that is intended to run only for a week or so probably does not have the processing power to handle a key-exchange protocol. In addition, the raw data from individual nodes are of little value to any adversary. And since there are numerous sensors, forged data from a single node will not have a great affect on the overall "big picture" that is derived from the data. Alternatively, other applications need and can support protection schemes that use a key-exchange protocol. Typically, the problems with prekeying make it an undesirable approach unless the data sensitivity and node life spans are very limited. Section 9.1.4 discusses the problem of prekeying on microsensor networks.

12.4.3 Reconfiguring

Reconfiguring poses another problem in ad hoc networking. Because ad hoc networks are, by nature, dynamic, as nodes move they go out of radio contact with some nodes and come into contact with other nodes. The network topology itself changes over time. This means that a previous route from node X to node Y may no longer work. Ad hoc routing algorithms must be able to reconfigure the underlying view of the network dynamically. In addition, some segments of the network may become unreachable. Two nodes that were in contact previously now may lose contact, even when both nodes are stationary (if nodes along the path are mobile). Applications running in this environment must be fault-tolerant. And finally, as a node moves, it may establish contact with a completely different network. Again, the software must handle these changes, and the security components must provide protection rather than simply leaving nodes and the network "wide open."

12.4.4 Hostile environment

The mobile environment is often more hostile than the nonmobile one. In a non-mobile environment, physical boundaries and barriers have more meaning and change less frequently. In a mobile environment, eavesdropping is easier. Physical locations often are not secured, e.g., coffee shops and airports, and nodes go in and out of contact regularly. In some contexts, such as an ad hoc network of soldiers, even signal detection could cause them to be captured or worse.

12.5 Additional Issues: Commerce

Electronic commerce is a prime application of and domain for mobile computing. The vast commercial potential for this drives the development and deployment of the technology, making commonplace mobile computing a reality. Generally, security is at odds with convenience, and in a commercial market, convenience takes precedence. Businesses often regard the loss of revenue from fraud and theft as the "cost of doing business" and calculate it into their fees. For example, credit card companies lose billions of dollars a year on fraud; these losses are eventually passed on to the end customer.

Because of the importance of electronic commerce, we will discuss some of the security concerns that have the greatest impact on it.

12.5.1 Liability

Currently, liability issues relating to computer security are still being determined. Not only do computers contain potentially useful information, such as customer credit card numbers, but machines on the Internet also can be used as springboards to launch attacks on other Internet computers. Some businesses have been sued because of their lack of "due diligence" by not installing patches, antivirus software, and similar protection.

12.5.2 Fear, uncertainty, and doubt

Another aspect of electronic commerce is the "intangibles" from public perceptions. For example, a company that makes children's cartoons places a great deal of value on their wholesome, family-friendly image. Anything perceived to damage that image, such as having the company's Web site defaced by pornography, would be of more concern to the company than, say, copyright infringement, which can be handled by a lawsuit.

Companies that suffer from break-ins often are reluctant to report them because of the fear that such a report will hurt their reputations. Often, the costs of these break-ins simply are included as the cost of doing business.

Often security is perceived as a one-shot application, if at all. Install virus protection or a firewall, and the system is protected—end of story. Many places do not even go that far and simply feel that security can be added later, if needed. This is in the commercial domain; in the home computer domain, security is even scarcer, even though many home sites have high-speed, continuous connections to the Internet. The result is that known bugs and old exploit scripts continue to work years after they are published; for example, Web sites still see attempted attacks by the "code red" worm from 2001, years after it was first released.

12.5.3 Fraud

One of the biggest problems with electronic commerce is fraud, typically purchases billed with stolen or faked credit card numbers. The credit card numbers can be stolen by breaking into machines and getting customer lists or through many other means, including physical means, such as "dumpster diving." Often purchases are made to fake addresses.

One scam involved placing orders to be delivered to legitimate addresses. These often were in wealthy neighborhoods, with houses that had long driveways with mailboxes at the end of the driveway—driveways long enough that the delivery drivers could drop off a package in the mailbox, the thief who had ordered it could later retrieve the package, and the owner of the house would never see it or know of any package.

This is not a new problem, just another form of nonrepudiation. Because the recipient of the merchandise legitimately *can* repudiate (i.e., deny) the claim that she purchased the merchandise, the online vendor must take the loss because it was a *card not present* (CNP) purchase, in which there is no signature of the card owner. If there *were* nonrepudiation, the true purchaser of the merchandise would not be able to fake the credentials of another person, such as an "unforgeable" signature.

12.5.4 Big bucks at stake

The amount of money lost is large—billions of dollars each year. The potential market, from hardware to infrastructure providers to online merchants and services, is vast. The boom of the dot-com companies of the late 1990s attests to this. While many companies with shaky business plans failed, Amazon, eBay, and other vendors have demonstrated that electronic commerce can be profitable. Many vendors faced DoS attacks in 2000 (Messamer and Pappalardo, 2001), including Yahoo, eBay, CNN, and eTrade. Since margins are often small, profit lost from fraud or other security-related problems can push a business over the edge into failure. Computer security is essential to electronic commerce.

12.6 Additional Types of Attacks

We conclude the chapter with a discussion of a few specific types of attacks, specifically "man in the middle" attacks, traffic analysis, replay attacks, and buffer-overflow attacks.

12.6.1 "Man in the middle" attacks

A *man in the middle* attack occurs when a malicious node inserts itself in the path between two nodes. The most obvious use is to eavesdrop on the conversation between the two nodes. In addition, the malicious node may choose to modify the data from the source before it forwards them to the recipient or simply to drop certain packets.

A common way for such an attack to occur is through "ARP cache poisoning," in which the malicious node replies to an Address Resolution Protocol (ARP) query claiming to be the IP address of the real next node on the route to the destination. This may require launching a brief DoS attack on the real node to prevent it from being able to answer the ARP query in a timely fashion. By redirecting all traffic through itself, the malicious node can even defeat the protections of isolating traffic provided by switched Ethernets over the nonswitched, shared networks.

12.6.2 Traffic analysis

Even without decrypting the data, attackers can gain insight about the meaning of the data by observing properties such as message sizes, communication pairs, and the sequence of encrypted back-and-forth conversations. This technique is called *traffic analysis*. There are stories about pizza deliveries to the Pentagon dramatically increasing the night before military actions; another, during World War II, describes numerous science fiction magazine subscriptions noted to places such as Oak Ridge and Los Alamos (Spencer, 1990). If true, even without knowing the content of what was occurring, an adversary could make guesses and gain more information than is desired. Similarly, an adversary can surmise a great deal about the meaning of different communication flows without knowing the actual content because of standard protocol headers at each communication layer. For example, a typical Simple Mail Transfer Protocol (SMTP) conversation includes "HELO," "MAIL," "RCPT," and "DATA" messages, in addition to TCP protocol headers. Without knowing the content of the message, an observer may infer that one site is sending mail to another by observing 3 brief messages of the appropriate size followed by a longer ("DATA") message with the appropriate timing between the messages.

12.6.3 Replay attacks

A *replay attack* involves reusing data in a packet observed by a malicious node. For example, if a malicious node observes an authentication sequence, it may be able to resend the reply sequence if the authenticator issues the same challenge. Often the attacker must build up a large repository of previously seen challenges and responses, but disk space is cheap. The simplest version of this attack is observing a username and password if they are sent as plain text.

Replay attacks can work even if the data are encrypted if the malicious node has observed the challenge before. Replay attacks can be defeated by including information that cannot be reused in the packets, called a *nonce*, such as a sequence numbers or timestamps.

12.6.4 Buffer-overflow attacks

Buffer-overflow attacks are not characteristic of mobile or wireless systems or even networking or routing per se. But they are such a common type of attack that they deserve a mention. A *buffer-overflow attack* occurs when a code segment that reads input does not perform any bounds checking on the amount of input it accepts. If the input exceeds the input buffer space allocated for it, it overwrites memory adjacent to the buffer. Typically, the calling stack or data for local variables gets corrupted. The buffer overflow is designed so that the extra data cause the program to execute different code by changing variable values, program flow, or similar. Commonly, the overflow code causes a command shell to execute. If these attacks are performed on a program running with root privileges, then the shell has root as well.

Some languages, such as Java, have built-in boundary checking to prevent buffer overflows. Others, such as Perl and Lisp, automatically increase buffer sizes as required. These techniques eliminate the effectiveness of many of the common buffer-overflow attacks. However, there is an enormous code base still out there with vulnerable code, much of it low-level operating system code. The effort to review, let alone fix, this code is equally enormous. Thus there will continue to be vulnerable code for quite some time.

12.7 Summary

In this chapter we presented four important security criteria and how certain attacks prevent a system from achieving these criteria. We also described how the mobile computing environment introduces further complications and risks. In Chap. 13 we focus on some general approaches to handling the security problems mentioned in this chapter.

12.8 References

Diffie, W., and M. Hellman, "New Directions in Cryptography," IEEE Transactions on Information Theory, Volume IT-23, number 6, pp. 644–654, 1976.

Gibson, S., "Distributed Reflection Denial of Service," Gibson Research Corp., *http://grc.com/dos/drdos.htm.*

Harkins, D., and D. Carrel, "The Internet Key Exchange (IKE)," November 1998, *http://www.ietf.org/rfc/rfc2409.txt.*

McClure, S., J. Scambray, and G. Kurtz, *Hacking Exposed.* Berkeley, CA: Osborne/McGraw-Hill, 1999.

Messamer, E., and D. Pappalardo, "One Year after DoS Attacks, Vulnerabilities Remain," CNN.com/SCI-TECH, February 8, 2001, *http://www.cnn.com/2001/TECH/internet/02/08/ddos.anniversary.idg/index.html.*

Moore, D., V. Paxson, S. Savage, C. Shannon, S. Staniford, and N. Weaver, "The Spread of the Sapphire/Slammer Worm," February 2003, *http://www.cs.berkeley.edu/~nweaver/sapphire/.*

Spencer, H., "Re: Pentagon Pizza," comp.risks 10(16), July 31, 1990, *http://catless.ncl.ac.uk/Risks/10.16.html#subj11.1.*

Approaches to Security

In Chap. 12 we discussed some of the problems associated with providing security to mobile computing. This chapter presents, at a high level, several approaches taken to protecting data. We will see these methods employed in subsequent chapters.

13.1 Limit the Signal

The first way to protect data is to limit the accessibility of the signal. At the lowest levels, this means restricting who can see the signal on the wire.

13.1.1 Wire integrity and tapping

In a traditional Ethernet, signals do not radiate far beyond the wire, so eavesdropping requires physical proximity to the network cable and physically tapping the line. This is done by splicing into the line (old "thicknet" Ethernet used "vampire" taps to add stations to the network that actually tapped into the wire), stealing a line from a hub (which splits a network into a star topology), bridging the network, or similar techniques. Tapping can be detected by the change in electrical resistance in the wire. In addition, wires can be placed inside protective conduit, making it harder to tap. To increase the protection, the conduits can be pressurized with an inert gas (e.g., N_2) so that any physical taps cause a pressure drop, allowing the location (within a segment) of the tap to be detected.

13.1.2 Physical limitation

The problem then becomes physically limiting who has access to the machines or the endpoints once the problem of tapping the wire between endpoints has been addressed. Machines need to run secure (enough) operating systems and have accounts and secure passwords. Users need to log out or lock screens if they are not present to prevent "lunchtime hacking," in which an unauthorized user sneaks into an authorized user's office to use his computer while the authorized user is gone. Another approach to deal with this is to use a behavior-based intrusion detection system that detects changes in the usage pattern.

The problem in public computer areas or labs still exists, in which an unauthorized user can unplug the cable to a protected computer and plug it into her own. Routers and switches can filter *outbound* packets and block access to an unknown Ethernet card. However, a user can reboot a legitimate machine using her own floppy disk or CD-ROM, running her own operating system. To prevent this, machines can be configured so that they cannot be booted from these devices.

If the information is important enough to protect, additional protection can be added to any computer area, such as surveillance cameras or human guards. It becomes a game of cat-and-mouse, in which the weak link in the system moves elsewhere to a different (and stronger) component of the system, making it more difficult to subvert the system security. The operating expense also increases with protection.

To increase protection further, data on the network should be encrypted so that even if there is a physical breach of the network and someone can eavesdrop, the stolen data will be meaningless. We discuss the basics of encryption in the next section.

13.2   Encryption

One of the most common and useful techniques to protect data is encryption. Encryption has been used for thousands of years; in fact, a simple technique called the *Caesar cipher* was said to be used by Julius Caesar. Since modern computer speeds have rendered many techniques insecure, we will focus on the essential components of the methods that are used currently. As computers get faster, key lengths may need to increase; however, since each bit added to a key doubles the time required to break the algorithm via a brute-force attack, the algorithms are scalable, assuming that no fundamental flaws are discovered. Encryption methods can be divided into two classes: public key and private key. Both of these are discussed below.

13.2.1   Public and private key encryption

Private key or symmetric encryption.   In *symmetric* or *private key encryption*, the same key is used to encrypt as well as decrypt the data. Both parties share the key, which must be kept secret. The encryption and decryption functions can be described as: $C = E_k(P)$ $P = D_k(C)$.

The encryption and decryption functions, E and D, respectively, may be the same function or may be different, depending on the algorithm used. The encryption function takes the key k and plaintext P and produces ciphertext C as output. The decryption function takes the key k and ciphertext C and produces plaintext P as output.

Public key or asymmetric encryption.   In *asymmetric* or *public key encryption*, there are two keys, a *public key* and a *private key*. As the names imply, the private key is known only by the owner, whereas the public key is made publicly available. The public and private key encryption and decryption can be described as:

$$C = E_{\text{private}}(P) \quad P = D_{\text{public}}(C)$$

$$C = E_{\text{public}}(P) \quad P = D_{\text{private}}(C)$$

Text that is encrypted with the private key can be decrypted only with the public key, and text that is encrypted with the public key can be decrypted only with the private key.

13.2.2   Computational and data overhead

Encryption is an effective way to protect information. However, there are costs associated with using encryption. In this section we discuss the computational and data overhead associated with using encryption (and decryption).

Private key algorithms generally run significantly faster than public key ones. This computational overhead becomes a factor when significant amounts of data must be encrypted (or decrypted) or if the processing power of the computer is limited, as in palmtop devices or sensor networks.

Block-oriented algorithms might pad the plaintext (e.g., with spaces or nulls) to fill up a block, but the ciphertext is generally the same size as the plaintext. Thus the data overhead from padding is minimal, especially considering modern data transfer rates and the size of primary and secondary storage. The keys themselves are not generally a significant overhead, unless there are many of them; a 1024-bit key consumes only 128 bytes. However, because a message encrypted using a public key can only be decrypted by a single recipient, i.e., the owner of the private key, a message sent to several recipients must be encrypted separately, once for each recipient, thus significantly increasing the size of the encrypted message. The same message could be encrypted only once with the sender's private key, but that will not provide privacy because anyone could obtain and use the sender's public key to decrypt the message. Each recipient causes the message size to increase by the size of the message to encrypt. In this approach, for n recipients and a message length of $|M|$, the message overhead is $n \times |M|$.

Hybrid approaches reduce these costs. For example, Pretty Good Privacy (PGP) (RSA, 2004; http://www.pgp.com) combines public and private key encryption. The data are encrypted using a symmetric key approach and a randomly generated one-time key (a *session key*). The session key itself is encrypted using public key encryption. This method allows the text of the message to be encrypted using the faster symmetric key approach. If the sender wants to include additional recipients, only the symmetric key itself must be encrypted using a different public key rather than the entire message. By reducing this overhead, PGP scales well for multiple recipients. In this approach, for n recipients, a key length of $|K|$, and a message length of $|M|$, the message overhead is $n \times |K| + |M|$. A strong session key is on the order of 128–256 bits (16–32 bytes); generally, $|K| \ll |M|$, therefore the message overhead of this approach is significantly less than that of a public key approach.

13.3   Integrity Codes

Integrity codes are used to detect changes to data from either accidental or deliberate errors, such as during transmission or from a compromised middleman. We describe the principles behind the techniques and discuss how they can be used to help protect data.

13.3.1   Checksum versus cryptographic hash

Checksums are a common type of basic integrity code and are transmitted with the data. The receiver applies a simple formula to the data and compares the result with the integrity code that was transmitted to ensure with high probability that he received the message intact. Parity bits and cyclic redundancy checks (CRCs) are examples of checksums.

Parity bits, used by (very) old modems, are a simple example of a checksum. In even parity, all the bits in a byte added together must be even, and if not, the parity bit is set to 1 (to make it even). Odd parity is similar, but the bits must sum to an odd number. This method can detect single-bit errors.

A *cyclic redundancy check* (CRC) is more sophisticated checksum that uses polynomial math to compute the parity bits. Thus the input is divided by a polynomial term, and the remainder represents the CRC. This operation is equivalent to binary arithmetic with no carries. The polynomial used for CRC-32 is $G(x) = x^{32} + x^{26} + x^{23} + x^{22} + x^{16} + x^{12} + x^{11} + x^{10} + x^8 + x^7 + x^5 + x^4 + x^2 + x + 1$ and was selected because its properties allow it to detect single-bit errors, two-bit errors, errors with an odd number of bits, and burst errors. CRC-32 is a 32-bit CRC used in many applications, including Ethernet and file compression. For a detailed explanation of the mathematics behind CRC computation and implementations of CRC algorithms in C, refer to Williams (1993).

The main problem with checksums is that they are easily predicted and forged. If a bit is changed, it is easy to determine how the checksum will change. Similarly, for every bit change, it is easy to determine an additional bit in the stream to change to counteract the effect of the first change, thus leaving the unchanged checksum still valid. Thus, while checksums can detect simple transmission and other accidental errors, they cannot detect malicious errors in which the data are modified *intentionally* (see Section 15.3.6). Protection from malicious changes requires a cryptographically secure hash.

A *hash function* is a function that takes an input of arbitrary size and produces a fixed-size output. A *one-way hash function* is a function that is computationally infeasible to invert; i.e., the source cannot be easily determined given the hash (RSA, 2004; Schneier, 1996).

A *cryptographically secure hash* is a one-way hash function in which a small change in the input results in a very large change in the hash. Typically, a single-bit change in the source changes approximately half the bits in the hash. MD2, MD5, and SHA (RSA, 2004; Schneier, 1996) are examples of cryptographically secure hash functions. These functions protect the integrity of data, when used in a message authentication code (MACs), discussed in the next section.

13.3.2 Message authentication code (MAC)

A *message authentication code* (MAC) is a one-way hash function plus a secret key. MACs can be performed in two ways, as shown in Fig. 13.1. In the first approach, the function computes the MAC by encrypting the hash of the message using the key; in the second approach, the function computes the MAC by taking the hash of the message and the key concatenated together. MACs protect the message's authenticity without secrecy. The MACs are attached to the message.

MACs serve as a type of tamperproof seal, allowing the easy detection of changes to the original message. Because the MAC uses a cryptographically secure hash, any change to the message results in a very different hash, and it is computationally infeasible to determine how to change the message so that it once again matches the MAC. Since a (secret) key is required to create the MAC, the MAC cannot be replaced with one that matches the modified message without access to the key (which, presumably, is not compromised). MACs are a relatively low-cost way to protect the integrity of a message. There are various ways to

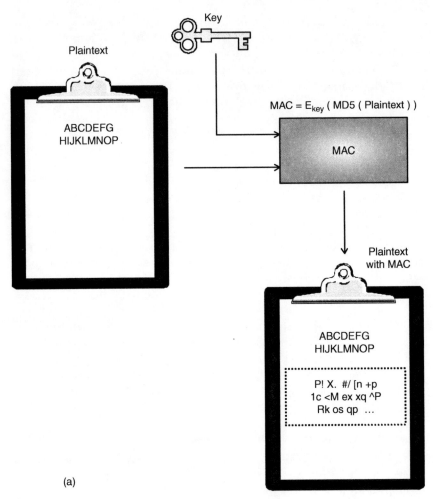

$$MAC = E_{key} \, (\, MD5 \, (\, Plaintext \,) \,)$$

(a)

Figure 13.1   Two methods to compute a MAC.

create a MAC, e.g., putting the key at the beginning or at the end of a message. In the book *Applied Cryptography*, Schneier (1996) presents several algorithms to compute MACs, such as MAA and RIPE-MAC, and one-way hash functions and stream ciphers.

13.3.3   Payload versus header

A packet consists of a *header* and a *payload*. The header contains source and destination address information and additional information such as the protocol, sequence number, and special flags. The payload is the

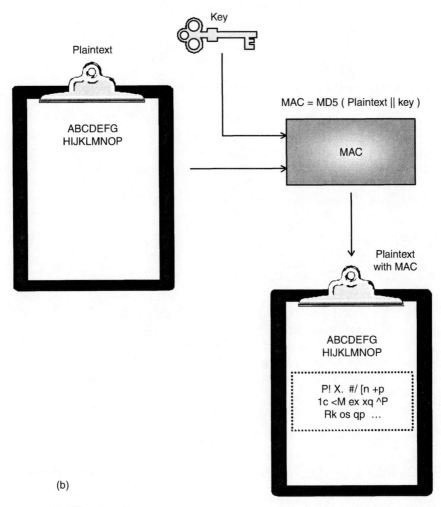

(b)

Figure 13.1   *(Continued)*

data. Protocols can be layered, so one layer's payload can be a complete packet (header and payload) from another higher layer. One question to consider when using encryption is whether to encrypt just the payload or the entire packet.

The simplest approach is to encrypt just the payload. This simplifies the process because only the endpoints need to have keys or perform any encryption or decryption. However, there is a lot of information in the packet headers that can be useful for attackers. Even exposing the sender and recipients can be more than is desired. Another alternative is to encrypt the packet header as well.

13.3.4   Traffic analysis

Even with encryption and integrity codes, packets are still vulnerable to a traffic analysis attack. This is particularly problematic if the header and payload are encrypted together because many parts of the header are predictable. And given enough samples of plaintext and the equivalent ciphertext, the key can sometimes be determined. An observer near either the source or the destination of the packets can infer quite a bit by just watching the patterns of data flow without knowing the contents of the flow. IP Security (IPSec) addresses some of the threats of traffic analysis, especially by using its *tunnel mode.*

13.4   IPSec

IP Security (IPSec) are protocols that provide security for Internet Protocol (IP) packets (IETF, 2004; Kent, 1998). There are three main components to IPSec. The *Internet key exchange* (IKE) defines a hybrid protocol to negotiate and provide authenticated keying material for security associations in a protected manner. The *authentication header* (AH) provides message integrity. The *encapsulating security payload* (ESP) provides confidentiality. We discuss AH and ESP in the following subsections.

13.4.1   Authentication header (AH)

The AH uses a MAC (see Section 13.3.2), referred to as an *integrity check value* (ICV), to guarantee the integrity of the data. It also can prevent replay attacks (see Section 12.6.3) by including a sequence number in the header. Similar to ESP (described below), it can operate in *transport* or *tunnel* mode. In transport mode, the AH information is added immediately following the IP header information; in tunnel mode, the entire original IP datagram becomes the payload of the IPSec packet, with the AH providing integrity for both its headers and the payload.

For point-to-point (i.e., nonmulticast) communication, AH algorithms include "keyed MACs based on symmetric encryption algorithms (e.g., DES) or on one-way hash functions (e.g., MD5 or SHA-1) (Kent, 1998a)."

On receiving a packet, the destination checks the packet's sequence number to make sure that the packet is not a duplicate and discards it if it is. The destination then computes the MAC for the packet and compares it with the ICV in the packet. If the values match, it accepts the packet. If not, it discards the packet and makes an entry in an audit log.

13.4.2 Encapsulating security payload (ESP)

The ESP is a mechanism to provide confidentiality and integrity to data by encrypting the payload. ESP operates in one of two modes, tunnel mode or transport mode, and the packet's payload consists of either the upper-layer protocol (e.g., TCP, UDP, ICMP, or IGMP) or the entire IP datagram, respectively. The ESP consists of the encrypted data plus additional data fields, such as initialization vectors, integrity check values, padding, etc., as needed by the particular encryption algorithm used. In *transport mode*, the ESP is added after the IP header, before any upper-layer protocols. The original IP headers are still visible. In *tunnel mode*, the *entire* IP datagram is encrypted within the ESP. Tunnel mode can be used by security gateways. The end-point hosts communicate with the security gateways (through a protected intranet), and the security gateways communicate with each other via IPSec. The only plaintext an eavesdropper on the Internet can see is the security gateway addresses. The eavesdropper will *not* see the true source and destination because they are encrypted, contained within the ESP. This technique helps to prevent traffic analysis attacks because the eavesdropper sees only a collection of packets and cannot tell individual flows or conversations that comprise the flow. Figure 13.2 illustrates how ESP works. RFC 2406 details ESP (Kent, 1998b).

13.5 Other Security-Related Mechanisms

This chapter concludes with a brief summary of additional methods used to secure systems.

13.5.1 Authentication protocols

Authentication protocols provide a mechanism to verify a user's claimed identity when he establishes a connection to a remote system. Protocols that allow a remote system login or code execution, such as telnet, ssh (http://www.openssh.com; http://www.ietf.org/html.charters/secsh-charter.html), and RPC, use the remote operating system's authentication mechanism (typically a username and password combination). Other protocols use different mechanisms, such as the Point-to-Point Protocol (PPP), which is used commonly to provide an IP connection over a serial (modem) line. Two mechanisms for authentication are the Challenge Handshake Authentication Protocol (CHAP) and the Lightweight Extensible Authentication Protocol (LEAP), which are based on the Extensible Authentication Protocol (EAP). We describe each below.

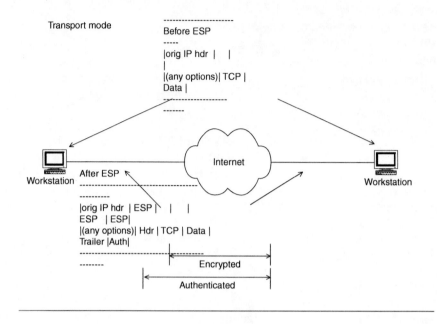

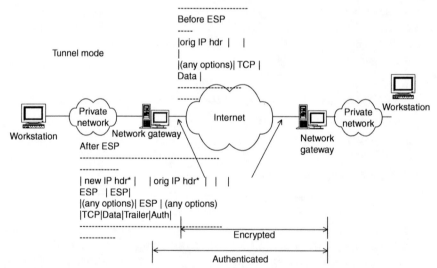

Figure 13.2 Tunnel and transport mode.

Challenge handshake authentication protocol (CHAP). The Challenge Handshake Authentication Protocol (CHAP) is an example of a popular protocol that authenticates users of the Point-to-Point Protocol (Simpson, 1996) based dial-up systems. CHAP allows one system to identify itself to another when they both have a shared secret. The *authenticator* is the system at the end of the link with resource to grant. It grants its resources to remote systems only after they have completed authentication. The *peer* is the system attempting to connect to the authenticator. If both systems require authentication, the authenticator and the peer can switch roles and use CHAP again in the other direction.

In CHAP, the authenticator issues a *challenge* to the peer. The peer sends a *response* back to the authenticator, and the authenticator then replies with "success" if the peer properly identified itself or "failure" otherwise. After success, the authenticator issues new challenges periodically. Figure 13.3 shows an example of the challenge/response conversation in CHAP.

The challenge consists of an identifier and a value. The response is generated by computing the MD5 hash of the identifier concatenated to the shared secret concatenated to the value. The authenticator computes this value too, compares it with the peer's response and issues a success message if they match and a failure message if not.

The algorithm is fast and simple. The identifier number in the challenge helps to prevent replay attacks because the identifiers are not reused. Since the authenticator rechallenges the peer again periodically, the time of exposure to a particular attack is limited. However, this

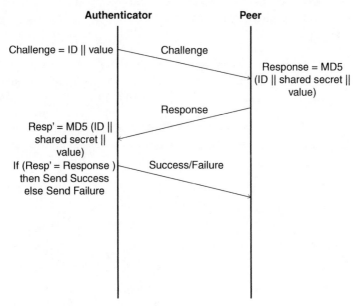

Figure 13.3 CHAP authentication.

protocol does require a shared secret between the two systems. Operating system passwords typically are stored as one-way hashes that cannot be reversed and thus cannot be used with CHAP. Therefore, the plaintext password must be available somewhere. If a third-party performs the authentication, then the shared secret must be transferred there by some mechanism, which creates a risk of exposure.

Extensible authentication protocol (EAP).   The Extensible Authentication Protocol (EAP), as the name suggests, is a generic authentication protocol that allows different authentication methods to be used and new protocols to be added easily. PPP supports EAP (Blunk and Vollbrecht, 1998; Aboba and Simon, 1999), as does WEP and 802.11i (see Sections 15.4 and 15.5).

There are four different types of EAP packets: *request, response, success,* and *failure.* The request packet, shown in Fig. 13.4, is sent by the authenticator to the peer (or "supplicant" in 802.1x terminology (see Section 15.5.2)) and contains a "type" field to indicate what is being requested. The "identifier" field allows the peer to differentiate between retransmissions of old requests and new requests. The "type" field specifies what type of authentication mechanism is being used. A value of 13 specifies EAP Transport Layer Security (TLS) (Aboba et al., 2004). The "type-data" field contains data specific to the authentication mechanism used. The response packet shares the same format as the request and is sent from the peer to the authenticator in response to a request. The code indicates the type of packet, 1 for request and 2 for response. The "flags" field allows specifying message fragmentation, start messages, and the presence of the TLS "message length" field.

EAP initially defines several types of request/response types and allows for additional ones to be added to the standard, thus making it an extensible protocol. The protocol allows the peer to negotiate with the authenticator as to the type of authentication used. All implementations must support the "MD5-challenge" type, which is similar to CHAP with MD5 specified as the algorithm. The authenticator sends a "challenge" to the peer as a request, and the peer sends a "response" that is calculated by

```
0                   1                   2                   3
0 1 2 3 4 5 6 7 8 9 0 1 2 3 4 5 6 7 8 9 0 1 2 3 4 5 6 7 8 9 0 1
+-+-+-+-+-+-+-+-+-+-+-+-+-+-+-+-+-+-+-+-+-+-+-+-+-+-+-+-+-+-+-+-+
|      Code     |   Identifier  |             Length            |
+-+-+-+-+-+-+-+-+-+-+-+-+-+-+-+-+-+-+-+-+-+-+-+-+-+-+-+-+-+-+-+-+
|      Type     |     Flags     |       TLS Message Length      |
+-+-+-+-+-+-+-+-+-+-+-+-+-+-+-+-+-+-+-+-+-+-+-+-+-+-+-+-+-+-+-+-+
|        TLS Message Length     |          TLS Data ...         |
+-+-+-+-+-+-+-+-+-+-+-+-+-+-+-+-+-+-+-+-+-+-+-+-+-+-+-+-+-+-+-+-+
```

Figure 13.4   An EAP request/response packet.

taking the MD5 hash of the concatenation of the identifier, a shared secret, and the challenge. The authenticator then performs the same calculation and compares its value with that of the response and sends a success or failure packet as appropriate. The success and failure packets are sent by the authenticator to the peer to indicate successful or unsuccessful authentication, respectively.

EAP provides a simple mechanism to allow one endpoint to authenticate itself to another. It specifies a simple protocol to allow the exchange of data. EAP defines a basic mechanism (MD5) to be used for authentication and allows additional mechanisms to be added, extending the protocol.

Variants of EAP have been proposed to address some of its limitations, including the lack of protection of the user's identity, lack of a key exchange mechanism, and lack of support for packet fragmentation and reassembly. Protected EAP (PEAP) was designed by Microsoft to extend EAP by wrapping EAP within TLS (Andersson et al., 2002). Similarly, the Extensible Authentication Protocol–Tunneling, Trusted Layer Security (EAP-TTLS) (Aboba and Simon, 1999) is a competing standard. Cisco has created the Lightweight EAP (LEAP), a proprietary protocol based on EAP. Note that this is not to be confused with the similarly named Lightweight and Efficient Application Protocol (LEAP; see *http://www.leapforum.org/*).

EAP is, as the name suggests, an extensible framework that is used as a base for various other authentication protocols (Aboba et al., 2004; Zorn and Cobb, 1998; Aboba and Simon, 1999).

Remote authentication dial In user service (RADIUS). The Remote Authentication Dial In User Service (RADIUS) is a protocol to convey authentication, authorization, and configuration information between a network access server and an authentication server and authenticate the links along the path to the node requesting authentication. This protocol was intended originally to provide centralized management of dispersed serial lines and modem pools. It uses a single "database" of users that includes authentication data, specifically user names and passwords, and configuration information, which includes the types of services to provide the user, such as SLIP, PPP, telnet, etc. RADIUS is used as a general-purpose third-party authentication server for network access, as part of 802.1x, and 802.11i (see Sections. 15.5 and 15.5.2). RADIUS uses the challenge/response model described earlier and supports CHAP as well as other protocols (Rigney et al., 2000).

13.5.2 AAA

Authentication, authorization, and *auditing* (AAA) help to maintain the security of systems. An IETF Working Group focuses on developing the requirements for AAA and protocols that implement them (IETF, 2004a). Authentication answers the question, "Are they who they claim to be?" and is defined as follows: "The act of verifying a claimed identity, in the form of a preexisting label from a mutually known name space, as the originator of a message (message authentication) or as the endpoint of a channel (entity authentication) (Aboba and Wood, 2003)." Digital signatures, MACs, authentication headers, passwords, and biometrics are all mechanisms that provide authentication.

One form of passwords is the *one-time password*, in which a different password (or phrase) is used each time for authentication. SKEY is an example of a one-time password scheme, in which a function is applied repeatedly to a random number. Each result, $x_1, x_2, \ldots, x_n$, is saved in a list for the user, and the final result is saved by the system. The user authenticates by providing x_{n-1}, which the system uses to compute x_n and compares with the stored value. If successful, the system then stores x_{n-1}, and the user must provide x_{n-2} to authenticate. Passwords are useless once they have been used because no further computations depend on them (Haller, 1994; Haller, 1995).

Authorization answers the question, "Do they have permission to do it?" and is defined as "The act of determining if a particular right, such as access to some resource, can be granted to the presenter of a particular credential (IETF, 2004a)." The authorization process involves checking privileges, access control lists, and possibly roles (http://csrc.nist.rist.gov/rbac; NIST, 1995; Ferraiolo and Kuhn, 1992).

Accounting is the process of maintaining an audit log or history of what happened and is defined as "The collection of resource consumption data for the purposes of capacity and trend analysis, cost allocation, auditing, and billing. Accounting management requires that resource consumption be measured, rated, assigned, and communicated between appropriate parties (IETF, 2004a)." Audit logs do not provide protection on their own, and in fact, many are used solely for billing purposes. However, by maintaining a record of users, transactions, times, and other similar information, the logs can indicate what information was altered, deleted, or otherwise compromised; when it occurred, exactly or within a time window of a certain granularity; and who was responsible.

Audit logs indicate if it was a legitimate, authorized user (performing illegal operations), and if it was an external user, it can provide some information as to the source. Note that external users attacking through the Internet typically use compromised machines (*zombies*) to perform the attacks, often employing a number of zombies as stepping stones. The IP address in an audit log will point only to the last zombie in the chain; typically zombies are (re)configured so that their auditing mechanisms are disabled.

13.5.3 Special hardware

In addition to the techniques discussed in this chapter, special hardware can be used. One type of hardware is smartcards that contain cryptographic tokens or run algorithms to generate one-time passwords.

Another type of hardware is a transmitter that makes it difficult to receive transmissions. Personal and local area networks, as well as cell phones (see Chaps. 14 through 17), transmit with only enough power for the corresponding receiver to detect the signal successfully. In addition, they use *frequency-hopping* and *spread-spectrum technology* to increase the effective bandwidth that also has the effect of making it more difficult to reconstruct the signal. Of course, anyone with the right hardware, such as an 802.11b card, can receive 802.11b signals. Military communications use similar techniques, but the hardware is restricted.

Finally, another security risk is the electromagnetic radiation generated by the computer itself. With the right hardware, it is possible to reconstruct the source of the signals (a screen image, keys pressed by a wireless keyboard, etc.). The Department of Defense has developed a set of standards and procedures to reduce the emanations to a sufficiently low level, known as Transient ElectroMagnetic Pulse Emantation STandard (TEMPEST) (Caloyannides, 2001). In addition, it is possible to reconstruct a screen image by observing the glow on a wall opposite a CRT monitor using special hardware and applying signal processing techniques (Huhn, 2002). By restricting physical access to the area, this threat can be mitigated. "Physical access" in this sense means direct contact with the computer, as well as any indirect contact, such as radio and light waves. The room must be electrically shielded, to block any electromagnetic radiation, and no one outside the room is permitted to look inside. Unfortunately for those working in such places, no windows are permitted.

13.6 Summary

In this chapter we introduced techniques to provide data security, including public and private key encryption and message integrity codes. We described IPSec and some authentication protocols used for dial-up protocols such as PPP. In the next several chapters we will see how some of these techniques are used to protect personal, local, and wide area networks.

13.7 References

Aboba, B., and D. Simon, RFC 2716, "PPP EAP TLS Authentication Protocol," October 1999, http://www.ietf.org/rfc/rfc2716.txt

Aboba, B., D. Simon, RFC 2716, "PPP EAP TLS Authentication, Authentication Protocol," October 1999 ftp://ftp.rfc-editor.org/in-notes/rfc2716.txt

Aboba, B., L. Blunk, J. Vollbrecht, J. Carlson, and H. Levkowetz (eds.). "Extensible Authentication Protocol (EAP)," Request for Comment 3748, June 2004, ftp://ftp.rfc-editor.org/in-notes/rfc3748.txt

Aboba, B., and J. Wood, "Authentication, Authorization, and Accounting (AAA) Transport Profile," RFC 3539, June 2003, http://www.ietf.org/rfc/rfc3539.txt

Andersson, H., S. Josefsson, G. Zorn, D. Simon, and A. Palekar, "Protected EAP Protocop (PEAP)," Internet Draft, PPPEXT Working Group, February, 2002, http://www. globecom. net/ietf/draft/draft-josefsson-pppext-eap-tls-eap-02.html

"Authentication, Authorization and Accounting (AAA)," Internet Engineering Working Group, http://www.ietf.org/html.charters/aaa-charter.html

Blunk, L., and J. Vollbrecht, RFC 2284, "PPP Extensible Authentication Protocol (EAP)," March 1998, http://www.ietf.org/rfc/rfc2284.txt

Caloyannides, M., "Computer Forensics and Privacy," Artech House, 2001.

Ferraiolo and Kuhn, "Role Based Access Control," 15th National Computer Security Conference, 1992.

Funk, P., and S. Blake-Wilson, "EAP Tunneled TLS Authentication Protocl (EAP-TTLS)," Internet-Draft, PPPEXT Working Group, February, 2002, http://www. ietf.org/proceedings/ 02jul/I-D/draft-ietf-pppext-eap-ttls-01.txt

Haller, N.M., "The S/KEY TM One-Time Password System," In: Proceedings of the Internet Society Symposium on Network and Distributed Systems, 1994.

Haller, N.M., RFC 1760, "The S/KEY One-time Password System." February, 1995, http://www.ietf.org/rfc/rfc1760.txt

Huhn, M.G., "Optical Time-Domain Eavesdropping Risks of CRT Displays," In: Proceedings of IEEE Symposium on Security and Privacy, 2002.

IETF Internet Draft, "Authentication, Authorization and Accounting (AAA) Transport Profile," October 2002, http://www.ietf.org/internet-drafts/draft-ietf-aaa-transport-08.txt

"IP Security Protocol (IPSec)," Internet Engineering Task Force Working Group, http://www.ietf.org/html.charters/ipsec-charter.html

Kent, S., and R. Atkinson, RFC 2402, "IP Authentication Header (RFC 2402)," http://www.ietf.org/rfc/rfc2402.txt

Kent, S., and R. Atkinson, RFC 2406, "IP Encapsulating Security Payload (ESP) (RFC 2406)," http://www.ietf.org/rfc/rfc2406.txt

RBAC, "An Introduction to Role-Based Access Control," NIST CSL Bulletin on RBAC, December 1995, http://csrc.nist.gov/publications/nistbul/csl95-12.txt

Rigmey, C., A. Rubens, W. Simpson, and S. Willens, RFC 2865, "Remote Authentication Dial in User Service (RADIUS)," June 2000, http://www.ietf.org/rfc/ rfc2865.txt

RSA Online FAQs, "RSA Laboratories' Frequently Asked Questions About Today's Cryptography, Version 4.1," http://www.rsasecurity.com/rsalabs/faq/index.html

Schneier, B., Applied Cryptography, 2nd ed. New York, NY: John Wiley & Sons, 1996, pp. 455–459.

Simpson, W., RFC 1994, "PPP Challenge Handshake Authentication Protocol (CHAP)," August 1996, http://www.ietf.org/rfc/rfc1994.txt

Williams, R., "A Painless Guide to CRC Error Detection Algorithms," August 19, 1993, ftp://ftp.rocksoft.com/papers/crc_v3.txt

Zorn, G., and S. Cobb, "Microsoft PPP CHAP Extensions," RFC 2433, October 1998, http://www.ietf.org/rfc/rfc2433.txt

Security in Wireless Personal Area Networks

This chapter presents security techniques used in personal area networks. We will see how the ideas presented in Chaps. 12 and 13 are applied on the Bluetooth wireless personal area network (WPAN), specifically how data are protected via encryption, how authentication is performed (including key exchange), and what limitations and problems are present in the design.

14.1 Basic Idea

Wireless personal area networks (WPANs) provide connectivity between nodes that are relatively close, within 10 m. WPANs function as an alternative to a cable, e.g., linking a wireless headphone to a portable CD player. Bluetooth is an industry standard protocol for WPANs and is now developed as one of the IEEE 802 standards (IEEE, 2004; Bluetooth SIG, 2001).

Bluetooth can link a headphone and CD player together. It also can link two personal digital assistants (PDAs) together so that they can exchange business cards. Or a printer can use Bluetooth to let different devices send print jobs to it. While Bluetooth was intended originally to be a "wireless cable" to link devices together, its uses have expanded to include networking to support Internet protocol (IP). Bluetooth can serve as the physical layer link for IP services between wireless computers.

14.1.1 Bluetooth specifications

Since Bluetooth was designed to be a sort of "wireless cable" that connects devices, such as a computer and a printer or a CD player and headphones, it has a small coverage area, typically targeted at approximately 10 m, although the coverage can extend to 100 m; however, it adjusts its transmission power to what the receiver needs. There are three classes of Bluetooth transmission:

- Class 1: 100 mW (+20 dBm) max (100 m range)
- Class 2: 2.5 mW (+4 dBm) max
- Class 3:1 mW (0 dBm) max (10 m range)

Bluetooth supports point-to-point (unicast) and point-to-multipoint (multicast) transmission with a variety of data rates. The initial Bluetooth standard specified a transmission rate of 1 Mbps. At the low end, Task Group 4 in IEEE 802.15 supports low data rates, with low complexity and long battery life (lasting from months to years), of 20, 40, and 250 kbps. At the high end, Task Group 3 supports high data rates of 11 to 55 Mbps. TG3A uses an alternative physical layer, including ultra wideband (UWB) radio, with data rates beyond 100 Mbps.

Bluetooth networks transmit on the 2.4-GHz band, which is an unlicensed frequency range. It uses 79 channels between 2.402 and 2.480 GHz (in the United States and most of Europe). Because the band is unlicensed, Bluetooth can conflict with 802.11 and metropolitan area networks, such as 802.16; however, these groups are working together to ensure "harmonization" in order to minimize conflicts with each other. In addition, microwave ovens operate on that frequency. While they generally are shielded properly, testing shows that poorly shielded ovens can jam radios and reduce throughput by 75 percent (although an owner of such a leaky oven probably has bigger problems to worry about than reduced throughput).

Bluetooth uses a fast-frequency-hopping algorithm, which switches frequencies at a rate of 1600 times per second. Packets are short, typically around 350 bytes, and forward error correction (FEC) provides data integrity.

14.1.2 Bluetooth network terms

Master.   Each Bluetooth network, called a *piconet*, (see below) has one device serving as the *master*, and the other devices serve as *slaves*. The master determines the frequency-hopping sequence of the piconet, as well as performing other functions, such as swapping *parked* nodes into the piconet. A node can be in multiple piconets but serves as the master of at most one piconet.

Slave.   All nonmaster nodes in a piconet are *slaves*.

Piconet. The smallest collection of nodes connected together by Bluetooth is called a *piconet*. A piconet is a collection of two to eight Bluetooth active devices. One unit acts as *master*, and the remaining devices act as *slaves*. Piconet addresses are 3-bit values, thus supporting up to seven slaves and one master. Piconets are ad hoc networks, forming dynamically. Each piconet has a different frequency-hopping sequence determined by the device address of the master.

Scatternet (multipoint). A *scatternet* is a collection of piconets joined by a device that is a master in one piconet and a slave in another.

Bluetooth allows multipoint transmissions. Within a piconet, the master can send to multiple recipients. There is no specification for scatternet multipoint, and different vendors implement multipoint transmissions differently with no interoperability guarantees.

Bluetooth node states. Bluetooth nodes are in one of several states. Devices not connected are in *standby mode* and monitor 32 frequencies once every 1.28 s. Devices connected to a piconet are in the *active state*. Devices wishing to connect to another device send out a *page* if they know the device address of the destination. If the destination address is unknown, they send out an *inquiry*, and a piconet master that receives the inquiry sends a *page* on 16 frequencies and then on the other 16 frequencies if no response is received (Monson, 1999; Pahlavan and Krishnamurthy, 2002).

In addition to these states, Bluetooth devices can be in one of three power saving states: *hold, park,* and *sniff.*

Hold. Devices placed in *hold mode* are completely inactive, except for an internal timer that determines when the device leaves hold mode and resumes normal operation. The master places a slave in hold mode; a slave also can request that its master place it in hold mode. This is used for low-power nodes.

Park. Devices in *park mode* remain synchronized to the piconet but have no 3-bit piconet address and do not participate in the piconet. They listen to traffic from the master in order to resynchronize with the piconet and receive broadcast messages. The master places nodes in and out of the park state. Parked nodes *can* request the master to remove them from that state during a certain time slot. Also, parking allows the master to support more than seven devices—the master parks devices to free up the 3-bit piconet address, so other devices can participate.

Sniff. Devices in *sniff mode* listen to the piconet at a reduced rate, thus saving power.

14.1.3 Bluetooth security mechanisms

Bluetooth has several mechanisms to provide security. The security mechanisms are defined in the Bluetooth specification (Bluetooth, 2001); Vainio (2000) and others (Fleming et al., 2000) provide descriptions of the mechanism. First, the transmitters use the lowest power required for their data to be received. This means that intruder nodes farther away from the transmitter than the receiver is find the signal to be weak at best. This requires the intruder to be in very close physical proximity, which may expose the intruder to a higher risk of detection than he is willing to accept. In truth, this mechanism was designed to save power and increase the life of the device, but it also has the fortunate side effect of increasing physical security.

Channel hopping provides additional protection, making it difficult to snoop on the data stream. The fast rate of hopping makes it hard for a casual observer to "sniff" the data stream off of one channel or guess the hopping sequence. Again, this mechanism was not designed for security but instead was designed to allow many nodes to share the same frequency bands without interfering with each other.

Data are protected by the optional use of encryption. The encryption algorithm is, essentially, a stream cipher that XORs the data stream with a stream of numbers from a pseudorandom-number generator (PRNG) seeded by an encryption key. The keys are created and distributed by a key exchange algorithm, so keys are not sent as plaintext. Finally, nodes can perform authentication and authorization to verify the identity and access of both parties that are communicating. We discuss the keys and encryption mechanisms in the next sections.

14.2 Bluetooth Security Modes

Bluetooth defines three modes of security for devices: nonsecure, service-level enforced security, and link-level enforced security.

- *Nonsecure.* A device in the *nonsecure mode* does not initiate any security procedure. This is intended for public use devices, such as a walk-up printer.

- *Service-level enforced security.* A device in the *service-level enforced security mode* permits access to itself depending on the service request. For example, a PC may allow a user to download files to it but does not allow its own files to be read.

- *Link-level enforced security.* A device in the *link-level enforced security mode* requires authentication and authorization for use, e.g., cell phones.

The Bluetooth standard defines security levels for devices and services. It defines two security levels for devices: *trusted* and *untrusted.*

Trusted devices allow unrestricted access to all services, whereas untrusted devices do not. Services can be in one of three security levels: *open, authentication*, and *authentication and authorization*. In the first, the services are open to all devices. The second requires authentication from the devices, whereas the third requires both authentication *and* authorization for the device.

14.3 Basic Security Mechanisms

Bluetooth uses encryption and link-layer keys. Encryption keys protect the data in a session, whereas link-layer keys provide authentication and serve as a parameter when deriving the encryption keys. Link-layer key lifetimes are either *semipermanent* or *temporary*. Semipermanent keys can be used after the current session to authenticate Bluetooth devices. Temporary keys can be used only during the current session; they are often used in point-to-multipoint communication in which the same information is transmitted to several recipients.

Four entities are used for link-layer security:

- A 48-bit publicly available *device address*, fixed and unique for each device
- A 128-bit pseudorandom private key for authentication
- An 8- to 128-bit private key for encryption (The variable length accommodates different countries' export restrictions.)
- A 128-bit pseudorandom number generated by the device

Note that the device address is the only publicly available value. Link-layer keys are used in authentication and serve as a parameter in deriving the encryption key. The pseudorandom private key can be one of four basic types. Generally, it is fixed for each Bluetooth unit, although it can be derived from the private key from each of a pair of communicating devices.

The four basic types of link-layer keys used in Bluetooth security are

- The *initialization key* is used as a link-layer key when there are not yet any unit or combination keys. This key is used only during installation and typically requires the user to enter a personal identification number (PIN) on the unit.
- The *unit key* is generated in each device when the device is installed and is stored in nonvolatile memory and (almost) never changed.
- The *combination key* is derived from information from two devices that communicate with each other. A different combination key is generated for each pair of communicating devices.

■ The *master key* is a temporary key that replaces the current link-layer key. It can be used when the master device wants to transmit to multiple recipients at once.

The keys are used to create the (shared) encryption key (see Section 13.2). When two units need to communicate with each other, they can use the unit key of one of the devices, or they can use a combination of both unit keys for more security, if desired. If more than one unit is to share a transmission, i.e., multipoint, then nodes in the multipoint transmission can use a master key. Finally, a node uses an initialization key if no link-layer key is available. Each of the keys is discussed below.

14.3.1 Initialization key

The security layer uses the initialization key to form a secure channel to exchange other link-layer keys. It creates the initialization key by using a combination of a PIN code, which varies from 1 to 16 octets (8 to 128 bits), with the (fixed) Bluetooth device address (48 bits), and a random 128-bit number using the E_{22} algorithm. The strength of the initialization key lies in the length of the PIN, which can be entered manually into each device or stored. The initialization key is used only for key exchange during the generation of the other link-layer keys and is discarded after the key exchange.

14.3.2 Unit key

The unit key is associated with the device. It is 128 bits and is generated with the E_{21} algorithm using the Bluetooth device address and a random number, both 128 bits long. The device creates the key the first time it is operated, stores the unit key in nonvolatile memory, and rarely changes it. Devices can choose whose unit key to use. This is particularly useful if one device has limited memory and cannot store extra keys.

14.3.3 Combination key

The combination key allows two devices to communicate with each other securely. This key is generated during the initialization process if it is needed by both devices concurrently. The two devices, A and B, each compute a number, LK_K$_A$ and LK_K$_B$, respectively. Each node computes this number using the E_{21} algorithm, which takes a random number and the Bluetooth device address (a fixed 48-bit number) as input. The two devices then securely exchange the random numbers they used by XORing the number with the current link-layer key (which is the initialization key) and transmitting the result to the other device. Each device extracts the random number by XORing it with the current link-layer key. Since A and

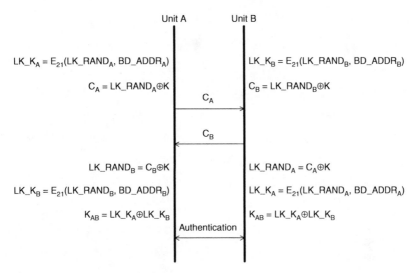

Figure 14.1 The combination key.

B know each other's Bluetooth device address, A now computes LK_K$_B$ with the E$_{21}$ algorithm using B's random number and B's device address. Similarly, B computes LK_K$_A$. The combination key is computed by XORing LK_K$_A$ and LK_K$_B$ (see Fig. 14.1).

14.3.4 Master key

The master key is the only temporary key. The master device generates it using the E$_{22}$ algorithm with two 128-bit random numbers. A random number is sent to slaves, which use it and the current link-layer key to generate an *overlay*. The master key is XORed with the overlay by the master and sent to the slaves, which can extract the master key from it. This procedure must be done for each slave (see Fig. 14.2).

14.4 Encryption

There are three modes for Bluetooth encryption:

- In the first mode, nothing is encrypted.
- In the second mode, broadcast traffic is not encrypted, but individually addressed traffic is encrypted with the master key.
- In the third mode, all traffic is encrypted with the master key.

The packet payload is encrypted when encryption is enabled. Encryption is performed with the E_0 *stream cipher* and is resynchronized for

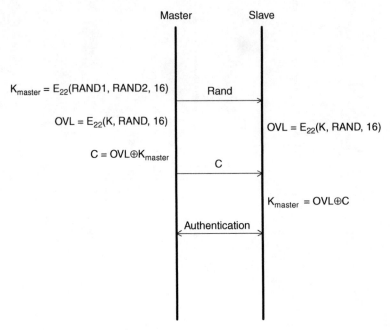

Figure 14.2 The master key.

each payload. The E_0 stream cipher consists of a payload key generator, a key stream generator, and the encryption/decryption part (see
Fig. 14.3).

Essentially, the algorithm consists of XORing the data payload stream
with a stream of pseudorandom numbers. The pseudorandom-number

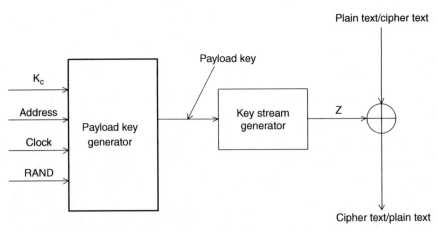

Figure 14.3 Bluetooth encryption.

generator is initialized with an encryption key. This key is generated from the current link-layer key, a 96-bit ciphering offset number (COF), and a 128-bit random number. The COF is based on the authenticated ciphering offset (ACO), which is generated during the authentication process.

When the link manager (LM) activates encryption, it generates the encryption key. This key automatically changes every time the Bluetooth device enters the encryption mode.

14.5 Authentication

Bluetooth authentication uses a challenge/response protocol with symmetric encryption and shared secrets. If unit A wants to verify unit B's claim of identity, unit A sends a challenge to unit B. Unit B encrypts the challenge and sends the result back as a response. Unit A also encrypts the challenge and then compares it with the response it receives from unit B. If they match, then they both share the same secret key, and thus unit A has authenticated unit B. If unit B wants to authenticate unit A, then the process repeats, with unit B sending a (different) challenge to unit A (see Fig. 14.4).

The challenge is a random number that becomes an input to the encryption algorithm E_1. The algorithm takes two other inputs, the Bluetooth device address of the "claimant" (unit B) and the link-layer key, which is the shared secret. The E_1 function produces a value, the *authenticated ciphering offset* (ACO), that is used as a parameter for the encryption key generation if authentication is successful (see Fig. 14.5).

14.6 Limitations and Problems

Since Bluetooth uses the unlicensed 2.4-GHz radio band, it is susceptible to both unintentional and intentional jamming. Other devices that operate in the same band, including some 802.11 and home/RF devices, as well as microwave ovens, could interfere with signal reception. The signals also could be jammed intentionally by transmitters that are nearby or significantly stronger than the Bluetooth transmitters. However, the low power requirements and signal hopping make Bluetooth more resilient to casual jamming.

Other problems include key management. Until a secure link is established, the keys and data used to derive them are sent in the clear. The only mitigation is to use a PIN code entered into each device, but this can be cumbersome. Bluetooth supports only device authentication, not user authentication. Thus, a stolen device, such as a PDA, could gain unauthorized access to data and resources unless the applications themselves provide an additional layer of security.

Finally, like any wireless device, care must be used when connecting it to an existing network because it might be unintentionally exposing

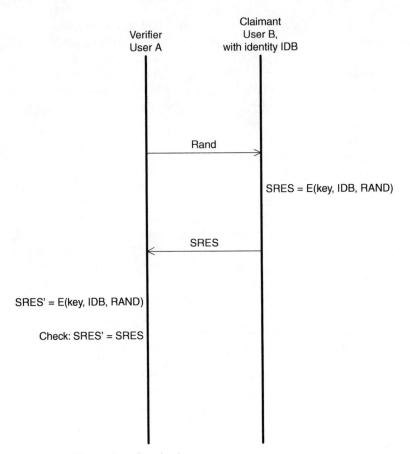

Figure 14.4 Bluetooth authentication.

part of the internal network that had been protected behind a firewall (Nichols and Lekkas, 2002).

Bluetooth Attacks. In addition to the previously discussed limitations, Bluetooth is vulnerable to a number of attacks (Anand, 2001; Sutherland, 2000; Jakobsson and Wetzel, 2001). If two sites, A and B, communicate with each other using A's unit key (K_A) because of limited memory on A, then afterwards site B can impersonate A as well as eavesdrop on A's communications because B knows that the key that will be used. Variations of a man-in-the-middle attack are possible. If an attacker can synchronize with the frequency-hopping sequence, then it can eavesdrop.

The secrecy of the combination key depends on the PIN, which is 1 to 16 characters long and typically has a default value of zero. Many PINs

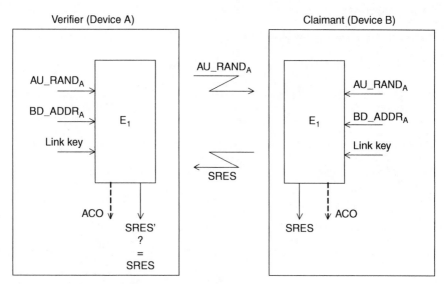

Figure 14.5 Bluetooth authentication in detail.

are only four characters long, so brute-force attacks against the PIN are quite feasible. It should be noted that short PINs are a problem in the implementation and use, not in the Bluetooth specification.

Another area of vulnerability is the cipher algorithms used. Jakobsson and Wetzel described a cipher attack on a 256-bit key that requires $O(2^{66})$ operations. This attack is not practical, but it does show that the search space, which has a size of $O(2^{256})$ for a brute-force attack on a 256-bit key, can be vastly reduced. This casts doubt on the overall strength of the encryption, and the question remains open as to how much more efficient the attack can be made.

These authors also describe another type of attack in which the location and movements of the victim are tracked rather than the content of the data. For example, tracker devices could be installed at airports and train stations and record the IDs of all Bluetooth devices that pass. When fed to a central repository, attackers would build up a picture of people's movements as well as their current location.

14.7 Summary

In this chapter we presented Bluetooth or IEEE 802.15, a protocol for WPANs, describing how it establishes the various link keys it uses to perform encryption and how it authenticates devices. We also described some limitations of the security mechanisms. Chapter 15 covers the IEEE 802.11 WLAN protocol and its security mechanisms.

14.8   References

Anand, N., "An Overview of Bluetooth Security," February 22, 2001, SANS web site, *http://www.giac.org/practical/gsec/Nikhil_Anand_GSEC.pdf.*

Bluetooth Special Interest Group, "Chapter 14 Bluetooth Security," Specification of the Bluetooth System, Core, Volume 1, Version 1.1, February 22, 2001, The Official Bluetooth web site, *http://bluetooth.com/pdf/Bluetooth_11_Specifications_Book.pdf,* pp. 148–177.

Fliming, K., U. Gadamsetty, R. Hunter, S. Kambhatla, S. Rajagopal, and S. Ramakesavan, "Architectural Overview of Intel's Bluetooth Software Stack," Intel Technology Jouranl, 2nd Quarter, 2000, *http://www.intel.com/technology/itj/q22000/pdf/art_2.pdf.*

http://grouper.ieee.org/groups/802/15/.

Jakobsson, M., and S. Wetzel, "Security Weaknesses in Bluetooth," *Lecture Notes in Computer Science*, vol. 2020. Berlin: Springer-Verlag, 2001.

Monson, H., "Bluetooth technology and implications," Dec. 14, 1999, *http://www.sysopt.com/articles/bluetooth/index2.html.*

Nichols, R.K., and P.C. Lekkas, *Wireless Security, Models, Threats, and Solutions.* New York, NY: McGraw-Hill, 2002.

Pahlavan, K., and P. Krishnamurthy, *Principles of Wireless Networks, A Unified Approach.* New York, NY: Prentice-Hall, 2002.

Sutherland, E., "Despite the Hype, Bluetooth has Security Issues that cannot be ignored," Nov. 28, 2000, *http://alllinuxdevices.com/news_story.php3?ltsn=2000-11-29-002-03-PS-HH-WL.*

Vainio, J., "Bluetooth Security," May 25, 2000, *http://www.niksula.cs.hut.fi/~jiitv/bluesec.html.*

15

Security in Wireless Local Area Networks

This chapter presents the security techniques used in local area networks, specifically the IEEE 802.11 standard (IEEE, 1999). After an introduction to the IEEE 802.11 wireless local area network (WLAN) standard, we discuss its security mechanisms and focus on the flawed wired-equivalent privacy (WEP), covering its intentions and shortcomings, as well as the ways to get the best protection given limited coverage; WiFi Protected Access (WPA), an interim protocol to fix the shortcomings of WEP; and 802.11i, the IEEE standard to provide strong encryption, key management, and support for authentication. We conclude by discussing virtual private networks (VPNs) and firewall protections in the context of WLANs.

15.1 Basic Idea

WLAN coverage has a radius of around 100 m typically. This covers several rooms or a small company with a few offices. Of course, actual coverage depends on where it is deployed, the material in the walls, the frequency range, other nearby radio sources, etc. WLANs offer a cheap alternative to running a wire to every office, allowing fast installation.

Many wireless access points (see below) work directly "out of the box," requiring no configuration. The user simply plugs them into the network and a power outlet, and they work. The downside is that most devices default to being very open, with most security features disabled; these features often are overlooked for an "out of the box" installation. In addition, the users may not know that some of the security features are either limited or flawed.

IEEE 802 is the LAN/metropolitan area network (MAN) standard committee that has numerous subgroups within it. IEEE 802.11 is the WLAN standard subcommittee. And within it, there are several committees for different 802.11 standards. Each of the standards subcommittees represents different technologies or protocols. Some use different frequencies, and some use different protocols. Below we define some basic architecture concepts common to all WLANs.

The *wireless station* (WS) is the remote or mobile unit. The *access point* (AP) or *base station* is the nonmobile unit that connects the wireless network into a wire-based network. The AP acts as a bridge or router and usually has some protection mechanisms built in. 802.11 networks can be organized in two different ways: infrastructure or ad hoc.

A *basic service set* (BSS), identified by a 6-byte string, is a network formed by an AP and the wireless stations that are associated with it.

An *extended service set* (ESS) is two or more BSSs that form a single logical network. As they move, wireless stations can switch seamlessly from one AP to another with no disruption of service. The APs coordinate the *handoff* among themselves, generally via an Ethernet connection. Figure 15.1 shows an example of two BSSs forming an ESS.

When wireless stations communicate through an AP, it is called *infrastructure mode*.

An *independent basic service set* (IBSS) is a set of wireless stations that communicate with each other directly without using an AP. In contrast to infrastructure mode, this type of communication is called *ad hoc mode*, and such a network is referred to as an *ad hoc network*. Typically, ad hoc networks form when a group of wireless stations wish to share

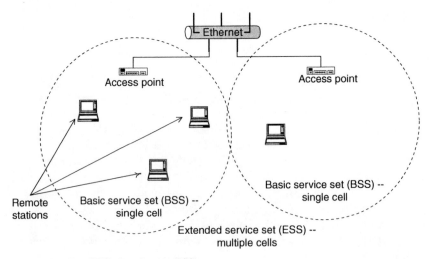

Figure 15.1 Two BSSs forming an ESS.

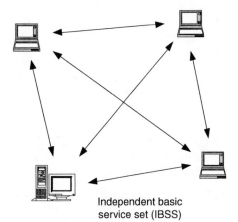

Independent basic
service set (IBSS)

Figure 15.2   An IBSS.

information with each other. If all the ad hoc nodes are within range of each other, routing is trivial (i.e., broadcast). The situation becomes more complex when the ad hoc network extends beyond the reception radius of the nodes, and complex ad hoc routing protocols must be used (Perkins et al., 2003; Park and Corson, 2000; Clausen and Jacquet, 2003; Moy, 1998). Figure 15.2 shows an example of an IBSS. Wireless ad hoc routing is discussed in Chap. 12.

A *service set identifier* (SSID) is a 32 byte string that identifies the name of the network, either the IBSS or the ESS.

15.2 Wireless Alphabet Soup

IEEE 802.11 was the initial protocol created in 1997 using the 2.4 GHz frequency range and supporting 1 and 2 Mbps via frequency-hopping spread spectrum (FHSS) and direct-sequence spread spectrum (DSSS). Different task groups within the 802.11 working group created and revised the 802.11 standard, e.g., support for higher data rates and the use of the 5-GHz frequency range. Each new or revised standard receives a new suffix letter. Currently, there is an "alphabet soup" of IEEE 802.11 protocols (Geier, 2002). Before discussing the most relevant protocols relating to security, we provide a brief overview of the various "802.11-something" protocols.

- *802.11a* is a physical-layer standard that uses *orthogonal frequency division multiplexing* (OFDM) in the 5-GHz band, supporting speeds from 6 to 54 Mbps. 802.11a offers the highest speeds currently, although the range for the highest speeds is limited, and transmission rates drop to slower speeds beyond a short distance. 802.11a has leap-frogged over

802.11b as the fastest 802.11 technology available, having a maximum speed of 54 Mbps. However, it faces competition from 802.11g, which provides similar speeds but with better signal propagation than 802.11a and is compatible with (soon to be legacy) 802.11b cards.

- *802.11b* uses DSSS in the 2.4-GHz range to achieve faster speeds of 5.5 and 11 Mbps using complementary code keying (CCK) and is widely deployed. *Wired-equivalent privacy* (WEP) is the scheme to provide data protection and is described later in this chapter. 802.11b is currently the most widely deployed version of 802.11 cards for home and business but this will change rapidly as 802.11a and 802.11g become more widely available and less expensive.

- *802.11c* provides required information to ensure proper bridging operations and is used when developing APs.

- *802.11d* provides "global harmonization." It defines physical-layer requirements to satisfy the different regulatory organizations in different parts of the world, e.g., United States, Japan, and Europe. This includes both the 2.4- and 5-GHz bands and only affects those developing 802.11 products.

- *802.11e* extends the MAC layer of 802.11 to provide quality-of-service (QoS) support for audio and video applications. These MAC-level changes will affect all 802.11 operating frequencies (i.e., 2.4 and 5 GHz) and will be backwards-compatible with the existing protocol.

- *802.11f* defines a standard so that different APs can communicate with each other. This "inter access point protocol" will allow wireless stations to "roam" from one AP to another. Currently, 802.11 defines no standard, so each vendor can create its own incompatible means to implement roaming.

- *802.11g* specifies a higher-speed extension to the 2.4-GHz band. 802.11g extends 802.11b to support up to 54 Mbps. 802.11g uses OFDM rather than DSSS. Essentially, 802.11g is designed to make 802.11b compete with the bandwidth of 802.11a.

- *802.11h* provides "spectrum-managed 802.11a" to address the requirements in Europe for use of the 5-GHz band. The functions provided include dynamic channel selection (DCS) and transmit power control (TPC), which will help to prevent any interference with satellite communications. 802.11h eventually will replace 802.11a.

- *802.11i* standardizes MAC enhancements for 802.11 security. It is designed to address the problems and shortcomings of WEP, incorporating 802.1x and stronger encryption techniques, such as the advanced encryption standard (AES), the follow-on to DES. 802.11i updates the MAC layer to provide security for all 802.11 protocols. In the meantime, many vendors are using WPA, which incorporates many features that are in the proposed 802.11i specification, even though the standard is still being developed. Section 15.5 discusses the 802.11i protocol in more detail.

- *802.11j* addresses 4.9- to 5.0-GHz operation in Japan (group formed on November 2002).

- *802.11k* defines and exposes radio and network information to facilitate the management and maintenance of a wireless and mobile LAN. Also, it will enable new applications to be created based on this radio information, such as location-enabled services.

In the following sections we discuss WEP, WPA, and 802.11i, three protocols to provide protection for 802.11.

15.3 Wired-Equivalent Privacy (WEP)

WEP is the security scheme provided with 802.11b. Since wireless communication presents an easy target for casual eavesdropping, WEP was designed to raise the baseline security level to be comparable with standard wired Ethernet. Sniffing packets off a wired network requires a user to physically tap into the network; the WEP designers wanted to make sniffers go through a similar level of effort to get similar information from a wireless network. However, several severe design flaws rendered WEP virtually useless against a skilled, knowledgeable attacker. In this section we present the design goals of WEP, its data frame, and how encryption, authentication, and decryption work, and then we discuss the flaws and potential remedies. Even though WEP is considered passé, especially now that WPA and 802.11i exist, it is useful to understand what it was designed to do and what its problems are.

15.3.1 WEP goals

WEP was designed originally to support a few criteria. First, it had to be "reasonably strong." Of course, this is a debatable point, but the goal was to raise the bar on security so that some effort must be spent to break the protection. Second, it had to be self-synchronizing. Stations must be able to resynchronize with the AP without requiring user intervention, such as a password, because the stations may go in and out of coverage frequently. Third, it must be computationally efficient so that it can be performed in either hardware or software because some processors may be low-power, low-speed devices. Fourth, it had to be exportable. Although the United States relaxed some of the encryption restrictions in January of 2000 as part of the "Wassenaar arrangement," (Wassenaar, 2003)

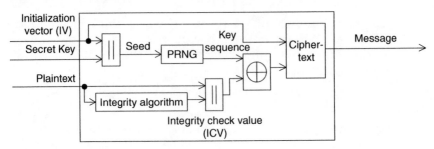

Figure 15.3 Block diagram of WEP encryption.

other countries still tightly restrict encryption technology. On the other hand, no country restricts the strength of protection used for authentication. And finally, WEP must be optional.

WEP consists of a secret key of either 40 or 104 bits (5 or 13 bytes) and an initialization vector (IV) of 24 bits. Thus the total *protection,* as it is sometimes called, is 64 or 128 bits (often mistakenly referred to as 64- or 128-bit "keys" even though the keys are 40 or 104 bits). The key plus the IV is used to seed an RC4-based pseudorandom-number generator (PRNG). This sends a stream of pseudorandom numbers that is XORed with the data stream to produce the ciphertext. In addition, an integrity check value (ICV) indicates if the data stream was corrupted. The ICV is a simple CRC-32 checksum. Figure 15.3 shows a block diagram of WEP encryption.

15.3.2 WEP data frame

The WEP data frame, shown in Fig. 15.4, consists of an IV of 4 bytes, the data or protocol data unit (PDU) of 1 or more bytes, and the ICV of 4 bytes. The IV can be further divided into 3 bytes (24 bits) of the actual initialization vector plus 1 byte that uses 2 bits to specify a key and 6 bits of padding. With the 2 bits, the device can store up to four different secret keys (recall that the keys are not transmitted but are local to the device).

15.3.3 WEP encryption

The encryption process is shown in the block diagram in Fig. 15.3. It takes the plaintext message, the IV, and the secret key as input and produces as output a message consisting of the ciphertext message and the IV by performing the following steps:

1. Compute the ICV using CRC-32 over the plaintext message.

2. Concatenate the ICV to the plaintext message.

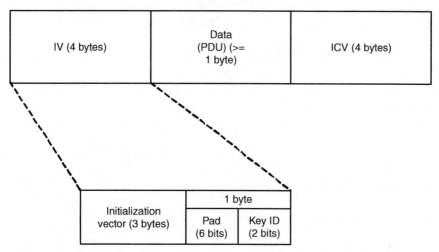

Figure 15.4 WEP data frame.

3. Choose a random IV and concatenate it to the secret key, and use it as input to the RC4 PRNG to produce the pseudorandom key sequence.

4. Encrypt the plaintext and the ICV by doing a bitwise XOR with the key sequence from the PRNG to produce the ciphertext.

5. Append the IV to the front of ciphertext.

15.3.4 WEP decryption

Decryption of WEP data is, more or less, just the reverse of the encryption. The algorithm takes the secret key and the message consisting of the ciphertext and ICV as input and produces the plaintext message and an error flag as output by performing the following steps:

1. Generate the key sequence k using the IV of the message.

2. Decrypt the ciphertext message by doing a bitwise XOR with k to generate the original plaintext and ICV.

3. Verify the integrity of the message by computing the ICV on plaintext, ICV', and comparing it with the recovered ICV from step 2.

4. Trap errors, if ICV ≠ ICV', by sending an error to the MAC management layer and back to the sending station.

15.3.5 WEP authentication

APs perform an optional challenge/response style of authentication to the wireless stations, as shown in Fig. 15.5, as follows:

Wireless Station   **Access Point**

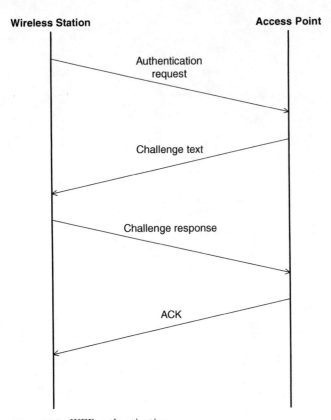

Figure 15.5 WEP authentication.

1. The wireless station (WS) sends an authentication request to the AP.

2. The AP sends a (random) challenge text T back to the WS.

3. The WS sends the challenge response, which is text T, encrypted with a shared secret key.

4. The AP sends an acknowledgment (ACK) if the response is valid and a NACK if it is invalid.

15.3.6 WEP flaws

There are a number of problems with WEP. First, from the start, it was designed to be "as good as" wired Ethernet, as opposed to providing "strong" security. WEP has been broken by various groups (Walker, 2000; Borisov et al., 2001; Fluhrer et al., 2001).

WEP provides no automated key management, and the IEEE standard does not specify any distribution mechanism. All keys must be entered manually, and typically, all wireless stations in one network use the same password. The risk that a device inadvertently shares the password increases with the number of devices sharing this password. In addition, rekeying an entire network can be an administrative nightmare. All users must be informed that the passwords (i.e., keys) are changing as of a certain date, after which they will be locked out of the network until they get the new code.

Often the same key is used for both encryption and authentication, which then ties the two functions together (i.e., if one is compromised, then both are). Another weakness in the design is the limit of the number of keys that can be used, i.e., four. Finally, using a single key for the whole network increases the chance of keystream reuse.

Collision attacks. Keystream reuse is another problem. The more often a key is used, the easier it is to crack it. The three factors in WEP are the encryption key (IV plus secret key), the plaintext, and the ciphertext. If any two are known, the third can (eventually) be derived. The encryption key consists of the IV, which is transmitted in the clear with each packet, and the secret key, which is not transmitted.

While changing the IV with every packet is recommended, the IEEE does not specify a standard as to how often the IV must change. Reusing the IV values increases the risk of an attacker being able to subvert WEP's protection.

The IV often is initialized to 0. This happens when the wireless card is initialized, such as when the computer is rebooted or the card is removed and inserted or possibly if the wireless station is brought out of a sleep mode. Typically, the IV is incremented by one each time, so there is a low-value bias to the IVs. If a random increment is used, a 50 percent chance of collision exists after approximately 5000 packets. If sequential increments are used, then a 24-bit IV can roll over in less than half of a day in a busy net (note that it is possible to *make* a net busy by inducing traffic on it, even if the encrypted data cannot be decrypted).

A keystream attack requires ciphertext from a reused keystream and partial knowledge of the plaintext. In essence, since $P \text{ XOR } K = C,$ if P (plaintext) and C (ciphertext) are known, K (key) can be computed. The known plaintext can be from predicted data (e.g., the "password:" prompt at a login) or generated by an attacker (e.g., junk mail sent to the remote machine), or in certain cases an AP may broadcast *both* encrypted and unencrypted data such as if the same data must be sent to a wireless station *and* a station on the wired network to which the AP connects. Attackers can build decryption dictionaries by observing the traffic on the network. This requires time and space, although given that tens of gigabytes of disk space are commonplace, space is no longer a serious constraint. If a known keystream *is* reused, then the attacker can use it to create arbitrary messages.

Forgery attacks. Another weakness of WEP is that the ICV uses the CRC-32 as a checksum rather than a cryptographically secure hash function (see Section 13.3.1). CRC-32 is an unkeyed linear function of the message. An attacker can alter bits in the encrypted message *and* then alter the checksum to match the modified encrypted message without having any knowledge of the plaintext message. This means that WEP cannot (adequately) ensure message integrity.

While the RC4 algorithm generally is strong, an attacker can use the APs to decrypt the traffic. If the AP can be convinced to route traffic through the wired network, the data is sent in the clear, which provides plaintext to compare with the ciphertext transmitted by the AP. ARP cache poisoning is one method to do this, although it is not a flaw in WEP. IP redirection is another, in which the attacker modifies the encrypted packet's address so that it is delivered to the attacker's machine in the clear. Another attack type is reaction attacks, in which the attacker cleverly flips a few bits in a message and observes any TCP ACKs (which have a known, fixed packet size), which serves as an oracle to decrypt 1 bit of the plaintext.

Weak key attacks. While RC4 generally is a strong algorithm, there are certain initialization vectors that produce poor results and reveal some of the bits of the plaintext, and the pattern for these IVs is well known. Many programs, such as WEPCrack (2001), can analyze a packet stream taken from AirSnort (2004) looking for the bad IVs and use it to crack the secret key. It takes around 7 GB of data, on average, to produce enough bad IVs to crack the secret key. This may be a lot of data for a home network but not that much for an active network (less than 2 hours at 11 Mbps). And an attacker could induce the traffic on a network if not enough is present. This is one of the most serious problems with WEP, because the tools to crack WEP are widely available.

Replay attacks. Finally, WEP is vulnerable to replay attacks, in which an attacker can eavesdrop and record a sequence of packets from an authorized user and then replay the recorded sequence back to the AP, impersonating the authorized user. This type of attack could be used to authenticate an AP or even to replay a traffic stream such as an Internet purchase (imagine discovering that you bought 50 copies of a DVD instead of one).

15.3.7 WEP fixes

Before discussing the protocols to replace WEP, we briefly discuss how some of WEP's weaknesses can be addressed.

First, to prevent reaction attacks, cryptographically secure hashes, such as MD5 and SHA-1, should be used as integrity check codes rather than a CRC. Second, better key management can increase security. Individual keys should be used per wireless station rather than per network. And changing of keys should be done, like voting in Chicago, early and often. These, of course, require changes to the protocol.

Some simple network configuration changes also can increase the protection. First, the wireless networks should be *outside* the firewall or at least *not inside* the network, possibly blocked by another firewall. In this way, if the wireless network is compromised, it does not bypass any of the security mechanisms used to protect the wired network from machines on the Internet. A VPN tunnel should be used to get from the wireless network into the protected network. Finally, a router or firewall can be configured to block *outbound* traffic from the wireless network to the Internet so that attackers cannot use the wireless network as a launching point for attacks on the Internet. Figure 15.6 shows an example of such a configuration. The best answer at this point is to upgrade to a better protocol. WPA is currently available and provides

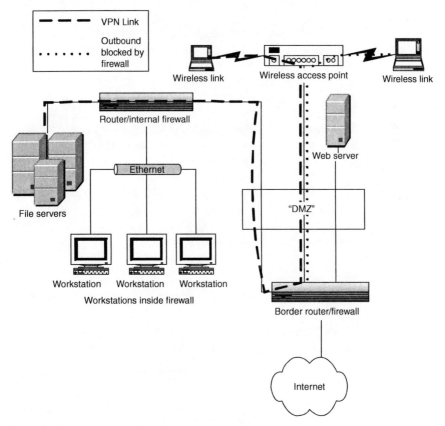

Figure 15.6 A protected wireless network configuration.

stronger protection than WEP, and 802.11i was ratified as a standard in late June 2004 and will be available as "WPA2."

15.4 WPA

Wi-Fi protected access (WPA) was created as an interim measure to increase the security of 802.11b networks. Recognizing that WEP has too many flaws but that it will still be some time before the IEEE adopts the 802.11i protocol for security, WPA was created in 2002 by a vendor alliance (the Wi-Fi Alliance) to provide stronger protection until 802.11i arrived.

Many of the ideas in 802.11i are in WPA, and in fact, vendors expect that the upgrades will involve driver and firmware changes only.

Instead of 40-bit keys, as used in WEP, WPA uses 128-bit keys for encryption and hashing to generate new "random" keys for each use. This key protocol is called the *Temporal Key Integrity Protocol* (TKIP).

The Extensible Authentication Protocol (EAP) allows network administrators to select the method to use for authentication, such as biometric. Also, authentication is performed both ways, by the client and by the server. WEP provides only client authentication via a static password, which must be shared by all users of a network.

WPA also provides automatic key management to generate, configure, and distribute keys (Paulson, 2003).

15.5 802.11i

The 802.11i Working Group is tasked with providing security for 802.11 and addressing the weaknesses and shortcomings of WEP. 802.11i provides two layers, the lower for encryption and the higher for access control. The lower layer supports two encryption protocols, TKIP for legacy equipment and CCMP for future equipment. 802.1x, an IEEE standard for port-based network access control, provides the authentication and key management (IEEE, 2004). We discuss the two layers below (Cam-Winget et al., 2003; Walker, 2002; Eaton, 2002).

15.5.1 Encryption protocols

The Temporal Key Integrity Protocol (TKIP) and the Counter Mode with CBC-MAC Protocol (CCMP) are two encryption algorithms supported by the 802.11i standard. The standard is designed to be extensible, so new algorithms could be added, such as the Advanced Encryption Standard (AES).

TKIP. is a short-term fix for the weaknesses of WEP that maintains compatibility with existing hardware. TKIP requires four new algorithms: a message integrity code (MIC) called "Michael"; IV sequencing, a new per-packet key construction; and a key distribution. TKIP was designed to fix the biggest flaws in WEP and provide protection against collision, weak key, forgery, and replay attacks.

Michael is a keyed hash that was designed to be a computationally low-cost MIC to run on low-power processors. It uses a 64-bit key, the source and destination address, and the plaintext data of the 802.11 frame. It partitions packets into 32-bit blocks and computes the result using shifts, XORs, and addition to create the 64-bit result. TKIP requires the keys to be changed at least once per minute and whenever there is an MIC validation error. The MIC prevents forgery attacks.

TKIP extends the 24-bit IV to 48 bits, referred to as the *TKIP sequence counter* (TSC). While WEP never specified how often the IV should change, TKIP requires that the TSC be updated with every packet. The TSC is constructed from the first and second bytes of the WEP IV and adds 4 extra bytes as the *extended IV*. The initialization vectors are now required to be a strictly increasing sequence that starts at 0 when the base key is set. When the IV reaches its maximum value, data traffic halts, and the protocol must generate a new base key and restart the IV. Because of this, any out-of-sequence packet is discarded, which prevents replay attacks.

TKIP extends the MAC protocol data unit (MPDU) by 12 bytes total, 4 for the extended IV and 8 for the MIC, and is 20 bytes longer than an unencrypted frame. WEP frames extend the frame by 8 bytes compared with an unencrypted frame.

The per-packet encryption key is generated by combining the temporal key, the transmitter address, and the TSC in a nonlinear two-phase key mixing function. The first phase uses the temporal key, the transmitter MAC address, and the 4 most significant bytes of the TSC to create an intermediate value that can be cached and used for up to 2^{16} packets. Note that since the transmitter address is used to create the key, different hosts generate different values even using the same temporal key. The second phase takes the intermediate value and mixes it with the 2 least significant bytes of the TSC to produce the per-packet key. The second phase decorrelates the packet sequence numbers from the per-packet key, blocking weak key attacks.

TKIP uses two keys. One is the 64-bit key Michael uses, and the other is a 128-bit key used by the mixing function to create the per-packet encryption key.

802.1x authenticates the remote station after it associates with the AP. It then gets a fresh master key and distributes it.

TKIP is designed so that an attack that changes the packet sequence number also changes the per-packet encryption key so that either the traditional WEP ICV or the TKIP MIC can catch the error. The MIC makes it computationally infeasible to create an attack that alters the data in a packet. Since the MIC uses the source and destination address, packets cannot be redirected to unauthorized destinations or fake a source address.

CCMP. The Counter Mode with CBC-MAC Protocol (CCMP) is a new encryption method defined by 802.11i and designed to provide a long-term solution to the problems in WEP without TKIP's constraint of using only existing hardware. CBC-MAC is a method to make a Message Authentication Protocol (MAC) using cipher block chaining (CBC) (see Section 13.3.2) (NIST, 1985). CBC-MAC uses the advanced encryption standard (AES), which can support a number of different modes or algorithms. The counter mode provides privacy, whereas the CBC-MAC provides authentication and data integrity. AES is a symmetric, iterated block-mode cipher using 128-bit blocks for encryption. The encryption key length for 802.11i is set at 128 bits as well.

CCMP adds 16 bytes to the frame size compared with unencrypted packets and is identical to a TKIP frame except that there is no ICV; thus it is 4 bytes shorter than a TKIP frame.

CCMP uses a 48-bit initialization vector called the *packet number* (PN). Similar to TKIP, the CCMP PN is much longer than WEP and allows the same AES key to be used for the lifetime of the association. The PN serves as a sequence number that CCMP uses to prevent replay attacks.

AES, unlike TKIP, does not need any per-packet keys and thus has no per-packet key-derivation function. The same AES key is used to provide confidentiality and integrity for all the data sent during a single association. Like TKIP, CCMP uses an MIC to protect the integrity of the data. The MIC length ranges from 2 to 16 bytes and is significantly stronger than TKIP's Michael. CCMP does *not* require any ICV, unlike TKIP and WEP (Cam-Wignet et al., 2002).

15.5.2   Access control via 802.1x

IEEE 802.1x is a standard for "port-based network access control" and is used by 802.11i as the mechanism to provide user authentication and encryption key distribution, both features that WEP did not provide. 802.1x is designed for both wired and wireless networks and provides a framework in which different upper-layer authentication protocols can be used. A *port* is any sort of controlled access and can be a router or a switch for a wired Ethernet, a modem line for a dial-up network, or a wireless AP.

802.1x defines three roles. An *authenticator* is the endpoint that enforces the authentication process for the other endpoint of the connection. The *supplicant* is the endpoint requesting access from the authenticator at the other end of the link. An *authentication server* (AS) is the entity that decides, based on the credentials provided by the supplicant, if the supplicant is authorized to access the services provided by the authenticator.

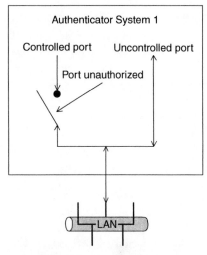

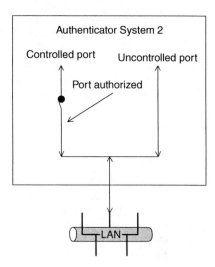

Figure 15.7   802.1x authentication.

The AP acts as a bridge and forwards the credentials from the supplicant to the AS. Typically, RADIUS is the server used as the AS.

The authenticator defines two types of ports. An *uncontrolled port* allows uncontrolled exchange of data between the two ends regardless of the authorization state. A *controlled port* allows the exchange of data only if the current state of the port is "authorized."

Figure 15.7 shows an example of how the authentication works. Initially, the supplicant can communicate with only the authentication server, via the authenticator's uncontrolled port. Once the supplicant has established its authentication, the controlled port's state is changed from "unauthorized" to "authorized," and the supplicant then has access to the services provided by the authenticator—in this case, access to the Internet.

The actual authentication protocol and credentials used depend on the upper-level authentication protocol used. 802.1x merely provides the framework for the exchange of data for these higher-layer protocols.

For 802.11i, two types of keys are generated. *Session* (or *pairwise*) keys are unique to each association between a client and an AP and create a private virtual port between the two. *Group* (or *groupwise*) keys are shared by all the clients connected to an AP and are used for multicast traffic.

Unlike WEP, the keys are generated dynamically, without any intervention of a network administrator. By using an authentication server, the system supports a centralized security management model as opposed to having credentials spread over many APs on a network. Since each individual remote station uses a different primary key for encryption, it becomes much harder to eavesdrop on a network or build up a dictionary of keys.

A master key, called the *pairwise master key* (PMK), is used to generate the lower-level keys employed by the MAC-layer encryption (for TKIP and CCMP). A RADIUS server can serve as the AS, if present. Alternatively, the network can be configured to use shared keys, preloaded into every device, and not use dynamic key management. This would be more typical for home use. If no AS is present, then the PMK must be entered manually into each device (the AP and the remote station). The PMK is still used to generate the session keys that are used for the actual MAC-layer data encryption.

15.6   Fixes and "Best Practices"

The 802.11 area is very dynamic, with new standards that are evolving rapidly. In fact, the area is changing so fast that any detailed technical advice we present will, in all likelihood, be out of date 6 months after it is published. And, in fact, in the time between writing and editing this book, 802.11i was ratified. Instead, we present some of the broader aspects of how to protect a system, rather than focusing on specific, ephemeral technical details.

15.6.1   Anything is better than nothing

There are various flaws in the different protection schemes and mechanisms available. However, unless the goal is to provide open access to everyone, WLANs should not be run "wide open" with no protections at all. Anything that makes it more difficult for an adversary to gain access to your network increases the level of protection.

Many access points provide MAC address—based protection. The administrator can specify a set of MAC addresses that are allowed to use the AP. This is another way to "raise the bar" for security. However, several problems exist with this approach. The size of the list of MAC addresses in the AP may be limited, and thus this might not be usable on a network with a large number of users. If the user base is relatively dynamic, it might be impractical to continually update each AP with a new list of good or bad MACs. Finally, some wireless cards allow the user to specify the MAC address, so they can be faked and are not a reliable way to block intruder access.

15.6.2   Know thine enemy

Having a realistic assessment of the capabilities and motivations of an adversary allows administrators to make informed decisions about the types and levels of protections needed. Is she a casual adversary who is looking for Internet access from the first easily available system? Or is she targeting your system in particular? How much time and money is she willing to spend trying to get at your resources? Does she care if you detect her activities? Geier (2002a) suggests several security policies for wireless networks.

15.6.3 Use whatever wireless security mechanisms are present

The best answer is to use 802.11i with 802.1x for authentication and have a separate authentication server. Smaller residential setups may not have all these components, and some installations may be stuck with using legacy hardware (even though individual wireless cards and APs are relatively inexpensive). If 802.11i, called "WPA2," is not available, then use WPA and, literally better than nothing, WEP.

Using WEP forces the adversary to spend some, albeit a small, amount of time to break into the system. Some manufacturers have fixes, such as WEPplus, that fix some of WEP's worse problems by not using known bad initialization vector values.

IEEE 802.1x, discussed in Section 15.5.2, provides "port-based network access control" using existing standards, including the Extensible Authentication Protocol (EAP), the Challenge Handshake Protocol (CHAP), and RADIUS (see Section 13.5.1).

WPA was the bridge to provide better security than WEP until 802.11i was finalized and products that support it become available. As of this writing, only one product supports 802.11i, but more are sure to follow shortly.

15.6.4 End-to-end VPN

Virtual private network (VPN) software allows a remote system to be part of another subnet by encapsulating the data, encrypting it, and then sending the packet to a firewall (*tunneling*). The firewall decrypts the data, unencapsulates them, and then retransmits them on the private network, making it appear as if the data originated locally. Similarly, data bound for the remote system from inside the private network are sent to the firewall and similarly tunneled to the remote system, at which point they are delivered as if they were a local packet. Since the data are encrypted as they pass between the nodes in the "virtual" network, eavesdroppers cannot compromise the data confidentiality.

By using VPN software or hardware, the wireless network can sit "outside" the trusted network and tunnel out of only that subnet. VPNs generally use strong encryption.

15.6.5 Firewall protection

A firewall protects networks, both wired and wireless, by serving as the delineation between a private network sitting "behind" the firewall and the public network exposed to arbitrary Internet traffic. Only certain, specified traffic enters the private network. Traffic may be permitted to, say, the mail and Web servers, whereas all other traffic is blocked. For a home installation that has no public servers, no unsolicited inbound traffic may be permitted. Usually, inbound traffic is permitted if it was initiated by an outbound connection from behind the firewall. Firewalls also can prevent inside traffic from getting out, preserving data confidentiality, as well as preventing internal machines from being used as the source of an attack against other machines on the Internet. A firewall may be a separate computer or hardware device, or it can be some form of "personal firewall" software running on the user's machine. Hardware firewalls typically are better than personal firewalls because they serve a single purpose and run on a simplified operating system that runs only the firewall software. There are a number of products that combine the functionality of a cable modem, router, and firewall using Network Address Translation (NAT) that allow numerous home computers to share a single IP address on the Internet using the built-in Dynamic Host Configuration Protocol (DHCP) to issue the internal IP addresses. These devices are inexpensive and very convenient and as a side effect protect the users' computers from hostile Internet traffic. Figure 15.8 shows an example of such a configuration. They may not, however, protect the wireless component of the network adequately by default. Also, some compatibility problems exist between NAT and IPsec, which is used in some VPN implementations.

15.6.6   Use whatever else is available

SSID.   A simple first step to protect a wireless AP is to change the SSID to something other than the default. For example, it is well known that *linksys* and *tsunami* are the default SSIDs for LinkSys and Cisco APs, respectively. The less an attacker knows about the system, such as the AP hardware, the better.

Use good passwords.   All default passwords for hardware are public knowledge, so it is important to at least change the password so that it is not "default" or blank. The next step is to use a good password that is not guessed or cracked easily, e.g., avoid dictionary words.

Closed network.   In addition, the AP can be set so that it does not broadcast the SSID in the beacon frame; i.e., it is not announcing its presence continuously. This is known as a *closed network* (see Section 12.2.5). A wireless station must know the SSID before it can associate with the AP, as opposed to casually observing the SSID in the beacon. However, this merely obscures the SSID because it is broadcast in the clear by the wireless station when it associates with the AP and can be detected easily by a wireless packet sniffer. However, it does raise the level of protection.

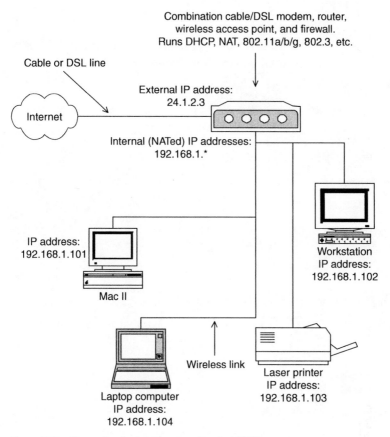

Figure 15.8   Example of a home network using NAT.

Limit the radio wave propagation.   While shielding your house or living in a Faraday cage may eliminate stray radio transmissions, it is a bit extreme for home use. Moving the AP toward the center of the desired coverage area is a simple step that forces the signals to pass through more walls before reaching outside attenuating them more than if the AP sat right next to a window. Commercial wireless networks require more planning in terms of physical layout than home networks (assuming that a single AP is insufficient). Directional antennas also limit the signal propagation.

IDS.   Running an intrusion detection system (IDS) provides an additional layer of security. While an IDS does not prevent an attack, it allows the user to know that one occurred, often giving enough details to help stop the attack or prevent it from recurring. There are many freely available IDSs, such as snort (http://www.snort.org).

15.7 Summary

In this chapter we presented the basic components and operation of WLANs using the IEEE 802.11 protocol. We described the numerous 802.11 subgroups and WEP, the original security mechanism of 802.11b, specifically, its goals and functions, as well as the problems in its design. We described 802.11i, the follow-on security protocol for 802.11, as well as "best practices" for use in concert with the mechanisms in the 802.11 protocol. Chapter 16 covers wireless metropolitan area networks (WMANs), which provide Internet connectivity *to* the local area networks.

15.8 References

AirSnort Homepage, *http://airsnort.shmoo.com/*

Borisov, N., I. Goldberg, and D. Wagner, "Intercepting Mobile Communications: The Insecurity of 802.11," in Proceedings of the 7th Annual International Conference on Mobile Computing and Networking, July 2001, Rome, Italy, *http://www.isaac.cs.berkeley.edu/isaac/mobicom.pdf.*

Cam-Winget N., R. Housley, D. Wagner, and J. Walker, "Security Flaws in 802.11 Data Link Protocols," Communications of the ACM, May 2003, 46(5), pp. 35–39.

Cam-Wignet N., T. Moore, D. Stanley, and J. Walker, "IEEE 802.11i Overview," presented at the NIST 802.11 Wireless LAN Security Workshop, Falls Church, Virginia, December 4-5, 2002, *http://csrc.nist.gov/wireless/ S10_802.11i%20Overview-jw1.pdf.*

Clausen, T., P. Jacquet, IETF RFC 3626, "Optimized Link State Routing Protocol (OLSR)," October 2003, *http://www.ietf.org/rfc/rfc3626.txt.*

Dailey Paulson L., "Vendors Push Wireless LAN Security," Computer, January, 2003, p. 28.

Eaton, D., "Diving into the 802.11i Spec: A Tutorial," CommsDesign, November 26, 2002, *http:// www.commsdesign.com/design_corner/OEG20021126S0003.*

Fluhrer, S., I. Mantin, and A. Shamir, "Weaknesses in the Key Scheduling Algorithm of RC4," Lecture Notes in Computer Science, Volume 2256, 2001.

Geier J., "802.11 Alphabet Soup," Tutorial on 80211planet web site, August 5, 2002 *http://www.80211-planet.com/tutorials/article.php/1439551.*

Geier J., "The Guts of WLAN Security Policy," November 12, 2002, *http://www.80211-planet.com/tutorials/article.php/1499151.*

IEEE Standard 802.11, "Wireless LAN Medium Access Control (MAC) Sublayer," 1999.

Moy, J., IETF RFC 2328, "OSPF Version 2," April 1998, *http://www.ietf.org/rfc/rfc2328.txt.*

National Institute of Standards and Technology. FIPS Pub 113: "Computer Data Authentication," May 30, 1985.

Park, V., and M.S. Corson, IETF MANET Internet Draft, draft-ietf-MANET-tora-spec-03.txt," November 2000.

Perkins, C.E., E.M. Belding-Royer, and Samir Das, IETF RFC 3562, "Ad Hoc On Demand Routing Distance Vector (AODV) Routing," July 2003, *http://www.ietf.org/rfc/rfc3561.txt.*

Walker, J., "Unsafe at any key size: an analysis of WEP encapsulation," Tech. Rep. 03628E, IEEE 802.11 committee, March 2000, *http://grouper.ieee.org/groups/802/11/Documents/DocumentHolder/0-362.zip.*

Walker J., "802.11 Security Considerations and Solutions," Intel Developer Forum, Spring 2002, *http://developer.intel.com/idf.*

"WEPCrack, and 802.11 Key Breaker," *http://wepcrack.sourceforge.net/.*

"802.1x—Port Based Network Access Control", *http://grouper.ieee.org/ groups/802/1/pages/802.1x.html.*

16

Security in Wireless Metropolitan Area Networks (802.16)

Metropolitan area networks (MANs) link local area networks (LANs) to larger networks, such as the Internet, through a high-speed connection. This chapter presents the security mechanisms in the newly approved IEEE 802.16 standard for wireless MANs, which provides the interface to high-speed wireless links using licensed and unlicensed radio bands to fixed locations, such as rooftop antennas. We describe the protocol, the security mechanisms, and potential limitations.

16.1 Broadband Wireless Access

MANs link commercial and residential buildings to the Internet through a high-speed connection. Typically, an Internet service provider (ISP) supplies the MAN. The links can be dedicated high-speed lines, such as T1 (1.54 Mbps), T3 (45 Mbps), OC3 (155 Mbps), OC12 (622 Mbps), and beyond, or they can be a broadband link, i.e., a link (wire) that carries more than one channel at once, such as cable modem or a digital subscriber line (DSL). Dedicated lines provide more reliable connections and higher speeds but generally are very expensive.

Note that the answer to the question, "What counts as high speed?" changes faster than a speeding packet (on a "high-speed" connection).

Cable modem and DSL, two popular, inexpensive options for Internet service for residential users and smaller businesses, currently provide around 3 Mbps (i.e., twice T1 speed) at a fraction of the cost of a dedicated line. However, a third option has become available recently and is growing in popularity: broadband wireless access (BWA).

16.2 IEEE 802.16

In late January 2003, the IEEE approved IEEE 802.16 as a standard. The first certified products are planned to be available by the end of 2004 (WiMAX Forum, 2004). 802.16 is the Working Group on broadband wireless access standards for MANs (IEEE, 2004). The goal of the group is to provide fixed BWA to large areas. They are, in essence, competing with DSLs and cable modems.

802.16 is related to 802.11 (see Chap. 15), the wireless LAN standard, in that they are both under the same 802 LAN/MAN standards committee. However, 802.16 provides fixed wireless connectivity; i.e., the source and destination do not move, typically using line-of-sight antennas. 802.16 supports communication from 2- to 66-GHz bands in both licensed and unlicensed bands. 802.16 is the only 802 protocol that supports transmission on licensed bands. Task Group A covers the 2- to 11-GHz bands, and Task Group C covers the 10- to 66-GHz bands. Task Group 2a covers the coexistence of 802.16 and 802.11 protocols on the same unlicensed frequencies.

802.16 connects the base station (BS) at the ISP to the subscriber station (SS) at the business or residence and supports speeds up to 268 Mbps.

802.16 supports three types of transmission methods at the physical layer: single-carrier modulation (SC), orthogonal frequency division multiplex (OFDM), and orthogonal frequency division multiple access (OFDMA). The unlicensed bands use OFDM, which 802.11a uses as well. Security for 802.16 is provided by a privacy sublayer, described in the next section (Marks et al., 2001; Marks, 2002).

16.3 802.16 Security

The privacy sublayer of 802.16, operating below the common-part sublayer, protects the transmitted data by providing security mechanisms for authentication and data encryption. Since MANs provide Internet connectivity service from a provider to a paying subscriber, the provider must verify the identify of the subscriber. A key management protocol, Privacy Key Management (PKM), allows the BS to control access to the network and the SS and BS to exchange keys. The encapsulation protocol encrypts the packet data that are transmitted. The packet data (the MAC PDU payload) is encrypted; the header information is not. MAC management messages also are not encrypted (IEEE, 2004).

The privacy sublayer is based on the Baseline Privacy Interface Plus (BPI+) specification for Data Over Cable System Interface Specification (DOCSIS). Each customer transceiver, the SS, has its own digital certificate, which it uses for authentication and key exchange.

16.3.1 Key management

The PKM protocol uses X.509 digital certificates, the RSA (Housley et al., 1999) public key encryption algorithm (Kaliski et al., 1998), and strong symmetric encryption for key exchange between the BS and the SS. As in many other systems, it uses a hybrid approach of public and symmetric encryption.

The public key encryption establishes a shared secret, the *authorization key* (AK), between the SS and BS. Then the AK is used with symmetric encryption to exchange the traffic encryption keys.

In the initial authorization exchange, the BS (server) authenticates the SS (client) through use of the X.509 certificate. Each SS is issued a certificate by the manufacturer, which contains the SS's public key and MAC address. When the BS receives a certificate from an SS, it verifies the certificate and then uses the SS's public key to encrypt the authorization key. By using the certificate for authentication, the BS prevents an attacker from using a *cloned SS* to masquerade as a legitimate subscriber and steal service from the BS. An attacker that does not have the SS's private key from the SS's certificate could not decrypt the authorization key sent by the BS and therefore not steal service successfully from the BS. Refer to the "Key management protocol" in Section 7.1.2 of the IEEE 802.16 standard (IEEE, 2001). Note that IEEE has just released a revised version of the 802.16 standard (IEEE, 2004). Figure 16.1 shows an example of key management.

16.3.2 Security associations

A *security association* (SA) is the set of security information shared by the BS and one or more SSs to support secure communications. 802.16 defines three types of SAs:

- *Primary.* A primary security association is established during the initialization process of the SS.
- *Static.* Static SAs are provisioned within the BS.
- *Dynamic.* Dynamic SAs are created and destroyed as specific service flows are created and destroyed.

Static and dynamic SAs can be shared by multiple SSs.

The SA's shared information may include the cryptographic suite employed, as well as the traffic encryption key (TEK) and initialization vectors (IVs). An SAID identifies each SA. A cryptographic suite is the SA's set of methods for data encryption, data authentication, and TEK exchange.

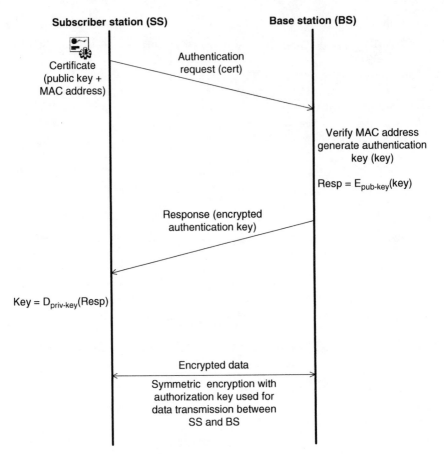

Figure 16.1 Key management in 802.16.

Currently, 802.16 supports the following two cryptographic suites:

- No encryption, no authentication, and 3-DES with a 128-bit key
- CBC mode, 56-bit DES, no authentication, and 3-DES with a 128-bit key

Additional codes are reserved for additional suites.

Each SS establishes an exclusive primary SA with its BS. Using PKM, the SS requests the keying material from the BS. The BS ensures that each client has access only to SAs that it is authorized to access.

16.3.3 Keying material lifetime

The SA's keying material includes the data encryption standard (DES) key and the CBC IV and has a limited lifetime. The BS informs the SS of the remaining lifetime of the keying material when it is delivered to the SS. The SS must request new keying material from the BS *before* the current one expires. If the current one does expire before a new one is received, the SS must perform a network entry (i.e., reinitialize and start over).

The AK lifetime is 7 days, and the grace time timer is 1 hour (Marks et al., 2001a). The grace time provides the SS with enough time to reauthorize, allowing for delays, before the current authorization key expires. The grace time specifies the time before the AK expires when reauthorization is scheduled to begin.

16.3.4 Subscriber station (SS) authorization

The SS first sends an authentication information message to the BS containing the manufacturer's X.509 certificate. The SS then sends an authorization request to the BS containing:

- The SS's X.509 certificate issued by the manufacturer
- A description of supported cryptographic algorithms
- A connection identifier (CID)

After the BS has validated the request, it creates an authorization key and encrypts it with the SS's public key. The reply includes:

- The encrypted AK
- A 4-bit sequence number to distinguish different AKs
- The key's lifetime
- The identifier for the SA (SAID)

After initial authorization, an SS reauthorize itself periodically with the BS. Successive generations of AKs have overlapping lifetimes to avoid service interruptions. The SS and BS can have two simultaneously active AKs. Once the BS has an AK, it can obtain a TEK.

16.3.5 Encryption

For each SA, the SS requests a key from the BS. The SS sends a key request message, and the BS sends a key reply containing the keying material. The TEK in the message is triple DES encrypted using a two-key triple DES key encryption key derived from the AK. The reply also contains the CBC IV and the lifetime of the key. Similar to the authorization key, the SS maintains two overlapping keys. The second key becomes active halfway through the life of the first key, and the first expires halfway through the life of the second, causing the SS to send a new key request to the BS. Each successive key maintains this half-step synchronization.

The BS is able to choose what cryptographic suite to use, based on the list of available suites the SS provided.

16.4 Problems and Limitations

Johnson and Walker (2003) point out several limitations to and potential problems with the current 802.16 specification and suggest enhancements to improve security. Since this is a rapidly evolving standard, we will cover just some of the highlights.

First, 802.11i (see Section 13.5) has now set the standard for current best practices in wireless security, and 802.16 falls short in a few places. Authentication occurs only in one direction; specifically, the base authenticates the subscriber, but not vice versa.

Authentication is based on X.509 certificates, which are difficult to administer. The RSA algorithm is used for key establishment which may be too compute intensive and slow for some devices or require more expensive hardware. DES is used for one of the encryption suites, which is not regarded as secure. There is no data authentication, which is regarded as mandatory in the wireless environment. In addition, there is no data replay protection.

Johnson and Walker make several recommendations. AES-128 should be added to the cryptography suites to enhance the protection. Data authentication needs to be added. This would be in the form of a MAC using the AES block cipher, following what is done in 802.11i. IVs should be sequential in order to maximize the time between rekeying. For sequential IVs, reuse occurs after N IVs; for randomized IVs, reuse occurs after $\sqrt{N}$ IVs. Currently, the same TEK is used in both directions, for the uplink and the downlink. This leads to a higher chance of initialization vector reuse. Instead, two different keys should be used.

Authorization should be bidirectional; i.e., users should perform authorization on the BS. Rather than relying solely on the factory-installed RSA key pair, adding an 802.1x-style authentication suite would add capabilities and bring it in line with the other 802 standards (802.3 and 802.11). In addition, rather than using a factory-installed key pair, a smart card system would allow more flexibility.

16.5 Summary

In this chapter we presented the IEEE 802.16 WMAN protocol and its security mechanisms. 802.16 provides support for MANs, providing the "last mile" of a high-speed wireless Internet link to a commercial subscriber and eventually residential subscribers. As mentioned earlier, 802.16 is still a new, evolving standard, and as of this writing, it will be some time before commercial hardware is available. Thus our discussion focused on the mechanisms in the standard, as well as changes that would provide protection similar to 802.11i by using the encryption suites and authentication protocols in 802.11i and 802.1x. Chapter 17 covers WWANs.

16.6 References

IEEE Standard 802.16-2001, "Air Interface for Fixed Broadband Wireless Access Systems," April 2002, available through Get IEEE 802™ at: *http://standards.ieee.org/getieee802/download/802.16-2001.pdf*.

IEEE Standard 802.16-2004, "Air Interface for Fixed Broadband Wireless Access Systems," August 2004.

Housley, R., W. Ford, W. Polk, and D. Solo, "Internet X.509 Public Key Infrastructure Certificate and CRL Profile," RFC 2459, January 1999, *http://www.ietf.org/rfc/rfc2459.txt*.

http://grouper.ieee.org/groups/802/16/.

http://www.cablemodem.com/specifications/specifications11.html.

Johnson, D., and J. Walker, "802.16 Security Enhancements," IEEE 802.16 Presentation Submission, Contributed document IEEE C802.16d-03/60r1, September 9, 2003, *http://grouper.ieee.org/groups/802/16/tgd/contrib/C80216d-03_60r1. pdf*.

Kaliski, B., and J. Staddon, "PKCS #1: RSA Cryptography Specifications Version 2.0," October 1998, *http://www.ietf.org/rfc/rfc2437.txt*.

Marks, R.B., C. Eklund, K. Standwood, and S. Wang, "The 802.16 WirelessMAN MAC: It's Done, but What Is It?", 802 LMSC Plenary Session, November 2001, *http://grouper.ieee.org/groups/802/16/docs/01/80216-01_58r1.pdf*.

Marks, R.B., I.C. Gifford, and B. O'Hara, "Standards from IEEE 802 Unleash the Wireless Internet," IEEE Microwave Magazine 2, pp. 46–56, June 2001, *http://grouper.ieee.org/groups/16/docs/01/80216c-01_10.pdf*.

Marks, R.B., "Consensus IEEE 802.16 Standard Marks Maturation of Broadband Wireless Access Industry," EE Times, April 1, 2002, *http://grouper.ieee.org/groups/802/16/docs/02/C80216-02_04.pdf*.

Worldwide Interoperability for Microwave Access Forum (WIMAX), *http://www.wimaxforum.org* (December 2003).

17

Security in Wide Area Networks

Wireless wide area networks (WWANs) are very large-scale wireless networks. 802.16 provided coverage in a metropolitan area, say, 10 miles, to a fixed location. A WWAN must provide continuous coverage across a much larger area, such as an entire state or country. And nodes in a WWAN can move and must remain connected. Currently, the only technology that provides this type of coverage is satellite and cell phones. Although satellite technology provides truly global coverage, the current cost makes it impractical for common use. Cell phone technology, on the other hand, is practically ubiquitous and much more affordable. Therefore, we focus on cell-based WWANs in this chapter.

17.1 Basic Idea

Low-power cells share frequencies and use spread-spectrum technology allowing multiple users per channel per cell. High redundancy allows voice quality to be maintained. Handoffs from one cell to another occur as the mobile unit passes out of one cell's coverage area and into another's. (Actually, multiple base stations are received at once, and the strongest is used.) While many locations have poor or no coverage, cellular wireless networks provide the closest approximation to ubiquitous connectivity this side of satellite phones. And in large metropolitan areas, cell coverage *is* ubiquitous, including base stations located in underground public transportation stations, so connectivity is maintained even on a moving train.

Initially, cellular networks just carried voice. The next step was to use these networks to carry data, say, to provide Internet connectivity through the cell phone. The first incarnation was cellular modems, which converted a digital signal to an analog signal by modulating a carrier.

It then converted the carrier to a digital signal, sent it across the air and through the wires, and then converted it back to an analog signal that was input into the receiver's modem. The modem then converted the analog signal back to a digital signal. This connects to some sort of dial-up that presumably connects to the Internet. Figure 17.1 shows the process of hops and conversions. Obviously, the results are suboptimal because performance is lost with each conversion.

A more direct approach, now available, takes the digital information from the computer and transmits it directly over the air to the cellular

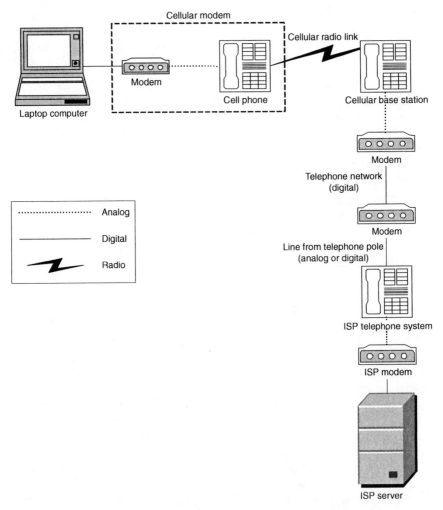

Figure 17.1   An inefficient way to send digital data over a cellular network.

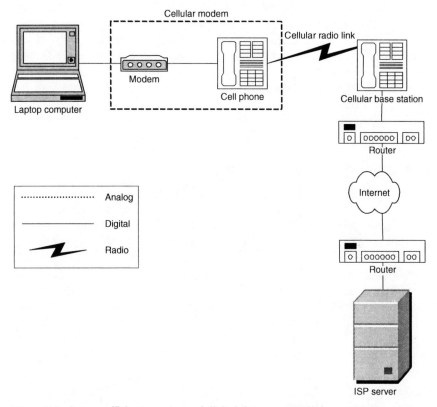

Figure 17.2 A more efficient way to send digital data over a cellular network.

provider. The cellular provider functions directly as an ISP and provides Internet connectivity. The cellular provider gives the device connecting to the phone a direct Internet protocol (IP) address. Figure 17.2 shows this approach.

The United States uses three wireless network standards: TDMA, CDMA, and GSM. Time domain multiple access (TDMA) uses differential quadrature phase shift keying (DQPSK) for time multiplexing information to archive a data rate of 48.6 kbps. Code division multiple access (CDMA) is a spread-spectrum technique similar to direct sequence spread spectrum (DSSS) used in 802.11b (Nichols and Lekkas, 2002). GSM uses a combination of TDMA/frequency division multiple access (FDMA).

17.2 CDMA

CMDA is a spread-spectrum technique. The transmitter uses a code, shared by both endpoints, to send each bit of data across a large frequency range. The receiver uses the code to reconstruct the original data from the spread-spectrum signal. This frequency-spreading technique makes it very difficult to intercept the signal unless the code is known. While CDMA had been developed originally for military applications, its commercial goal was its larger capacity over TDMA-based systems rather than security. CDMA's relatively strong security property comes from the low probability of interception (LPI) of the data because of the encoding used for spread spectrum as compared with GSM's weak(er) encryption of its data (Nichols and Lekkas, 2002).

17.3   GSM

Global systems for mobile communication (GSM) is one type of cellular phone network, and it has security mechanisms that provide authentication and encryption. GSM is based on TDMA; thus, intercepting the signals is much easier than in CDMA. Therefore, GSM has separate security mechanisms to encrypt the data it transmits. GSM security mechanisms are based on a shared secret between the home location register (HLR) and the subscriber identity module (SIM)—in other words, the security modules in the phone and the central station. A subscriber identity module is a removable hardware device that provides security, is managed by network operators, and is independent of the terminal device in which it resides. GSM provides mechanisms for authentication and encryption.

17.3.1   GSM authentication

The shared secret K_i is a 128-bit key. Authentication is performed when the HLR or base station sends a 128-bit random number called a *challenge* to the mobile station (MS), i.e., the phone. The MS calculates the response, a 32-bit signed response (SRES), by using the A_3 algorithm feeding the challenge and the shared secret as input. The base station then compares the SRES received from the MS with the expected value. Figure 17.3 shows the process (Pesonen, 1999).

17.3.2   GSM encryption

The MS and base station use a 64-bit session key K_c for data encryption of the over-the-air channel. They calculate K_c by using K_i and a 128-bit random number, which are the same numbers used to calculate the SRES. Instead of using the A_8 algorithm as was originally specified, however, most manufacturers use the A_3 algorithm to calculate K_c as well. This is done to reduce the number of cryptographic algorithms to encode in the telephone firmware.

The session key is *not* used to encrypt the data directly. Instead, it is used to generate the keystream that encrypts the data. In Chap. 13 we

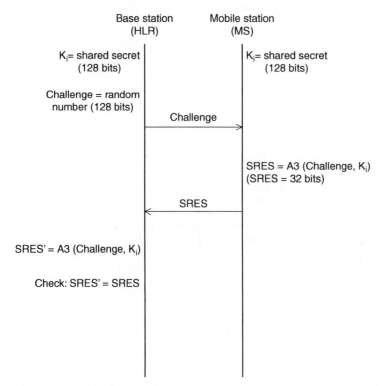

Base station
(HLR)

Mobile station
(MS)

K_i = shared secret
(128 bits)

K_i = shared secret
(128 bits)

Challenge = random
number (128 bits)

Challenge

SRES = A3 (Challenge, K_i)
(SRES = 32 bits)

SRES

SRES' = A3 (Challenge, K_i)

Check: SRES' = SRES

Figure 17.3 GSM authentication.

showed that a basic stream encryption algorithm works by XORing the datastream with a keystream generated by a pseudorandom-number generator (PRNG) provided with an initial seed. In this case, the seed is K_c, and the PRNG is the A_5 algorithm. Actually, the seed is K_c and the frame number (which is 22 bits). Figure 17.4 shows GSM encryption.

The frame numbers are generated implicitly and can be guessed. Thus, if an adversary determines K_c, he can decrypt the traffic. The mobile station can authenticate itself at the beginning of each call, but this is not generally done in practice. Thus an MS can retain the same K_c for days. In addition, once the base station (base transceiver station, BTS) receives the frames, it decrypts the data and sends them in plaintext to the backbone network.

17.4 Problems with GSM Security

GSM security has several shortcomings, including session life, weaknesses in the COMP-128 algorithm, as well as encryption used between only the MH and BS. We discuss these limitations and others below.

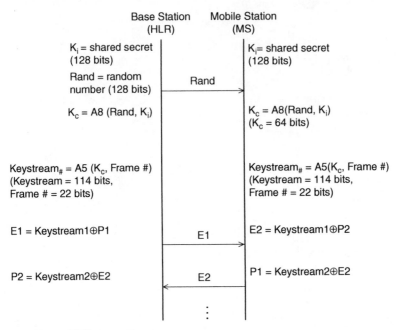

Figure 17.4 GSM encryption.

17.4.1 Session life

The first problem is the long life of authenticated sessions. While the mobile station *may* be requested to reauthenticate at the beginning of each call, typically this is not done. This means that the same session key K_c is used for days. The longer a session key is used, the weaker it becomes.

17.4.2 Weak encryption algorithm

Traffic is encrypted via the A_5 algorithm *only* over the air between the mobile and base stations. The data are decrypted when they arrive at the base station and are sent from the base station to the operator's backbone network in plaintext.

Almost all GSM implementations use the COMP-128 algorithm for both A_3 and A_8 algorithms. The session key K_c generated by COMP-128 is actually 54 bits, with 10 zero bits added to pad it to 64 bits. Obviously, this reduces the key space. Even non-COMP-128 algorithms use only 54 bits for K_c.

While real-time interception and decryption of over-the-air transmissions is (currently thought of to be) impossible, there are other attacks to GSM security. Pesonen (1999) calculates that a brute-force attack on a 2^{54} bit key using a 600-MHz Pentium III would take about 250 hours using one processor. "Modern" computers are easily five times faster, dropping the time to 50 hours. Pesonen also mentions a technique to reduce the time by one third, thus reducing the time to, roughly, 30 hours. By using multiple CPUs and distributed computing, the time can be reduced further. Therefore, what was infeasible a few years ago quickly becomes very possible and affordable. Additional approaches, such as divide and conquer, further reduce the search space. Nichols and Lekkas (2002a) describe the cryptanalysis of the A_5 algorithm. Golic (1997) published an attack on the A_5 algorithm with a complexity of $O(2^{40})$, as well as time and space trade-offs to further reduce the computation time.

17.4.3 Encryption between mobile host and base station only

Gaining access to the signaling network allows an eavesdropper to listen to unencrypted data traffic, as well as all the authentication data (RAND, SRES, and K_c). Networks commonly use microwave or satellite links, for which eavesdropping equipment does exist.

17.4.4 Limits to the secret key

If an attacker compromises the secret key K_i, the entire security scheme is compromised. While the GSM network detects if two phones with the same IDs are operating simultaneously and closes the account, the network cannot detect a passive eavesdropper silently decrypting the data. In addition, it is possible to retrieve K_i from a subscriber identity module (SIM) because of a flaw in COMP-128. By sending selected challenges, the COMP-128 algorithm responds in a way that reveals information about K_i. This attack takes hours to complete and requires physical access to the SIM.

Two variants of the A_5 algorithm exist: $A_{5/1}$ and $A_{5/2}$, with the former more secure than the latter. In 2000, approximately 230 million customers in Europe and elsewhere were using these two algorithms. Biryukov and colleagues (Biryukov et al., 2002) propose two attacks on $A_{5/1}$ that can be performed in real time on a PC after a one-time data initialization step is completed. The first attack requires eavesdropping on the output of the algorithm for the first 2 minutes of the conversation and computes the key in approximately 1 second. The second attack requires 2 seconds of the conversation and computes the key in several minutes.

The main problem with GSM security is that it relies on the secrecy of the A_5 algorithm, which was not publicly scrutinized. Once the algorithm leaked out in the middle to late 1990s, serious flaws were discovered. This is an example of the fallacy of "security by obscurity."

17.4.5 Other problems

There are several other GSM problems. No data integrity algorithm is used; therefore, data could be modified and the receiver could not detect it. Authentication is performed in only one direction, from the user to the network. No mechanism exists to identify the network to the user. Also, there is no indication to the user that encryption is being used.

17.5 The Four Generations of Wireless: 1G–4G

The first generation (1G) of wireless wide area communications, present in the 1970s and 1980s, used analog signals to transmit voice signals only. The second generation (2G) started in the 1990s and used digital signals for voice and data. GSM and TDMA are 2G. 2.5G represents technology improvements in the between 2G and 3G. The third generation (3G) supports higher data rates, from 144 kbps to 2 Mbps and beyond, and typically is packet-switched using CDMA. 3G examples include EDGE, GPRS, and W-CDMA. The fourth generation (4G) will provide much higher data rates, in excess of 20 Mbps, and is expected to be deployed around 2006 to 2010 (http://www.netmotionwireless.com/resource/glossary.asp). At this point, it is still unclear what will be in 4G versus 3G, but mostly likely 4G will be integrated with WPANs and WLANs.

17.6 3G

3G security is based on GSM (Myagmar and Gupta, 2001) but is designed to fix its shortcomings. The security mechanisms of 3G provide authentication, confidentiality, and encryption, and are described below.

Authentication. GSM authentication provides protection from unauthorized service access and is based on the A_3 algorithm, which is known to have limitations. Encryption is used to protect both the user data and the signaling data. The A_8 and A_5 algorithms are used but are not strong enough.

In the authentication and key agreement (AKA) phase, the user and network authenticate each other and agree on a cipher key (CK) and an integrity key (IK). The keys expire after a specified time limit.

Confidentiality. Confidentiality is provided by identifying users with a permanent identity, called the *international mobile subscriber identity* (IMSI) and a *temporary mobile subscriber identity* (TMSI). Transmission of the IMSI is not protected; it is sent as plaintext. Therefore, a more secure mechanism is needed. The user and network agree on the cipher key and algorithm during the AKA phase.

Encryption.   Recall that a *subscriber identity module* (SIM) is a removable hardware device connected to the phone that provides security. In addition, the phones indicate when encryption is being used and what level is available (2G or 3G).

The Third Generation Partnership Project (3GPP) is a standards group focused on 3G technology, including GSM, General Packet Radio Service (GPRS), Enhanced Data rates for GSM Evolution (EDGE), and Wideband Code Division Multiple Access (WCDMA). GGPP security provides some changes and enhancements of GSM security. Security mechanisms include sequence numbers to defeat false base station attacks. Key lengths were increased so that stronger algorithms can be used for encryption and integrity. New mechanisms provide security support within and between networks. Links between the base station and switch are now protected because security is based within the switch. Integrity mechanisms for the terminal identity, the *international mobile equipment identity* (IMEI), were designed into the system from the beginning rather than added on as an afterthought.

The authentication algorithm is not defined, but guidance is given as to what to use. When roaming, only the level of protection supported by the smart card is used, so a GSM card is not protected against a false base station attack when using a 3GPP network (3GPP, 2000).

The IMEI, which identifies the phone, is not protected either, but it is not a security feature.

Users can unwittingly "camp" on a false base station and will no longer receive paging signals from the serving network (SN).

If encryption is disabled, an intruder can hijack incoming and outgoing calls by posing as a "man in the middle" and then taking over once the call is connected.

Attacker Capabilities.   In order to perform an attack, an attacker must have one or more of the following capabilities: eavesdropping, impersonation of a user, impersonation of a network, "man in the middle," or compromising authentication vectors in the network. We list the capabilities in increasing order of effort to obtain and complexity. Therefore, a given capability implies possessing all capabilities listed above it. We describe each capability in more detail below.

Eavesdropping.   This capability allows the intruder to receive signaling, data, and control information associated with other users. This requires a modified mobile station.

Impersonation of a user.   This capability allows the intruder to send signaling, control, and data information such that it appears to originate from a different user. This requires a modified mobile station.

Impersonation of a network. This capability allows the intruder to send signaling, control, and data information such that it appears to originate from a different network or system component. This requires a modified base station.

"Man in the middle." This capability allows the intruder to place himself between the target user and the network. Being a "man in the middle" allows the intruder to eavesdrop, modify, delete, reorder, replay, and fake signaling, control, and data messages between the user and the network. This requires a modified base station in conjunction with a modified mobile station.

Compromising authentication vectors in the network. This capability allows an intruder to possess a "compromised authentication vector" including challenge/response pairs and cipher and integrity keys. The intruder obtains this information by compromising network nodes or eavesdropping on signaling messages on network links.

In addition, there are several types of denial-of-service (DoS) attacks. An attacker can send a fake (spoofed) *user deregistration request,* rendering the victim unreachable. An attacker can fake a *location update request,* which causes the network to register the victim in a new (wrong) location, causing the victim to be unreachable because she will be paged in the wrong location. An attacker with a modified base station can entice a user to "camp" on (connect to) the false base station, rendering the victim out of reach of paging signals of the real network.

17.7 Limitations

2G data rates are on the order of 9.6 to 28.8 kbps. 3G currently goes up to 140 kbps. Within a few years, it should support speeds in the megabit per second range, with plans to support speeds in the tens, if not hundreds, of megabits per second. While the envisioned rates are, relatively speaking, "fast," they will always be orders of magnitude slower than other network types. The 802 wireless networks (802.11, 802.15, 802.16) support speeds from 50 to 200 Mbps now. Ethernet (802.3) supports gigabit and faster rates now. There always are trade-offs to consider when using the cellular networks. The speed is slower than other types of wireless, or wired, networks, but the mobility and ubiquity of coverage of the network supports highly mobile applications and platforms.

17.8 Summary

In this chapter we presented the security features of and the threats to WWANs based on cellular technology, specifically, GSM and 3G. 3G is based on CDMA, which has a low probability of interception that provides data security. 3G security extends features in GSM, has identified weaknesses in GSM and attempts to address them. As part of this effort, 3G has identified the major threats it faces in terms of security.

Cellular WWANs are evolving rapidly, with bandwidth moving from kilobits to megabits per second. Although these networks have limitations, they offer continuous wide area coverage across large areas. While total global coverage is unlikely for terrestrial-based technology, 3G, and 4G in the next few years, will play a major role in providing the infrastructure for ubiquitous mobile computing.

17.9 References

Biryukov, A., A. Shamir, and D. Wagner, "Real Time Cryptanalysis of A5/1 on a PC," *Lecture Notes in Computer Science*, vol. 1978. Berlin: Springer-Verlag, 2000.

Jovan, Dj., Golic J., "Cryptanalysis of Alleged A5 Stream Cipher," Advances in Cryptology—EUROCrypt '97, May 1997, pp. 239–255, *http://jya.com/a5-hack.htm*.

Lauri, Pesonen L., "GSM Interception," Department of Computer Science and Engineering, Helsinki University of Technology, November 21, 1999, *http://www.dia.unisa.it/ads.dir/corso-security/www/CORSO-9900/a5/Netsec/netsec.html*.

Nichols, R., and P. Lekkas, *Wireless Security Models, Threats, and Solutions*, McGraw-Hill, 2002, pp. 16–28, 494–495.

Nichols, R., and P. Lekkas, *Wireless Security Models, Threats, and Solutions*. New York, NY: McGraw-Hill, 2002, pp. 324–325.

Suvada, Myagmar S., and V. K. Gupta "3G Security Principals," 2001, *http://choices.cs.uiuc.edu/MobilSec/posted_docs/3G_Security_Overview.ppt*.

3GPP3G TR 33.900, "A Guide to 3rd Generation Security," January 2000, *ftp://ftp.3gpp.org/TSG_SA/WG3_Security/_Specs/33900-120.pdf*.

Brief Introduction to Wireless Communication and Networking

A.1 Wireless Communication Basics

Similar to wired communication, a one-way (simplex) wireless communication requires a *transmitter* (transmitter electronics plus antenna) generating (electromagnetic) signals that can be properly received and deciphered by a *receiver* (receiving electronics plus antenna). Bidirectional communication requires a pair of *transceivers*—each transceiver consists of a transmitter and a receiver. Although it is possible for a transceiver to be designed such that it can transmit and receive at the same time, for various reasons such as the cost of the unit and its weight, a transceiver is designed with shared components (such as antenna) between its transmitter and receiver—leading it to either send or receive signals at any given time. Hence, bidirectional communication in wireless networks is usually half duplex. Various factors such as the power used for amplification of the wireless signal, properties of the medium, sensitivity of the receiver, and the signal interference from various other sources affect the distance at which a wireless transceiver's signal can be received. Directionality of the transceiver's antenna also plays a crucial role in this. An *omnidirectional antenna* radiates signals with almost equal strength in all the directions, whereas a *directional antenna* radiates signals with substantially more strength in some directions than others. The region around a transceiver within which the signal of a transceiver can be satisfactorily received is called the *coverage area* of the transceiver. Conceptually, a bidirectional wireless link exists between a pair of transceivers within the coverage area of each other since each can receive signals sent by the other. All other factors remaining constant, the coverage area of a transceiver can be increased or decreased by adjusting the power used for amplifying the signal.

Suppose that the receiver and the transmitter are separated by a distance R. The average power P_r of the signal received by the receiver is proportional to the (average) power P_t of the transmitted signal. Under some simplifying assumptions (which we will not go into here) the relationship between P_t and P_r can be expressed as follows:

$$P_r = \frac{P_t K}{R^a} \tag{1}$$

where a is the propagation loss exponent and K is the proportionality constant, which depends on the characteristics of the antennas. The value of the propagation loss exponent is usually between 2 and 5. It should be noted here that the above relationship holds when the receiving antenna is in the far-field region (as opposed to near-field region), i.e., R is greater than some minimum distance that marks the boundary of the near-field region around the transmitting antenna.

Depending on the sensitivity of the receiver, the received power has to be greater than a certain minimum threshold value $P_{r\text{-th}}$ for the receiver to be able to detect the transmitted signal. Hence the maximum distance $R_{\max}$ that a receiver can be from the transmitter and still detect the signal from it can be determined from Eq. (1):

$$R_{\max} = \left(\frac{P_t K}{P_{r\text{-th}}} \right)^{1/a} \tag{2}$$

As can be seen from Eq. (2), as P_t increases $R_{\max}$ increases, although very slowly.

Conceptually, $R_{\max}$ is the radius of the region around the transmitter within which a mobile unit can communicate with a transmitter, called the *cell* (assuming an omnidirectional transmitting antenna). In reality, the cell boundary in a given direction may depend on many factors such as the directionality of the antenna and the terrain. So, although we denote the cells as circular regions, in reality a cell may have very irregular geometry. In fact there are two power thresholds that may be associated with the receiver:

1. $P_{r\text{-th}}$-detect: the threshold above which the transmitter detects that the channel is busy and refrains from transmission.

2. $P_{r\text{-th}}$-recv: the threshold above which the transmitter is able to correctly receive a packet.

Obviously, $P_{r\text{-th}}$-detect $\leq P_{r\text{-th}}$-recv $\rightarrow R_{max}$-detect $\geq R_{max}$-recv.

Wireless transceivers are usually designed to operate within a certain frequency range called a *frequency band*. Frequency bands are consistent with the frequency bands allocated by the FCC. Most of these bands require a license issued by the FCC to use them. The exceptions are the Industrial Scientific and Medical (ISM) bands that do not require licenses for using them, and hence they can be freely used as long as the user abides by the FCC guidelines of not using transmission power greater than 1 W.

The signal propagation properties vary with the frequency. As the frequency of the signal increases, its penetrability decreases and the distance to which it can propagate within a medium (say air) decreases. Hence, at higher frequencies (such as infrared) the transceivers have to be in the line-of-sight of each other. At lower frequencies (such as radio frequencies) the line-of-sight is not required, since signal can be reflected from various objects and reach the receiver from various indirect paths. Wireless signals experience fading. There are two types of fading: short-term fading and long-term fading. This results in severe variability of the available bandwidth of a wireless link. Fading in a wireless medium results in much higher bit error rates than in wired communication. For more in-depth coverage, interested readers can consult many excellent books available on wireless communication (Rappaport, 1996).

A.2 Wireless Network Architectures

There are several types of wireless networks being currently used. Figure A.1 shows three different types of mobile network architectures. Figure A.1 (1) shows a wireless LAN and mobile IP-based architecture, Fig. A.1 (2) shows a mobile cellular architecture, and Fig. A.1 (3) shows a mobile ad hoc network architecture. Each of these architectures is described below.

A.2.1 Wireless local area network (WLANs) and mobile IP

An access point (AP) is usually a (mostly static) wireless transceiver connected to the wired network. A mobile station can establish a bidirectional link with an access point using a common protocol such as IEEE 802.11 or Bluetooth. Of course, the mobile station has to have the appropriate protocol stack and hardware (e.g., PCMCIA card or built-in transceiver) to communicate with an IEEE 802.11 or a Bluetooth access point. Wireless LANs are being used to provide wireless access called "hot spots" or "oases" in places such as airports, coffee shops, hotels, shopping malls, and homes. A wireless LAN permits mobility within the

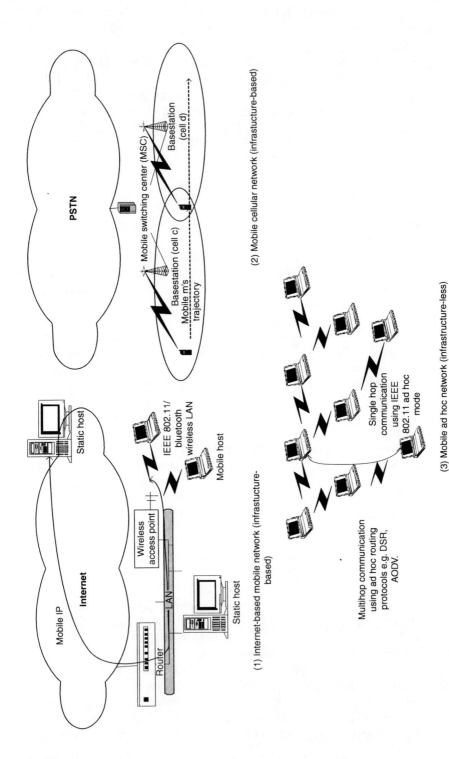

Figure A.1 Different types of mobile networks.

(1) Internet-based mobile network (infrastucture-based)

Mobile IP

Internet

Static host

Wireless access point

IEEE 802.11/ bluetooth wireless LAN

Mobile host

Router

LAN

Static host

(2) Mobile cellular network (infrastucture-based)

PSTN

Mobile switching center (MSC)

Basestation (cell d)

Basestation (cell c)

Mobile m's trajectory

(3) Mobile ad hoc network (infrastructure-less)

Single hop communication using IEEE 802.11 ad hoc mode

Multihop communication using ad hoc routing protocols e.g. DSR, AODV.

coverage area of the access point. For short-range technology, the radius of the coverage area can be a few meters (e.g., around few meters for RFIDS, around 10 m for Bluetooth), whereas for long-range technology this can range from several hundred to several thousand meters (e.g., around 300 m for IEEE 802.11 Wireless LANs, several kilometers for satellite cells). Protocols such as mobile IP provide wide-area mobility. Mobile IP is an extension of the Internet Protocol (IP) to support seamless mobility across IP networks.

A.2.2 Wireless cellular networks

In a wireless cellular network, several access points, called base stations (BS), are deployed such that each location in the service area is within the range of at least one base station. These base stations can communicate with each other and other static computers via a wired (or another wireless) network. The coverage area of a base station is called its (wireless) cell.

When an active mobile station (a mobile unit that is actively involved in communication) moves out of the coverage area of one base station, it (or the system) tries to establish a connection to another base station in order to continue communicating this process is called a *hand-off*. In a cellular network, cells of adjoining base stations usually overlap to allow a smooth hand-off of a mobile's connections from one base station to another. Further, in wireless telephony the available frequency range (bandwidth) is divided into wireless communication channels. A channel is wide enough (minimum of 64 kbps) to accommodate one voice circuit. Adjoining cells use different (sets of) channels to reduce (co-channel) interference.

Mobile ad hoc networks. A mobile ad hoc network is a network established by an ad hoc group of mobile computers with wireless transmission capabilities. IEEE 802.11 has an ad hoc mode that permits establishing such networks. Since the transmission range of each mobile computer is limited, multihop packet transmission may be needed to deliver a packet to its destination mobile computer. Several multihop routing protocols have been developed, for example, dynamic source routing (DSR) and ad hoc distance vector (AODV). Multihop routing in mobile ad hoc networks poses several challenges. For one, the topology of the network changes as the nodes move around. Further, the nodes in the network are mostly battery powered and so an intermediate node's energy is depleted in forwarding a packet for someone else. Hence, the development of energy-efficient routing protocols for mobile ad hoc networks is an active area of research.

For more in-depth coverage, interested readers can consult many excellent books available on wireless and mobile networks (Pahlavan and Krishnamurthy, 2002).

A.3 References

T. Rappaport, *Wireless Communications and Principles*, Prentice Hall, Upper Saddle River, New Jersey, 1996.

K. Pahlavan and P. Krishnamurthy, *Priciples of Wireless Networks*, Prentice Hall, Upper Saddle River, New Jersey, 2002.

Questions

Chapter 1

1. What characteristics distinguish mobile computing from distributed computing?

2. Conduct a survey to find out about the state-of-the-art in mobile computing.

3. Compare the mobile computing applications supported on cellular phones with those supported on laptops and personal digital assistants (PDAs).

4. What are the most important challenges facing mobile computing today? Explain each of the challenges.

5. Adaptation requires a mobile client to sense changes in its environment (such as, change in the received signal strength, and roundtrip delays), guess the cause of these changes, and then react appropriately. However, a mobile client can only have local information about its environment. Give an example showing that two very different changes in the global environment can lead to the same changes in the local environment of a mobile client.

6. An adaptive mobile information system may trade off the fidelity of data delivered with the performance (e.g., it may choose to deliver lower fidelity data to achieve low data access latency). Describe how one could compare the "goodness" of adaptive mobile information systems.

7. Application-aware adaptation requires the underlying system to not only allocate resources to various concurrent applications (processes) running on the system but also do resource revocation when the available resources (such as bandwidth) become scarce. Discuss some of the issues that should be taken into account in developing a resource revocation scheme.

8. In which of the layer(s) in the entire system architecture including the networking protocol stack, the operating system, and the middleware should support for mobility be incorporated? Provide justification for your answer.

9. Give four different examples of adaptations from the computing domain. Identify the type of adaptation involved for each. Develop state-based models for these adaptation techniques.

10. In what ways (if any) does mobility impact the peer-to-peer computing model?

11. Explain all the different kinds of adaptations performed/supported by Odyssey, Rover, and WebExpress.

12. Read the abstracts of papers published in recent conferences on mobile computing, such as ACM Mobicom and MobiSys, and identify some of the current research challenges being addressed by researchers.

13. Identify some of the formal methods that can help in design and specification of adaptive software systems.

Chapter 2

1. What are the various components of a mobility management system?

2. What are the various issues involved in handover in a cellular system?

3. Explain the impact of size and shape of location area on the cost of registration-area-based location management schemes.

4. Compare the location management schemes for mobile phones in cellular networks with location management schemes developed for supporting Internet mobility.

5. List some of the techniques for improving scalability of location management.

6. What is the importance of call-to-mobility ratio in the context of location management schemes?

7. Is it possible in a location management scheme to reduce search (update) cost without impacting the update (search) cost? Explain your answer using examples.

8. Forwarding pointer chaining is an optimization technique suitable for low CMRs. Also, eager cache updating is an optimization technique suitable for high CMRs.

   a. Justify why we would want to have both techniques available in a mobile cellular network.

   b. Is it beneficial to maintain a forwarding pointer chain as well as do eager cache maintenance for a mobile user? Explain why or why not.

9. Would it make sense to simultaneously use lazy and eager cache updating for a mobile user? Explain.

10. Describe how domain name system (DNS) can be used in location management.

11. Cellular networks are typically modeled in two dimensions. Does it make sense to model them in three dimensions as overlapping spheres? Explain why or why not. Assuming that it makes sense to model them in three dimensions, explain how that can affect location management algorithms. Clearly state all assumptions you make.

12. How is mobility supported in IPv6?

13. Research various techniques proposed for location management in mobile ad hoc networks.

14. Study various routing schemes developed for mobile ad hoc networks, e.g., dynamic source routing (DSR) and ad hoc on-demand distance vector (AODV), from the perspective of what location management operations (search and update) are included in each.

15. Explain some of the security problems related to location management.

16. Explain some of the privacy concerns related to location management. Propose a scheme for privacy-preserving location management.

17. List some of the currently available location-based services.

Chapter 3

1. Conduct a survey on currently available push-based commercial data services for mobile users.

2. List advantages and disadvantages of the pull-based and push-based data access models, from the perspective of mobile computing.

3. Explain issues involved in the use of caching for improving data availability. Contrast how the caching techniques used for improving availability differ from those used for improving performance.

4. Traditionally, performance of a caching technique is measured using the metric of miss ratio (the percentage of data accesses which resulted in a cache miss). The underlying assumption of this metric is that all cache misses have the same cost.

    a. Explain why this is a valid assumption in current computer systems where cached and primary copies are strongly connected.

    b. Explain why this assumption does not hold in a mobile computing environment with frequent disconnections and variable quality links.

   c. Suggest a possible set of metrics for caching in mobile computing. State any assumptions you make about the mobile computing environment.

5. What are the different types of consistency maintenance schemes used for distributed data caching for Internet-based applications? Discuss how mobility impacts these techniques.

6. Can the two proxies (intercepts) used in the WebExpress architecture be combined into a single proxy? Why or why not?

7. Explain with example(s) how the use of cyclic redundancy codes (CRC) in WebExpress helps in reducing the traffic between the client and the server.

8. Develop a scheme for caching location-sensitive data? How is your scheme different from traditional caching schemes for non-location-sensitive data? What performance metrics would you use to evaluate your scheme? Does your scheme take into account the mobility pattern of a mobile?

9. How do the research distributed caching techniques developed for mobile ad hoc networks differ from those developed for infrastructure-based mobile systems?

10. What are some of the issues in data management for mobile databases?

11. How can caching help in saving energy on a mobile computing device?

Chapter 4

1. What is the relationship between pervasive or ubiquitous computing and mobile computing?

2. What is the importance of proactivity and self-tuning in the development of ubiquitous computing (ubicomp) systems?

3. List and give examples of different types of contexts relevant in the mobile computing domain.

4. Design a ubicomp object (e.g., a ubicomp pen). Develop usage scenarios of your ubicomp object, clearly identifying the contexts it handles. Identify the technologies that are needed to implement your ubicomp object. Are these technologies currently widely available?

5. Give an example to show that in some cases humans and technologies have to coevolve for the technology to become "invisible."

6. Design an e-mail system to minimize human distraction.

7. Investigate the notion of augmented reality. How is it different from virtual reality?

8. List some of the metrics for evaluating ubicomp objects.

9. Study the project Aura at CMU and summarize its key features in relation to ubiquitous computing.

10. Discuss some of the ways in which pervasive computing technology can impact important applications such as health monitoring and homeland security.

11. Describe the ubiquitous computing applications envisioned in some Hollywood movies such as *Minority Report*. What are some of the major advances that need to happen to make these applications a reality?

12. Play a devil's advocate role and make an argument in support of the claim that ubiquitous computing will never succeed.

13. What are some of the system management challenges associated with pervasive computing?

Chapter 6

1. Propose a creative mobile application that uses Odyssey for adaptation. Your solution should include a description of which resources must be monitored, the necessary warden(s), any additional servers, interactions with the viceroy, and the like. You should also develop a detailed diagram of the system.

2. What are the complications in providing a general definition of *fidelity*? Are fidelity levels between different data types comparable? Why or why not?

3. What problems arise when applications adapt independently to changing resource levels, rather than cooperating to perform adaptation?

4. Design a high-level mobile application using the Puppeteer model. Sketch a diagram showing where fidelity levels change in your adaptive application.

5. *Is* there a killer application for mobile agents? Is there a class of applications for which mobile agents very significantly reduce development effort?

6. The use of mobile agents raises some serious security risks. Identify and explain these risks. Can you think of a different computing model, which provides some (or all) of the benefits of mobile agents with significantly reduced risk?

7. Identify some circumstances in which temporal locality in an agent's communications paradigm is not a detriment.

8. What are the difficulties in supporting agent migration (at an implementation level)? What can be done to reduce the severity of these difficulties?

Chapter 7

1. How are service discovery frameworks and mobile agent systems similar? How are they different?

2. For this question, you should refer to RFCs 2608 and 2609, available at http://www.srvloc.org/srvloc-standards.html. A service template in the Service Location Protocol (SLP) defines a particular service type through the standardization of attributes (and their types) and the format of service URLs. After reviewing RFCs 2608 and 2609, and possibly the sample services available at http://www.isi.edu/in-notes/iana/assignments/svrloc-templates/ (e.g., printer), propose a robust standard for an SLP Blender. You may look at an actual blender, in an actual kitchen, to see how a blender behaves. Your service template should meet all the requirements specified in the RFCs.

3. Rework question 1, above, but use Universal Plug and Play (UPnP).

4. Rework question 1, above, but use Jini.

5. What are the primary benefits of service catalogs? Under what circumstances might it be better to choose a service discovery framework that doesn't use service catalogs?

6. UPnP and SLP can operate without service catalogs. What technical problems arise in trying to modify the Jini specification to work in a *directoryless* fashion (i.e., without lookup servers)? Ignore the fact that there are other centralized servers in a Jini setup (e.g., for RMI).

7. What are the significant barriers to interoperability for service discovery frameworks?

8. What benefits does the eventing service in a service discovery framework provide for an application developer?

9. What are the novel security features of the Ninja service discovery framework? Are these applicable to the other frameworks discussed in the chapter?

10. Research the MD5 and SHA-1 cryptographic hash algorithms. How can these kinds of algorithms be used to generate universally unique identifiers (UUIDs)?

Chapter 8

1. Select a potential application of wireless sensor networks. How do the limitations of wireless sensor nodes affect your design?

2. Review the latest version of the Zigbee specification to see what problems it addresses and what protocols for wireless sensor networks need to be built upon this platform.

3. Energy is severely limited in wireless sensor networks. Review the latest research on some passive energy source, such as solar or vibration energy. How much power can it supply and what duty cycle would this allow for a wireless sensor?

4. List the different constraints on a wireless sensor and the relationship among these constraints.

5. What are the differences between wireless sensor networks and other types of ad hoc networks?

6. Choose a critical wireless sensor node application and try to develop a complete list of possible security concerns.

7. Find a commercially available wireless sensor other than the MICA mote. What are the resources available on this device?

8. For some specific wireless sensor node application, characterize the traffic pattern for the application.

9. Consider an industrial or commercial application of wireless sensor nodes. Attempt to quantify the costs of the network versus the benefits. Include the entire life cycle of these sensors, including installation and maintenance, in your calculations.

10. What are the special requirements for space-based wireless sensor networks? How do these nodes differ from more typical wireless sensor nodes?

Chapter 9

1. Investigate cross-layer optimizations for wireless sensor networks. Review the current research literature and see what efficiencies are gained.

2. What are the advantages and disadvantages of caching sensor readings in a wireless sensor network?

3. Devise a communication protocol using a small number of LEDs to relay as much information as possible to an observer.

4. Download Tiny O/S or another publicly available operating system for wireless sensors. Implement a simple protocol and test it with the simulator.

5. How can the neighbors of a sensor (the local topology) change even if there is no mobility? What problems does this create?

6. What are the advantages and disadvantages of deploying wireless sensors in a random fashion, which leads to an arbitrary topology?

7. The book describes the long-term performance improvement of Carbon-Zinc batteries. Investigate the performance trend for some other battery technology or for small solar cells.

8. What special security requirements do biomedical sensors have?

9. Study the option of combining a rechargeable battery with a passive power source. Estimate the duty cycle using numbers for some existing commercial rechargeable batteries and passive energy sources.

10. Would a static storage device, such as a memory stick, alleviate the severe data storage limits of a wireless sensor? What is the cost for such an approach?

Chapter 10

1. Find out what the energy usage is for other wireless network cards. How does the energy usage vary with the different states? How would this affect the protocol design?

2. Carefully design a protocol for initializing, assigning, and incrementing counters for duplicate packet counters as described in the chapter.

3. What are the trade-offs with using a dual-radio solution for channel scheduling? What would be the advantages and disadvantages of using a speaker on each sensor for the same purpose?

4. What are the functions needed for autoconfiguration of a wireless sensor network?

5. Investigate what the resource requirements should be for a wireless sensor network to employ CDMA. How do these requirements compare with the capabilities of current wireless sensors?

6. For a real system, find the amount of energy used to power up the electronics versus the energy consumed for transmission. What is the optimal hop distance using this information, assuming a distance-squared model?

7. What are the advantages and disadvantages of using TDMA to schedule communications in a wireless sensor network?

8. How can mobile sensors be used to extend the lifetime of a wireless sensor network? Give an example.

9. What are the two general categories of medium access control (MAC) in a wireless network? Which is more suitable for a wireless sensor network? Justify your answer.

10. Investigate recent proposals for secure key distribution in wireless sensor networks.

Chapter 11

1. Propose different definitions of network lifetime. What are the advantages and disadvantages of each? Consider your answers from the perspective of specific applications.

2. Review some recent publications on time synchronization in wireless sensor networks. Compare these papers with the approaches described in this chapter.

3. Review some recent publications on location determination in wireless sensor networks. Compare these papers with the approaches described in this chapter.

4. What is meant by data-centric communication? Why is this a useful model for many sensor network applications?

5. What are the two models for communication in wireless sensor networks? Give an example application where each is an appropriate model.

6. Directional routing relies on a dense enough network to forward packets from the source to the destination. Assuming a random deployment, what density is required to ensure delivery to the destination with high probability?

7. How could trajectory-based routing be used for efficient broadcasting? What is the most efficient solution you can come up with?

8. What are the advantages and disadvantages of forming a group for consensus only after an event occurs?

9. Does data aggregation require time synchronization? Why or why not?

10. Give three examples where off-loading some of the work to the base station leads to better performance than distributing the problem to the sensor nodes throughout the network.

Chapter 12

1. Give three examples of real attacks against integrity, confidentiality, non-repudiation, and availability. Use news, bugtraq, securiteam, Slashdot, CERT, or other sites as sources for information on attacks.

2. What makes a distributed denial of service (DDoS) attack more difficult to detect and defend against than a normal denial of service (DoS) attack?

3. Create a DoS-resistant routing protocol (e.g., using dynamic rerouting, with pairs of nodes watching each other).

4. What is the difference between resource depletion versus resource exhaustion attacks? Give an example where depletion is worse than exhaustion. Give an example where exhaustion is worse than depletion.

5. Assume you are sitting in a coffee shop with a laptop computer equipped with a wireless card. Instead of connecting to the coffee shop's network, you accidentally connect to the network of the business next to the coffee shop. What are the legal implications for accidental theft of service? The answer may depend on the country and state and will depend on your actions.

6. Is war driving illegal? The answer will change with time, intent, and location.

7. Ad hoc wireless networks exacerbate many of the problems related to computer security. Are there any security problems they help mitigate?

8. Describe a method to avoid traffic analysis attacks.

9. Why is encryption not sufficient to defeat a replay attack? Provide an example. Show how it can be defeated by using a nonce. Give three different ways to generate the nonce.

10. Give an example of a buffer overflow attack.

Chapter 13

1. Explain how encryption provides security. Include more than just data confidentiality.

2. What are the benefits of using hybrid approaches for encryption?

3. Why does a MAC use a cryptographically secure hash function? What purpose does the key serve? Is it public or private?

4. What is the difference between tunnel mode and transport mode in IPSec and when would each be appropriate to use?

5. Why does the AH not encrypt the header? When would you want to encrypt the header? Would you ever not want to encrypt the payload? Why or why not?

6. Download the latest RFCs describing IPSec (hint: use www.rfc-editor.org) and describe the most recent developments and changes to the protocol.

7. IPSec and NAT were known to have compatibility problems. What work has been done to remedy the situation? Hint: look at www.ietf.org.

8. Describe the authentication mechanisms your computer uses when you connect to the Internet. What authorization methods are provided? What auditing?

Chapter 14

1. How many devices can be associated with a scatternet? How many devices can be associated with a scatternet at the same time? Why are these different?

2. Describe the three modes of security for Bluetooth devices.

3. Describe the four types of link keys used in Bluetooth. For what and how is each one used?

4. Why is the length of the PIN the key factor in the security of the keys?

5. The various algorithms used by Bluetooth, such as E_0, E_{21}, and E_{22}, are standard algorithms that have not been published (although many of them

have eventually been discovered). This is often referred to as *security by obscurity*. What are the risks? It seems like all security is based on hiding some information. How is this different?

6. Name four typical applications for a Bluetooth network. What are the security demands of each one? For each, is the basic Bluetooth security model sufficient? Why or why not?

Chapter 15

1. What is the point of an extended service set (ESS)? Why would you want multiple access points to share the same network name?

2. Give three examples of uses for an independent basic service set network.

3. No one has defined an independent extended service set network (IESS)— an ad hoc network that has multiple access points cooperating with each other, performing hand-offs and the like. Give an example of such a network (hint: two areas are military operation or disaster relief). And IESS has not been defined because many hard problems lurk when trying to make it work. Describe why an IESS is much more difficult to run than an ESS.

4. 802.11 is a very dynamic area. Pick any four of the subcommittees of the 802.11 group alphabet soup that has had recent activity (within the last year) and describe their accomplishments or their next goals. How many new subgroups have been created since this book was published?

5. Where did the designers of WEP go wrong? What did they get right?

6. Explain how WEP encryption and decryption works. Assuming the RC4 PRNG functions as advertised (and ignore the existence of bad IV values), on what does the security of the encryption depend? Is that reasonable or a flaw?

7. Could some other mechanisms, such as a network sniffer or intrusion detection system, be configured to watch for WEP attacks? If so, explain how. If not, explain why.

8. What is the difference between 802.11, 802.11i, and 802.1x?

9. Describe how an 802.11 system could be protected using four different mechanisms. This can include capabilities intrinsic to 802.11 hardware, as well as 802.11i or WEP, and additional layers.

Chapter 16

1. What are the differences between the needs of local and metropolitan area network users?

2. The destinations for WMANs are generally assumed to be fixed. What benefit is there to using wireless as opposed to just running a wire for "the last mile?"

3. What assumptions allow 802.16 to use certificates for key management?

4. How is 802.16 extensible?

5. What is the "grace time" and what happens if it is exceeded?

6. How is 802.16 encryption similar to 802.11i?

7. In 802.16, the BS authenticates the SS. What risks exist because the SS does not authenticate the BS?

8. Why are sequential IV values better than random ones? Why would it not be easy to guess the next IV if they are used sequentially?

Chapter 17

1. Why is WWANs an approximation of the notion of ubiquitous computing?

2. GSM uses 54 bits instead of 64 bits for the encryption key. By how many orders of magnitude is the search space reduced?

3. Calculate how long a brute-force attack on GSM encryption would take using Pesonen's approach using currently available technology, and using multiple processors? How much would it cost?

4. Research what the current state of the art is with respect to 4G systems. How close are they to wide deployment and what features do they have that 3G lack?

5. Show how several individual security technologies can be subverted (e.g., VPN, WEP, WPA, Bluetooth authentication, and firewalls). How are components vulnerable individually at the different layers?

6. Design a protocol to incorporate personal, local, and wide area networks for communication, depending on network availability, bandwidth requirements, and the like (e.g., virtual tourist kiosks).

7. Describe the different threats and requirements for security for a WPAN, WLAN, WMAN, and WWAN. Why does end-to-end encryption *not* solve all the problems?

8. Why can one wireless technology not support all potential uses (personal, local, metro, wide)? What are the strengths and weaknesses of each? Even if each *can* do it (e.g., internet over PAN or LAN supporting personal network), why are they not appropriate?

9. What are some of the weak points in 3G security?

10. 3G systems are vulnerable to what kind of attacks?

11. Would a combination of 3G/802.16/802.11/802.15 systems provide better or worse security? What kind of attacks could be mounted against this combination and what sort of resources would this involve? Why does the combination strengthen/weaken the security beyond just any one of the individual protocols?

Index

3G, 364, 367
3GPP (*see* Third Generation Partnership
 Project)
802.11a wireless networks, 116–117

AAA (*see* Authentication, authorization,
 and auditing)
AAFZ algorithm, 65, 67
ABs (*see* Authentication blocks)
Access points (APs), 27–30, 330, 332
Access time, 68
Access transparency, 4
ACPI (*see* Advanced Configuration and
 Power Interface)
Active Badge Location System,
 100–101, 106
Active context, 96
Actuators, 104
Ad hoc networks, 330–331
Ad hoc wireless sensor networks,
 171–188
 applications, 182, 185–188
 arbitrary topologies, 202–203
 autoconfiguration, 213–221
 changing group dynamics, 237–239
 changing membership, 201–202
 clustering, 232–235
 communication, 196
 communication scheduling, 225–226
 data loss, 206
 defined, 173
 dual-radio scheduling, 230–231
 duplicate message suppression,
 226–228
 energy-efficient communication,
 221–235
 example, 176
 fault tolerance/recovery, 273–279

 group communication, 207–208
 lack of centralized mechanisms, 178,
 192–193
 lack of preexisting
 infrastructure, 175
 limited access to base station,
 175, 177
 loss of connectivity, 205–206
 MAC layer protocols, 282–283
 maintaining consistent views,
 208–209
 message aggregation, 228–230
 mobility, 203–209, 235–240
 movement detection, 235–237
 multihop routing, 222–225
 power-limited devices, 177–178,
 193–196
 resource constraints, 191–200
 resynchronization, 239–240
 routing, 252–272
 security, 200–203, 293–295
 security protocols, 220–221,
 247–249
 self-organization, 247
 sleep-mode scheduling, 232
 uniform power dissipation,
 280–281
 unique features, 178–184
 (*See also* Smart sensors)
Adaptability, 3–4
 environmental constraints, 5–6
 transparency, 4–5
Adaptation(s), 110–111, 113–123
 agility, 116–117
 application-aware, 6–7, 19, 114,
 117–119
 changing functionality as mechanism
 for, 8–9

Adaptation(s) (*Cont.*):
    characterizing strategies for, 115–117
    as concept, 109
    conflicting, 122–123
    contextual, 98
    developing and incorporating, 11–12
    and fidelity, 115–116
    middleware for, 114–123
    Odyssey as application-aware, |
        117–119
    proxies for performing, 14–16
    and resource monitoring, 114–115
    site for performance of, 12–13
    types of, 114, 115
    varying data quality as mechanism for,
        9–11
Adaptation controller, 123
Adaptation policy language, 123
Address Resolution Protocol (ARP),
        294, 297
Advanced Configuration and Power
        Interface (ACPI), 114
Advanced Encryption Standard (AES),
        332, 340
Advertisement, service (*see* Service
        advertisement)
AES (*see* Advanced Encryption
        Standard)
Agents, mobile (*see* Mobile agents)
Aggregation, message, 228–230
Agility, 10–11, 116–117
Aglets, 131–132
AH (*see* Authentication header)
ALOHA, 59, 63, 218
Applets, 124
Application-aware adaptation, 6–7,
        19, 114
APs (*see* Access points)
Ara, 132–133
Architecture-based mobile wireless
        networks, 59
Architectureless mobile wireless
        networks, 59–60
Architectures:
    coordination of middleware, 122
    of mobile agent systems, 124
ARP (*see* Address Resolution Protocol)
ARP cache poisoning, 297
    (*See also* Attacks)
ARP spoofing, 294
    (*See also* Attacks)
ARQ (*see* Automatic Repeat
        Request)

AS (*see* Authentication server)
AS Scheme (*see* Asynchronous Stateful
        Scheme)
Asleep mode, 80
Asymmetric (public key) encryption, 303
    (*See also* Encryption)
Asymmetric links, 23, 59–60
Asynchronous Stateful Scheme
        (AS scheme), 73, 78–84
Asynchronous stateless invalidation-
        based caching strategies, 73
Atmel ATMEGA processor, 197
Attacks, 289–290, 297–299, 326–327,
        337–339, 365–366
Augmentation, contextual, 98
Authentication:
    in Bluetooth, 325
    end-to-end, 9
    in wired-equivalent privacy (WEP),
        335–336
Authentication, authorization, and
        auditing (AAA), 313–314
Authentication blocks (ABs), 165
Authentication header (AH), 308
    (*See also* IPsec)
Authentication protocols, 309–313
Authentication server (AS), 342–344
Authenticator, 342
    (*See also* Authentication)
Authorities (Telescript), 129
Autoconfiguration (ad hoc wireless sensor
        networks), 213–221
    and medium access control schedule
        construction, 216–220
    and neighborhood discovery,
        214–215
    and security protocol configuration,
        220–221
    and topology discovery, 215–216
Automatic contextual reconfiguration
        applications, 100
Automatic Repeat Request (ARQ), 274
Availability, 288–290
Awake mode, 80

Bandwidth, 61–62
    in ad hoc wireless sensor networks,
        216–217
    spatial reuse of, 223–224
Basic service set (BSS), 330
Basing operations, 18
Batteries, 5, 194–195
Battery technology, 2–3, 114

Bluetooth, 317–327
    authentication in, 325, 326
    encryption in, 323–325
    limitations of, 325–327
    network terminology for, 318–319
    security mechanisms in, 320
    security modes in, 320–323
    specifications, 317–318
Bluetooth attacks, 326–327
    (see also Attacks)
Bluetooth Service Discovery Protocol
        (SDP), 145, 167
Broadcast downlink channel, 62,
        64–65
Broadcast scheduling (broadcast disk
        scheduling), 65–67
Broadcast storm problem, 227
Broadcasting timestamp scheme
        (BT scheme), 75
Browsing, service, 141
BSS (see Basic service set), 330
    BT scheme (see Broadcasting
        timestamp scheme)
Buffer-overflow attacks, 298–299
    (see also Attacks)

Cache, defined, 67–68
Cache consistency maintenance, 69–70
Cache maintenance schemes, 41
    asynchronous stateful (AS), 78–84
    and deciding which data to cache,
        84–86
    and disconnected operation, 77–78
    invalidation-based, 73, 75–77
    polling-every-time, 72
    for push-based information
        dissemination, 74–75
    strong vs. weak consistency of, 72
    TTL-based, 72–73
Cache miss, 68
Cache timestamp, 81–83
Caching, 9, 60
    DNS, 50
    importance of, 67–68
    information, 57–59
    performance and architectural issues
        with, 70–72
    per-user location, 40–44
    in traditional distributed systems,
        68–69
    Web, 77, 86–88
    in WebExpress, 18
    (See also Cache maintenance schemes)

Caesar cipher, 302
Callback breaks, 70
Callbacks, 70, 78, 80
Call-to-mobility ratio (CMR), 39, 41
Card not present (CNP), 297
Care-of IP addresses, 51
Cartesian routing, 265, 267
Catalogs, service (see Service catalogs)
CBC-MAC Protocol (CCMP), 340, 342
CCK (see Complementary code keying)
CCMP (see CBC-MAC Protocol)
CDMA (see Code Division Multiple
        Access)
Cell, 28
Cell residency time, 32
CGI (see Common gateway interface)
Chain of forwarding pointers, 37–38
Challenge Handshake Authentication
        Protocol (CHAP), 310–312, 345
    (See also Authentication)
Channel allocation schemes, 29
Channel management, 29
CHAP (see Challenge Handshake
        Authentication Protocol)
Checkpointing processes, 130–131
Checksums, 304–305
Clear-to-Send (CTS) packets, 282
Clients, 8–9
    and service discovery, 139
    thin, 9
    weakly-connected, 59
Client-server (CS) model, 8–9, 19, 61
    as communication strategy, 132
    extended, 9
    and mobile agent systems, 124
    and service discovery, 111
    and system design, 138
Client-side intercepts (CSIs), 17
Closed networks, 346
    (See also War chalking)
Cluster heads, 247
Clustering, 232–235
CMR (see Call-to-mobility ratio)
CNP (see Card not present)
COBRA, 132
Coda, 9, 12, 59, 77–78, 114
Code Division Multiple Access (CDMA),
        29, 218, 359–360
Collision attacks, 337
    (See also Attacks)
Collision avoidance, 218–220
Combination key (Bluetooth), 322–323
Commerce (see Electronic commerce)

Common gateway interface (CGI), 18
Communication:
  among mobile agents, 131–133
  energy-efficient, 221–235
  group, 268–271
  mobile, 1
  multihop, 222–223
  wireless, 196
Communication scheduling, 225–226
Compiled computer languages, 130
Complementary code keying (CCK), 332
    (See also Encryption)
Computational capacity (sensor nodes),
    197–198, 201
Computing:
  disconnected, 77–78
  mobile (see Mobile computing)
  pervasive, 171
  ubiquitous, 92–93
    (See also Context-aware computing)
Computing context, 94
Concurrency, 117
Confidentiality, 288, 364
Conflicting adaptation, 122–123
Connections (Telescript), 129
Connectivity:
  in ad hoc wireless sensor networks,
    205–206
  strong, 12
  weak, 12, 59
Consistency, 10, 69–70, 208–209
Context, 94–97
Context history, 95
Context Toolkit, 104–105
Context-aware applications, 96–102
  adaptation type of, 99
  core capabilities needed in, 97–98
  development of, 100–102
  functional/service type of, 98
  initiation type of, 99
Context-aware computing, 2, 91–107
  and applications, 96–102
  example of, 91–92
  and meaning of context, 94–96
  middleware support for, 102–106
  as paradigm, 91
  and ubiquitous computing, 92–93
Context-dependent data, 60
Context-triggered actions, 100–101
Contextual commands, 100
Contextual information applications, 99
Contextual reminders, 101
Contextual selection, 99

Contextual selection applications, 99–100
Controlled flooding, 260
Controlled port, 343
Coordinate systems (for sensors),
    251–252
CRC (see Cyclic redundancy check)
Cryptographically secure hash, 305
CS model (see Client-server model)
CSIs (see Client-side intercepts)
CTS packets (see Clear-to-Send packets)
Cyclic redundancy check (CRC), 18, 87,
    304–305
    (See also Data integrity)

D'Agents, 132
DAs (See Directory agents)
Data:
  context-dependent, 60
  location-dependent, 60
  read-only, 68
  time-dependent, 60
Data adaptation, 9–11, 14
Data caching (see Caching)
Data encryption (see Encryption)
Data integrity, 287–288
Data loss, 206
Data Manipulation Interface (DMI), 121
Data Over Cable System Interface
    Specification (DOCSIS), 350
Data streams, fidelity of, 116
DCS (see Dynamic channel selection)
DDOS attacks (see Distributed denial-of-
    service attacks)
Dead sensor nodes, 278–279
Decryption, 335
Denial-of-service attacks (DOS attacks),
    289–290, 297, 366
    (See also Attacks)
Department of Defense, 186–187
Destination-Sequenced Distance-Vector
    (DSDV), 192
Detectability, 290–291
Development, application, 100–102
DHCP (see Dynamic Host Configuration
    Protocol)
Differencing (WebExpress), 18, 87
Differential quadrature phase shift
    keying (DQPSK), 359
Diffusion routing, 260–265
Digital Structure Algorithm (DSA), 165
Digital subscriber line (DSL), 289
Direct sequence spread spectrum (DSSS),
    331–332, 359

Directed diffusion, 261–262
Directional routing, 265–268
Directional source aware routing, 267
Directory agents (DAs), 156
Disconnected computing (disconnected
    operation), 77–78
Disconnections, 5–6
  and mobile agents, 126
  in Mowgli, 87
  voluntary vs. involuntary, 59
Discovery, service (*see* Service discovery)
Distributed denial-of-service (DDOS)
    attacks, 289
    (*See also* Attacks)
Distributed systems, 1, 68–69
DMI (*see* Data Manipulation Interface)
DNS (*see* Domain Name Server; Domain
    Name System)
DOCSIS (*see* Data Over Cable System
    Interface Specification)
Downlink, 78
Downlink channels, 29, 62
Doze mode, 57, 80
DQPSK (*see* Differential quadrature
    phase shift keying)
DSA (*see* Digital Structure Algorithm)
DSDV (*see* Destination-Sequenced
    Distance-Vector)
DSL (*see* Digital subscriber line), 289
DSR (*see* Dynamic Source Routing), 192
DSSS (*see* Direct sequence spread
    spectrum)
Dual-radio scheduling, 230–231
Duplicate message suppression,
    226–228
Dynamic allocation schemes, 85
Dynamic channel selection (DCS),
    332
Dynamic Host Configuration Protocol
    (DHCP), 52, 141, 154, 346
Dynamic Source Routing (DSR), 192
Dynamic update schemes, 33
Dynamic updates, 39–40, 141

Eager cache maintenance scheme, 41
EAP (*see* Extensible Authentication
    Protocol)
Electronic commerce, 295–297
Emulating state (Coda), 12
Encapsulating security payload (ESP),
    308–309
    (*See also* IPsec)
Encapsulation, 51

Encryption, 9, 220, 302–304, 353,
    363, 365
  in Bluetooth, 323–325
  in wired-equivalent privacy (WEP),
    334–335
End-to-end authentication, 9
Energy efficiency (ad hoc wireless
    networks), 221–235, 279–284
  performance and, 283–284
  via clustering, 232–235
  via communication scheduling,
    225–226
  via dual-radio scheduling, 230–231
  via duplicate message suppression,
    226–228
  via message aggregation, 228–230
  via multihop routing, 222–225
  via sleep-mode scheduling, 232
Energy requirements, 5, 57, 59
Enumeration-based context,
    94–96
Environment(s):
  hostile, 295
  mobile computing, 5
Environmental applications,
    187–188
Environmental context, 95
Environmental state, 11
ESP (*see* Encapsulating
    security payload)
ESS (*see* Extended service set)
Ethernet, 301
Event-driven routing, 253–254
Eventing, 141, 159–163
Expanded ring search, 33
Expensive links, 23
Export driver, 121
Exposed terminal problem,
    218–219
Extended CS model, 9
Extended service set (ESS), 330
Extensible Authentication Protocol (EAP),
    309, 312–313, 345
    (*See also* Authentication)

Failure transparency, 5
Failures, 3
FAs (*see* Foreign agents)
Fault tolerance, 130–131
Faulty sensor nodes, 278–279
FDMA (*see* Frequency division multiple
    access)
FEC (*see* Forward error correction)

FHSS (*see* Frequency-hopping spread spectrum)
Fidelity, 115-116
  defined, 9
  dimensions of, 10
File sharing, 58–59
Fire sensors, 188
Firewall protection, 345–346
Flat organization (registration area-based location management), 45–47
Flooding, controlled, 260
Foreign agents (FAs), 51
Foreign network, 51
Forgery attacks, 338
Forward Error Correction (FEC), 196, 274
Forwarding pointers, 36–38
Fraud, 296–297
Frequency division multiple access (FDMA), 216–218, 359
Frequency hopping, 315
Frequency-hopping spread spectrum (FHSS), 331
Functional context, 95
Functionality, adapting, 8–9

Garbage collection, by service discovery frameworks, 141–142, 156–159
General Event Notification Architecture (GENA), 159–160
Global Positioning System (GPS), 250, 252, 292
Global service discovery, 155
Global System for Mobile Communication, 28, 33–34, 360–364
GPS (*see* Global Positioning System)
Group communication, 268–271
Group keys, 343
GSM (*see* Global System for Mobile Communication)

HA (*see* Home agent)
Habitat monitoring, 173, 187–188
Hand-off, 29
Hard state information, 8
Hash functions, 305
Header, packet, 306, 308
Hidden terminal problem, 218–219
Hierarchical organization (registration area-based location management), 47–48

High-level context, 96
Hit ratio, 74–75
HLC (*see* Home location cache)
HLRs (*see* Home location registrars)
Hoard walking, 78
Hoarding (Coda), 12, 77–78
Hold mode (Bluetooth), 319
Home agent (HA), 51, 78, 80–83
Home location cache (HLC), 73, 78, 80–83
Home location registrars (HLRs), 30, 32–34, 36–37, 49–50
Home network, 51
Hot data items (hot items), 57, 62, 65
HTTP (*see* Hyper-Text Transfer Protocol)
Hyper-Text Transfer Protocol (HTTP), 17–19, 61, 87–88, 160, 163

IBSS (*see* Independent basic service set)
ICMP (*see* Internet Control Message Protocol)
ICP (*see* Internet Caching Protocol)
ICV (*see* Integrity check value)
IDS (*see* Intrusion detection system)
IEEE 802, 330
IEEE 802.1x, 342–344
IEEE 802.11, 329–348
IGMP (*see* Internet Group Management Protocol)
IKE (*see* Internet Key Exchange)
IMEI (*see* International mobile equipment identity)
Import driver, 121
Impulse-down strategy, 119
Impulse-up strategy, 119
IMSI (*see* International mobile subscriber identity)
Independent basic service set (IBSS), 330
Industrial applications, 186–187
Information caching, 57–59
Information sources, 55
Initialization key (Bluetooth), 322
Integrity, 287–288
Integrity check value (ICV), 308, 334–335, 338
  (*See also* WEP)
Integrity codes, 304–308
Intercept model, 18
Interceptor, 117
Intercepts, 17

International mobile equipment identity
(IMEI), 365
(*See also* GSM)
International mobile subscriber identity
(IMSI), 364
(*See also* GSM)
Internet Caching Protocol (ICP), 61
Internet Control Message Protocol
(ICMP), 289
Internet Group Management Protocol
(IGMP), 8
Internet Key Exchange (IKE), 308
(*See also* IPsec)
Internet Protocol (IP), 4, 28, 193, 289,
308, 359
Interoperability, 166–167
Intrusion detection system (IDS), 347
Invalidation reports, 75–77
Invalidation-based caching strategies,
73, 75–77
Involuntary disconnections, 59
IP (*see* Internet Protocol)
IP address, 50–53, 151
IP Security (IPSec), 308
IP tunneling, 51

Java, 130, 132–133, 143, 163, 165,
167, 299
Java Native Interface (JNI), 143
JavaSpaces, 133
Jini, 138, 143, 149–157, 163–164,
166–167
JNI (*see* Java Native Interface)

Kernel, OS, 117, 118
Key distribution protocols,
220–221
Keys:
  group vs. session, 343
  pairwise master, 344
  and security protocol configuration,
  248–249
  in wireless sensor nodes, 200–202
LANs (*see* Local area networks)
Laptops, 2
LAs (location areas), 34
Lazy cache maintenance scheme, 41
LEACH protocol, 247, 249, 256, 260,
280–281
LEAP (*see* Lightweight Extensible
Authentication Protocol)
Leasing (service discovery frameworks),
156–158

Least-recently used (LRU) algorithm,
18, 68
Lessee, 156
Lessor, 156
Liability, with electronic commerce, 296
Lightweight Extensible Authentication
Protocol (LEAP), 309
(*See also* Authentication)
Line Printer Daemon Protocol
(LPD), 145
Links:
  asymmetric, 23, 59–60
  expensive, 23
LIS (*see* Location information system)
Load balancing, 131
Local coordinate systems (for sensors),
251–252
Local location registrar, 34
Locality, temporal vs. spatial, 132
Location areas (LAs), 34
Location information system (LIS),
105–106
Location management:
  case studies, 48–53
  defined, 28
  registration area-based, 33–48
  search operation of, 28
  update operation of, 28
Location management optimization,
38–39
Location pointers, 36
Location registrars, 28, 30, 34, 49–50
Location transparency, 4
Location-dependent data, 60
Low-level context, 95–96
LPD Protocaol (Line Printer Daemon
Protocol)
LRU algorithm (*see* Least-recently used
algorithm)
MAC addresses (*see* Media Access Control
addresses)
MAC layer, 332
MAC layer protocols, 282–283
MACs (*see* Message authentication codes)
"Man in the middle" attacks, 297
MANETs (*see* Mobile ad hoc networks)
MANs (*see* Metropolitan area networks)
MARS (*see* Mobile Agent Reactive
Spaces)
Mars Exploration Rovers, 174
MASIF (*see* Mobile Agents System
Interoperability Facility)
Master (Bluetooth), 318

Master key (Bluetooth), 323
MCE (*see* Mobile computing
    environment)
MCHO (*see* Mobile-controlled handoff)
Media Access Control (MAC) addresses,
    143–144, 344
Medical applications, 185–186
Meeting places, 132–133
Meetings (Telescript), 129
Memory, shared, 68
Memory hierarchy, 68
Message aggregation, 228–230
Message authentication codes (MACs),
    305–308
Metropolitan area networks (MANs), 349
    security in, 350–354
    standards for, 350
MHTTP, 88
Mica, 197–198
Middleware:
    for adaptation, 114–123
    agility of, 116–117
    for context-aware applications,
        102–106
    and coordination of architectures,
        122
    defined, 109
    mobile agents as, 123–133
    service discovery (*see* Service discovery
        frameworks)
Middleware layer, 3
Migration of mobile agents, 130–131
Military applications, 182, 185
MIME (*see* Multipurpose Internet Mail
    Extensions)
Mobile ad hoc networks (MANETs),
    60–61, 173
Mobile Agent Reactive Spaces (MARS),
    133
Mobile agents, 109, 111, 123–133
    advantages of, 125–126
    applets vs., 124
    and application customization, 126
    communication strategies for,
        131–133
    disadvantages of, 126–127
    and dynamic CS architecture, 124
    and dynamic expansion of server
        functionality, 111
    meeting places for, 132–133
    migration strategies for, 130–131
    and process checkpointing, 130–131
    and security, 127

security with, 127
    for sets of applications, 125
    standardization issues with, 126
    in Telescript, 127–130
Mobile Agents System Interoperability
    Facility (MASIF), 126
Mobile communication, 1
Mobile computing, 1
    adaptability as key to, 3–7
    constraints of, 5–6
    mobile communications vs., 1
    scope of, 1–3
    vision of, 4
    environment (MCE), 5
    systems, 1
Mobile data caching (*see under* Caching)
Mobile Internet Protocol (Mobile IP), 28,
    50–53, 78
Mobile Web caching, 86–88
Mobile-controlled handoff (MCHO), 29
Mobility:
    of ad hoc wireless network nodes,
        235–240
    increases in, 27
    and security, 293
Mobility binding, 32, 51
Mobility management, 27–30
    (*See also* Location management)
Moore's law, 3, 5, 194
Motivating context, 95
Movement detection (ad hoc wireless
    sensor networks), 235–237
Mowgli, 86–88
MTCP, 87
Multicast discovery, 151–153
Multihop routing, 222–225, 294
Multipurpose Internet Mail Extensions
    (MIME), 19

Name transparency, 4–5
NAT (*see* Network Address Translation;
    Network address translator)
National Institute of Standards and
    Technology (NIST), 165
National Security Agency
    (NSA),165
NCHO (*see* Network-controlled
    handoff)
Neighborhood discovery (ad hoc wireless
    sensor networks), 214–215
Network Address Translation (NAT),
    346–347
    translator, 14

Network File System (NFS), 69
Network interface cards (NICs), 59
Network Time Protocol, 272
Network-controlled handoff (NCHO), 29
Networks, 1
    foreign, 51
    home, 51
NFS (*see* Network File System)
NICs (*see* Network interface cards)
Ninja, 145, 165
NIST (*see* National Institute of Standards
    and Technology)
NLANR, 61
Nonce, 298
Nonrepudiation, 288
NSA (*see* National Security Agency)

Odyssey, 10–11, 19–22, 115
    as application-aware adaptation,
        117–119
    sample video player application,
        119–120
OFDM (*see* Orthogonal frequency division
    multiplexing)
Offline algorithms, 84
On-demand downlink channel, 62–63
On-demand mode, 55, 57, 61
Online algorithms, 84
Open Systems Interconnect Standard
    (OSI standard), 195
Operating systems (OSs):
    and adaptation, 114
    for PDAs, 117
Optimization:
    location management, 38–39
    protocol, 87–88
Orthogonal frequency division
    multiplexing (OFDM),
    331–332
OSI standard (*see* Open Systems
    Interconnect standard)
OSs (*see* Operating systems)

P2P networks (*see* Peer-to-peer networks)
Paging, 28
Pairwise master key (PMK), 344
ParcTabs, 101
Parity bits, 304
    (*See also* Data integrity)
Park mode (Bluetooth), 319
Passive context, 96
Passive power sources, 194
Payload, packet, 306–308

PCS networks (*see* Personal
    communication service networks)
PDAs (*see* Personal digital assistants)
PDU (*see* Protocol data unit)
PEAP (*see* Protected EAP)
Peer-to-peer networks (P2P networks),
    61, 71
PEGASIS, 228–229, 256–257, 259–260,
    280–281
Permanent IP addresses, 51
Permits (Telescript), 129–130
Personal communication service (PCS)
    network(s), 33, 39
    location management scheme for,
        49–50
    service area of, 33–34
Personal digital assistants (PDAs), 2,
    101–102, 112, 317
    and concurrency, 117
    and mobile agents, 126
    mobility of, 203
Personalized mobile applications, 125
Per-user location caching, 40–44
Pervasive computing, 171
PGP (*see* Pretty Good Privacy)
Physical context, 94
Physical intercept (of signal), 291,
    301–302
Piconet (Bluetooth), 319
PIF (*see* Puppeteer Intermediate Format)
"Ping of death" attack, 289
    (*See also* Attacks)
PIX metric, 75
PKM (*see* Privacy Key Management)
Places (Telescript), 127–128
"Plug and play" technologies, 138, 141
PMK (*see* Pairwise master key)
Point-to-Point Protocol (PPP), 167, 309
Polling-every-time-based caching
    strategies, 72
Ports, controlled vs. uncontrolled, 343
Power conservation (*see* Energy
    efficiency)
PPP (*see* Point-to-Point Protocol)
Prefetching, 9, 60, 68
Prekeying, 294–295
Pretty Good Privacy (PGP), 304
    (*See also* Encryption)
Privacy Key Management (PKM),
    350–352
Private key (symmetric) encryption,
    302–303
    (*See also* Encryption)

Probe messages, 81
Process checkpointing, 130–131

Protected EAP (PEAP), 313
    (*See also* Authentication)
Protection, 334
Protocol data unit (PDU), 334
Protocol optimization, 87–88
Proxies, 13
    client-side, 17
    server-side, 17
    transcoding, 14–16
Proximity selection, 99
PSTN (*see* Public Switched Telephone
    Network)
Public key (asymmetric) encryption, 303
    (*see also* Encryption)
Public Switched Telephone Network
    (PSTN), 30
Publishing mode (*see* Push mode)
Publish-subscribe mode (*see* Push mode)
Pull mode, 55, 61
Puppeteer, 120–122
Puppeteer Intermediate Format (PIF),
    121–122
Push mode (publishing mode, publish-
    subscribe mode), 55–57
    advantages, 61–62
    bandwidth allocation, 63–65
    broadcast disk scheduling, 65–67
    cache maintenance schemes,
        74–75

QoS (*see* Quality of service)
QRPCs (*see* Queued remote procedure
    calls)
Quality of service (QoS), 2, 30, 110,
    206, 215
Queries (*see* Uplink requests)
Queued remote procedure calls (QRPCs),
    22–23

Radio frequency ID (RFID), 186–187
RADIUS, 313, 343
Random deployment of
    (of sensors), 246
RAs (*see* Registration areas)
RBS Protocol (*see* Reference Broadcast
    Synchronization protocol)
RCMR (*see* Regional call-to-mobility
    ratio)
RDOs (*see* Relocatable dynamic objects)
Read-only data, 68

Rebasing, 18
Reconfiguring, 295
Reference Broadcast Synchronization
    (RBS protocol), 272
Reference copy, 10, 115
Reference waveforms, 11
Regional call-to-mobility ratio
    (RCMR), 41
Registrars, single-location, 30
Registration area-based location
    management, 33–48
    dynamic updates, 39–40
    flat organization, 45–47
    forwarding pointers, 36–38
    hierarchical organization, 47–48
    location management optimization,
        38–39
    per-user location caching, 40–44
    replication of location information,
        44–45
Registration areas (RAs), 34
Registration operation, 28
Registry, 123
Reintegration state (Coda), 12
Relocatable dynamic objects (RDOs),
    22–23
Remote location registrar, 34
Remote Method Invocation (RMI), 132,
    149, 150
Remote procedure calls (RPCs),
    22–24, 309
Replay attacks, 298, 338
Reports, invalidation, 75–77
Request-to-Send (RTS packets), 218, 282
Resilient systems, 3
Resource discovery, contextual, 98
Resource monitoring, 114–115
Resource paucity (resource constraints),
    6, 59, 110
    in ad hoc wireless sensor networks,
        191–200
    and security, 291
    (*See also* Energy efficiency)
Resource Reservation Protocol (RSVP), 8
Resynchronization (ad hoc wireless sensor
    networks), 239–240
RFID (*see* Radio frequency ID)
Richness, 140
RMI (*see* Remote Method Invocation)
Role-based context, 96
Route aggregation, 50
Routing, 192, 220
    diffusion, 260–265

directional, 265–268
event-driven, 253–254
and group communication,
    268–271
multihop, 222–225
and periodic sensor readings,
    254–260
and security, 294
in wireless sensor networks, 252–272
Rover, 22–24
RPCs (*see* Remote procedure calls)
RSVP (*see* Resource Reservation Protocol)
RTS packets (*see* Request-to-Send
    packets)

SA-always, 84–86
Salutation, 138, 145, 167
SA-never, 84–86
Sapphire/Slammer SQL worm, 290
    (*See also* Attacks)
SAs (*see* Security associations; service
    agents)
Scalability, of sensors, 246
Scatternet (Bluetooth), 319
Scheduling:
    in ad hoc wireless sensor networks,
        216–217
    broadcast, 65–67
    communication, 225–226
    dual-radio, 230–231
    sleep-mode, 232
Scheme, 130
Scoping, by service discovery
    frameworks, 142
SCPD (*see* Service control protocol
    description)
SDP (*see* Service Discovery Protocol)
Searching, 28
Secure Hash Algorithm 1 (SHA-1), 165
    (*See also* MACs)
Security, 3, 5, 287–299, 301–315
    in ad hoc wireless sensor networks,
        200–203, 293–295
    and attacks, 297–299
    and availability, 288–290
    for commerce, 295–297
    and confidentiality, 288
    and detectability, 290–291
    general guidelines for, 344–347
    and integrity, 287–288
    with mobile agents, 127
    and mobility, 293
    and nonrepudiation, 288

and physical intercept of signal, 291
and resource depletion/exhaustion, 291
in service discovery frameworks,
    163–165
and theft of service, 291–292
via AAA, 313–314
via authentication protocols,
    309–313
via encryption, 302–304
via integrity codes, 304–308
via IPSec, 308–309
via signal limitation, 301–302
via special hardware, 315
and war driving/walking/chalking,
    292–293
in wide area networks (WANs),
    357–367
in wireless local area networks
    (WLANs), 329–348
in wireless metropolitan area networks
    (MANs), 349–354
in wireless personal area networks
    (WPANs), 317–327
Security associations (SAs), 351–353
Security protocols (ad hoc wireless
    networks), 220–221
SEKEN protocol, 248–249
Self-organization (ad hoc wireless sensor
    networks), 247–249
Sensing, contextual, 98
Sensors (*see* Smart sensors)
Serendipity, 106–107
Servers, 8–9, 68
Server-side intercepts (SSIs), 17
Service advertisement, 139–141, 153–155,
    158–159
Service agents (SAs), 164
Service area (of PCS network),
    33–34
Service browsing, 141
Service catalogs, 141, 155–156
Service control protocol description
    (SCPD), 147
Service discovery, 111–112
    defined, 113
    global, 155
Service discovery frameworks,
    137–168
    benefits, 137–138
    catalogs, service, 155–156
    common features, 138–142
    and CS system design, 138
    defined, 138

Service discovery frameworks (*Cont.*):
  eventing, 159–163
  garbage collection, 156–159
  interface definition, 149–150
  interoperability, 166–167
  leasing, 156–158
  multicast discovery, 151–155
  and security, 163–165
  security, 163–165
  and service advertisement, 139
  standardization process,
    144–145
  textual descriptions used by,
    145–149
  unicast discovery, 150–151
  universally unique identifiers,
    142–144
Service Discovery Protocol (SDP),
    145, 167
Service Location Protocol (SLP), 145, 148,
    154–156, 164–167
Service points (Ara), 132–133
Service set identifier (SSID), 331, 346
Service subtyping, 140
ServiceRegistrar, 151
Services:
  contextual, 103–104
  in service discovery frameworks,
    139–141
Session keys, 343
SHA-1 (*see* Secure Hash Algorithm 1)
Shared memory, 68
Signal limitation, 301–302
Simple Mail Transfer Protocol
    (SMTP), 298
Simple Object Access Protocol
    (SOAP), 147
Simple Service Discovery Protocol
    (SSDP), 159
Single-hop, 294–295
Single-location registrars, 30
Slave (Bluetooth), 318
Sleep-mode scheduling, 232
Sliding window, 43
Sliding-window dynamic data allocation
    scheme, 84–86
SLP (*see* Service Location Protocol)
S-MAC, 218
"Smart" batteries, 114
Smart Dust research group, 180
Smart sensors (sensor nodes, sensors),
    171–172, 179

agreement among, 274–278
and autoconfiguration,
    213–214
collaboration among, 271–272
dead/faulty, 278–279
deployment/configuration of,
    245–252
and group communication,
    207–208
limited computational capacity of,
    197–198, 201
limited input/output options with,
    199–200
limited resources available to, 180–181,
    193–196
limited storage capacity of, 198
location determination for, 249–252
loss of connectivity between,
    205–206
mobility of, 235–240
neighborhood discovery by,
    214–215
periodic readings of, 254–260
power management of, 281–282
random deployment of, 246
reconfiguration/redeployment
    of, 249
scalability of, 246
small size of, 181
as special-purpose devices, 179–180
Smartcards, 315
SMTP (*see* Simple Mail Transfer
    Protocol)
"Smurf" attack, 289
    (*See also* Attacks)
Sniff mode (Bluetooth), 319
SNR (*see* Signal-to-noise ratio)
SOAP (*see* Simple Object Access
    Protocol)
Social context, 95
Sockets, virtual, 18–19
Solar power, 194
Spatial locality, 132
Spatial reuse, 223–224
Split-phase operation, 22
Spoofing, 294
Spread-spectrum technology, 315
SSDP (*see* Simple Service Discovery
    Protocol)
SSID (*see* Service set identifier)
SSIs (*see* Server-side intercepts)
Staleness, 116

Standard Generalized Markup Language (SGML), 102
Standardization, of service discovery frameworks, 144–150
State information, 8
Stateful invalidation-based caching strategies, 73
Stateless invalidation-based caching strategies, 73
Static allocation schemes, 84
Static update schemes, 33
Step-down strategy, 119
Step-up strategy, 119
Stick-E Note, 101–102
Strong cache consistency, 69, 72
Strong connectivity, 12
Suboptimal system operation, 122
Subtyping, service, 140
Supplicant, 342
    (*See also* Authentication)
SWIM project, 187
Symmetric (private key) encryption, 302–303
    (*See also* Encryption)
"SYN flood" attack, 289
    (*See also* Attacks)
Synchronization, 271–272
Synchronous stateless invalidation-based caching strategies, 73
System operation, suboptimal, 122

Tacoma project, 126
Tapping, 301
TCP/IP (*see* Transmission Control Protocol/Internet Protocol)
TDMA (*see* Time division multiple access)
"Teardrop" attack, 289
    (*See also* Attacks)
Telescript, 127–130
TEMPEST, 315
Temporal context, 94–95
Temporal Key Integrity Protocol (TKIP), 340–342
Temporal locality, 132
Temporary mobile subscriber identity (TMSI), 364
Theft of service, 291–292
    (*See also* Attacks)
Thin clients, 9
Third Generation Partnership Project (3GPP), 365

Time division multiple access (TDMA), 217–218, 225–226, 250, 271, 282–283, 359
Time-based updates, 33, 39
Time-dependent data, 60
Timestamp, cache, 81–83
Time-to-live (ttl), 32, 51
Timing-Sync Protocol for Sensor Networks (TPSN), 272
TinyOS, 198
TKIP (*see* Temporal Key Integrity Protocol)
TLS (*see* Transport Layer Security)
TMSI (*see* Temporary mobile subscriber identity)
Topology discovery, 215–216
TPC (*see* Transmit power control)
TPSN (*see* Timing-Sync Protocol for Sensor Networks)
Traffic analysis, 298, 308
Trajectory-based forwarding, 267
Transcoding proxies, 14–16
Transcoding threshold, 14, 16
Transmission Control Protocol/Internet Protocol (TCP/IP), 11, 30, 87, 132, 148–149, 150–151, 160, 192, 196, 273, 289
Transparency, 4–5, 59
Transport Layer Security (TLS), 312
Transport mode, 309–310
    (*See also* IPsec)
Travel (Telescript), 128–129
ttl (*see* Time-to-live)
TTL-based caching strategies, 72–73, 86
Tunnel mode, 308–310
    (*see also* IPsec)
Tuple spaces, 133

UAs (*see* User agents)
UDP (*see* Unicast Discovery Protocol)
Uncontrolled port, 343
    (*See also* Authentication)
Unicast discovery, 150–151
Unicast Discovery Protocol (UDP), 152
Unit key (Bluetooth), 322
Universal Plug and Play (UPnP), 138, 145–148, 150, 155, 158–163, 166–167

Universally unique identifiers (UUIDs), 142–144, 152
Update lists, 41
Update schemes, 33
Updates:
    dynamic, 39–40, 141
    time-based, 33, 39
Updating, 28
Uplink channels, 29
Uplink request channel, 62–63
Uplink requests (queries), 78–79
UPnP (*see* Universal Plug and Play)
User agents (UAs), 164
User context, 94
User mobility, 5
User needs, prediction of, 123
UUIDs (*see* Universally unique identifiers)

Validity checks, 70
Venus (Coda file system), 12
VFS (*see* Virtual File System)
Vibration energy, 194
Viceroy (Odyssey), 20–22, 117, 119
Video, fidelity of, 115–116
Video streaming, 2, 4
Virtual File System (VFS), 117
Virtual private networks (VPNs), 345
Virtual sockets, 18–19
Visitor location registrars (VLRs), 36, 49
Voluntary disconnections, 59
VPNs (*see* Virtual private networks)

Wakeup, 80
WANs (*see* Wide area networks)
War chalking, 293
War dialing, 292
War driving, 292
War walking, 292
Wardens (Odyssey), 20–22, 117, 119
Waveforms, reference, 11
Weak cache consistency, 72
Weak connectivity, 12, 59
Weak key attacks, 338
        (*See also* Attacks)
Weakly-connected ad hoc networks, 60
Web caching, 77, 86–88
WebExpress, 17–19, 86

WEP (*see* Wired Equivalent Privacy)
WEPplus, 345
Wide area networks (WANs):
    about, 357–359
    security in, 359–367
Wi-Fi protected access (WPA), 340, 345
Wire integrity, 301
Wired Equivalent Privacy (WEP), 293, 332–340, 345
    authentication in, 335–336
    data frame, 334, 335
    decryption in, 335
    drawbacks of, 336–338
    encryption in, 334, 335
    fixes for, 338–340
    goals of, 333, 334
Wireless ad hoc sensor networks (*see* Ad hoc wireless sensor networks)
Wireless information systems, 56–57
Wireless local area networks (WLANs), 27, 329–331
    and 802.11 protocols, 331–333, 340–344
    security guidelines for, 344–347
    and WEP security scheme, 333–340
    and Wi-Fi protected access (WPA), 340
Wireless personal area networks (WPANs), 317–327 (*See also* Bluetooth)
Wireless station (WS), 330, 336
WLANs (*see* Wireless local area networks)
Working sets, 43
Working-set approach, 43
WPA (*see* Wi-Fi protected access)
WPANs (*see* Wireless personal area networks)
Write-disconnected state (Coda), 12
WS (*see* Wireless station)
WWW (*see* World Wide Web)
Xanim video player, 119–120
XML, 145–146, 152, 159–160, 165

ZebraNet project, 187
Zombies, 289–290